Springer Texts in Business and Economics

For further volumes:
http://www.springer.com/series/10099

Amiya K. Chakravarty

Supply Chain Transformation

Evolving with Emerging Business Paradigms

Springer

Amiya K. Chakravarty
DM School of Business
Northeastern University
Boston, MA
USA

ISSN 2192-4333 ISSN 2192-4341 (electronic)
ISBN 978-3-642-41910-2 ISBN 978-3-642-41911-9 (eBook)
DOI 10.1007/978-3-642-41911-9
Springer Heidelberg New York Dordrecht London

Library of Congress Control Number: 2014931758

Printed on acid-free paper

Springer is part of Springer Science+Business Media (www.springer.com)

This book is dedicated to my wife Indira,
my daughter and son, and their families

Foreword

The management of supply chains has been long recognized as a key requirement for the achievement of customer satisfaction in both product and service based industries. The business world, however, is changing in profound ways and it has become necessary for firms to adapt to prosper and even to survive. In order to do so, they must transform their supply chains in ways that are responsive to the wide range of changes in their environment which are driven by technology, government and market forces. This book provides a remarkable overview of how to drive such transformations.

The book covers an enormous range of topics that form the basics of modern supply chain management. The topics covered in the book include supply chain fundamentals such as the management of customers, products, processes, capacity, suppliers and logistics. It also considers newer areas where supply chains have had an impact, such as sustainability, health care, humanitarian relief, and emerging economies. In each chapter, the book reviews the state of the art by introducing current and emerging business paradigms and provides extensive coverage of specific case studies and examples. The approach taken is primarily conceptual and non-technical, and in so doing it provides the reader with actionable frameworks for navigating through the often confusing collection of choices that are available for the transformation of modern day supply chains. The book also contains many examples of model formulations, strategic frameworks and research results; all drawn from a wide range of sources. All of the material is consolidated, organized and explained in a clear and concise fashion that makes the content accessible to a wide range of readers, especially those looking for a strategic perspective.

Also included in the book is a chapter on how to transform the contemporary supply chain which contains descriptions of how companies have re-designed their supply chains in innovative ways in response to both external and internal drivers.

Both students and managers of supply chains will find this book to be of great value. It will be especially useful for those looking for an introduction to the many facets of the field as well those who are interested in an up-to-date overview of the field. The author has deep understanding of both the practical and academic aspects of supply chain management. Overall, the book provides the reader with a framework for using supply chain transformation to achieve competitive advantage.

Morris A. Cohen
The Wharton School, University of Pennsylvania
Philadelphia, PA, USA

August, 2013

Preface

Increasingly, companies motivated to create additional value for customers for competitive advantage are focusing on transforming their supply chains. While such values are shaped by the external drivers such as technology, market-volatility, preference for products and services, and disruptions, they can be stymied by the internal stresses arising from the need to minimize costs, limitations in redesigning processes, waste minimization, and the unavailability of knowledge-capital. It follows that for companies to survive and prosper, the type and extent of transformation must be specific to the business drivers. The relevant questions would be how might one identify the external forces that drive changes and the internal forces that act as constraints, and how to map the business drivers to the attributes of transformation.

Three dominant themes are discernible in the way the contemporary supply chains are being restructured: modification of the operations control system, leveraging of emerging technologies, and redesign of the supply chain architecture. Restructuring of operational control is aimed at the redesign of the flows in the system – material, information, and payments. It may include redesigning the inventory control system, modularizing products, modifying order processing system, implementing e-procurement, and enhancing shipment tracking.

Technology can impart enhanced value to the operational control infrastructure. The internet has helped transform supply chains from a sequential and linear relationship to a network of entities, and enabled customers to configure their own customized products. Other examples of technology include information sharing for transparency, radio frequency identification (RFID) for tracking shipments, mobile technology for real time decisions, cloud technology for better coordination and reduction of fixed cost, and the Web 2.0 for information exchange. A successful implementation of these technologies requires significant changes throughout the supply chain.

Architectural changes may segment the supply chain, incorporate flexible processes and equipment, move the customer order decoupling point (CODP) upstream, create logistics hubs and shared-service hubs, and develop country-specific customs duty and tax regimens. It enables companies tailor their supply chains through segment-specific supply chain processes, policies, and operational modes. It calls for a redesign of the supply chain architecture to create clusters of products, customers, and resources, based on shared sets of values. The intent is to design flexible and responsive supply chains that are capable of shifting back and forth between the complexity of customization and the simplicity of standardization. While the reengineering of operations is the most common form of transformation, architectural redesign is an undertaking that may require substantial new financial investments.

While the contemporary supply chain is well-structured, the evolving economic system may cause stresses to creep into this structure. The emergence of novel business paradigms – non-applicability of the traditional laws of supply and demand, dominance of negative externality effects, and anomalies of high growth rate coexisting with high supply side uncertainty – must be recognized in transforming supply chains. For example, healthcare delivery and humanitarian relief do not follow known supply/demand relationships; the negative externality effects are increasing sustainability concerns; and emerging economies, with dysfunctional business infrastructure, must manage high growth rates. How should the supply chain be transformed in such emerging scenarios?

The book delves into the transformation issues in contemporary supply chains, and extends the concepts to incorporate the emerging issues. It does so through ten chapters, divided into three sections. The first section builds the framework for transformation, while the second focuses on transforming the contemporary supply chain in terms of products, processes, supply base, procurement, logistics, and fulfillment. Section three is devoted to capturing the key issues in transforming supply chains for emerging economies, humanitarian relief, sustainability, and healthcare delivery.

The basic framework is discussed at length in Chap. 1. It outlines how the external and internal drivers should be mapped into generic supply chain capabilities: customer management, procurement, production, and fulfillment. The mapping identifies the capabilities that a company must strengthen based on the external and internal drivers it chooses to address.

Customer management is discussed in Chap. 2. It outlines how the process of mapping the customer's preferences becomes complex because of the presence of uncertainties in buying habits, pricing, lead times, and coalition formation. The supplier can respond by transforming his/her capabilities in order-decoupling, demand management (consolidation, allocation, shaping), bundling, flexibility, financial hedging, and revenue management. Examples from the airline, fashion goods, and computer industries are discussed.

Chapter 3 is focused at the interactions among products, processes, and the supply chain. Using product structure as the building block, the roles of product and process platforms in structuring transformation are highlighted through product modularity, component commonality, postponement, process structure, outsourcing, task sequencing, and task splitting.

In Chap. 4, the emphasis is on suppliers: building supplier's capability, and the interface between the supplier and the buyer. It is explained how the interface issues of fulfillment, collaboration, networking, and outsourcing are intricately related to the supplier's capabilities of procurement, quality, cost, and new knowledge creation. Increasingly, suppliers must manage complex technology portfolios, operate in uncertain environments, develop a sense of consumer needs, and manage risks. Concepts such as the supply-base, supplier development, supplier selection models, offshore outsourcing, and collaboration are elaborated.

In Chap. 5, the logistics functions are described – warehousing, transportation, packaging, and the customer interface, as it pertains to a contemporary supply chain. The strategic portfolio for logistics is established in terms of product characteristics and the corresponding logistics capability. Logistics needs of an evolving system are elaborated in terms of customization, globalization, outsourcing, sustainability, and new technology.

In Chap. 6, transformation approaches are discussed in view of the contemporary business drivers such as uncertainties, technology, and globalization. Three dominant themes of transformation are described: modification of operational control systems, leveraging of emerging technologies, and redesign of the supply chain architecture. Specific transformations include redesign of fulfillment, repositioning order decoupling point, creating flexibility, providing transparency, sharing services, managing tax burden, restructuring for RFID and mobile technology, incorporating cloud platform and Web 2.0, supply chain segmentation, and value chain restructuring.

The implications of market opportunities in emerging economies are discussed in Chap. 7. With emerging economies now too important to ignore, multinationals must operate in a dual economy – mature economies with orderly business environment but stymied growth rates, and emerging economies with high growth rates within a dysfunctional system. A number of business models that incorporate the characteristics of emerging economies are outlined, and the concept of global supply chain finance that decreases the suppliers' cost of borrowing in emerging economies is discussed.

The unique aspects of a humanitarian relief chain are carefully laid out in Chap. 8. While saving life creates social value, it generates profits for the participating retailers and logistics providers. However, the laws of supply and demand do not work well in a chaotic environment, as the consumer (aid recipient) neither places purchase orders nor pays for the delivered services. Conversely, the aid-donors who finance transactions have little control on what is delivered to the recipients. Coordination of prepositioned inventory with real time deliveries and the global flow of relief funds are discussed.

In Chap. 9, it is explained how the pursuit of sustainable practices may outweigh the costs incurred by individual companies while improving their business reputation. Novel sustainable products can create new opportunities. The supply chain provides the ideal platform for assessing impacts, as the decisions on resource use pattern and transportation may impact the entire business infrastructure. Supply chains need to be transformed in terms of reducing greenhouse gas emission from production processes, decreasing material consumption, reducing product recalls, facilitating recycling, and promoting humanitarian values. However, with a large number of stakeholders, the accountability of organizations becomes critical at every stage.

Chapter 10 delves into the business environment of the healthcare network, which is a supplier driven knowledge-intensive system. It is characterized by industry-fragmentation, market imperfections, payer's inability to control outcomes, defensive practice of medicine, dearth of investment in preventive healthcare, and risky investment in new drugs and treatments. Mutations in healthcare delivery are discussed in terms of extending the span of the delivery network to niche providers, serving segmented markets, leveraging connectivity, enabling patient-pulled services, offering multiple payment structures, improving the supply chains of drugs and devices, altering the ownership structure, building centers of excellence in treatments, and inculcating organizational responsibility.

Boston, MA, USA Amiya K. Chakravarty

Acknowledgement

The concepts and models discussed in this book have evolved over a number of years through the interactions with my students, colleagues, and business executives. It is the culmination of efforts made possible with the support of many individuals and organizations. Naturally, the list of individuals I am thankful to is extensive. Two names stand out – Nisha Chakravarty and Raj Beri – who made invaluable contributions throughout the writing of this book.

I appreciate the opportunity to "experiment" with several emerging concepts at doctoral seminars at the Vienna University of Business and Economics. I am thankful for the support of the DM School of Business at Northeastern University through the Philip McDonald Chair endowment.

This book could not have been completed without the continuous support of my family. I remain indebted forever.

Contents

Part I

The Concept

A framework that maps external and internal drivers into generic supply chain capabilities of customer management, procurement, production, and fulfillment. The map identifies the capabilities a company must strengthen based on the drivers it chooses to address.

A Framework for Transformation

1

1.1 Introduction

Transformation of supply chains can be driven by a company's internal problems or they can be externally driven. The type and extent of restructuring should be determined by the nature of these drivers. The relevant question to ask would be how one identifies the external/internal forces driving changes, and the mapping between the drivers and the supply chain capabilities.

Anna's Linens created a flexible, scalable, and efficient supply chain that would provide higher returns. They implemented a new enterprise resource planning system, and a warehouse management system. Simultaneously, they developed a logistics infrastructure centered on a regional distribution center, a consolidated corporate ordering process, and a set of metrics defined for the entire supply chain (RPE 2010). Anna's success lay in being able to control costs, better manage inventory, and eliminate dependence on third party providers.

Nokia's supply chain, relatively speaking, is very complex with global-manufacturing and procurement that involved 60 large suppliers, multiple factories, and 100 billion components. The pressure from new product introductions was intense, as every phone model offered on average 170 handset variations and 250 sales package variants (IBM 2006). Nokia created an innovative supplier network to meet customer expectations. The partnership was based upon up-to-date information, flexibility, and trust. In support, they developed a pull-driven supply chain linking suppliers, factories, telecom operators, contract manufacturers, and logistics service providers. They cut inventories but provided fast access for rapid ordering. Benefits include increased sales, reduced component inventories throughout the pipeline (including suppliers and customers), time-to-market, and risk management.

The two cases vividly reveal that companies operating in different economic environments require different approaches for their supply chains. Transformation of supply chains can be driven by a company's internal problems as in the case of Anna's Linens or they can be externally driven as in the case of Nokia. Clearly, the

A.K. Chakravarty, *Supply Chain Transformation*, Springer Texts in Business and Economics, DOI 10.1007/978-3-642-41911-9_1,

type and extent of restructuring should be determined by the nature of the internal and external drivers. The relevant question to ask would be how one identifies (i) the external/internal forces driving changes, and (ii) the mapping between the drivers and the possible supply chain modifications shown in Table 1.1.

While Anna's Linens and Nokia were able to restructure their supply chains successfully, others have failed. What differentiates a company in this context is the set of capabilities it possesses or can develop. These capabilities can be broadly categorized as customer management, procurement, production, outsourcing, and fulfillment. Each of these capabilities can be thought of as a grouping of the restructuring options similar to those shown in Table 1.1. A partial list of options for each category is shown in Table 1.2.

Table 1.1 Drivers for change and restructuring options

	Restructuring options					
Internal/external drivers	Ordering process	Outsource logistics	Central distribution center	Add features to products	Value chain splintering	Demand shaping
Capacity utilization			X		X	X
Cost minimization		X	X			
Product variety	X			X	X	X
Rapid response		X			X	
Emerging economies	X	X			X	

Table 1.2 Supply chain capability

Customer management	Procurement	Production	Fulfillment
Product design	Order management	Facilities	Delivery route planning
Demand shaping	Contract management	Flexibility, automation	Tracking of orders in transit
Preference mapping	Channel management	Outsourcing, quality assurance	Information technology
Pricing	Supplier relationship	Throughput maximization	Third party logistics

As capabilities can be expensive to build, companies may strategically focus on a subset of capabilities. As we would see later, the fit between the drivers (external/internal) and capabilities a firm possesses, helps determine whether the firm succeeds in its mission.

1.2 The Drivers of Change

In general, companies are subjected to a large number of external forces and internal stresses, and they need to find appropriate responses to survive and prosper. As we shall see later, these drivers may also define the opportunities a company may want to explore. External forces emanate from different sources such as customers, technology, locations, disruptions, and sustainability. Similarly the need for redesigning processes, reducing cost, and minimizing waste act as the internal drivers for change.

1.2.1 External Drivers

The major external drivers, mentioned earlier, include customer-preferences, technology, emerging economies, disruptions, and environmental sustainability, as shown in Fig. 1.1.

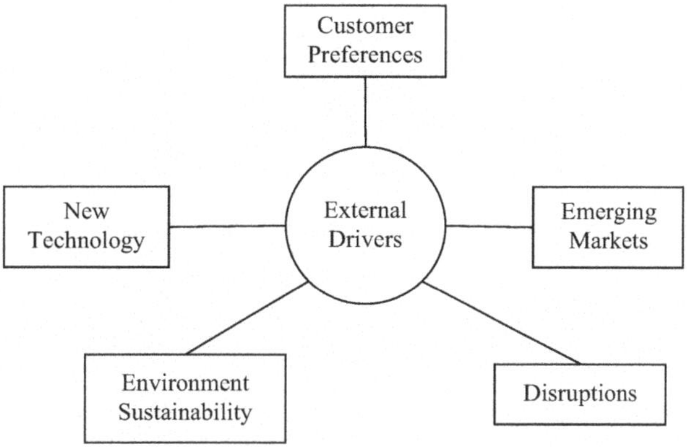

Fig. 1.1 External drivers

Customer preferences for product-variety, rapid response, and low cost, define the competitive landscape where supply chains must operate. As Chayet et al. (2011) point out, offering a variety of products can increase market share and it can lead to revenue growth; but it can also add complexity, reduce economies of scale, and result in increased costs. Therefore, the supply chain may need to be restructured to increase process flexibility. The front end marketing functions must be reengineered for creating a better brand image through product variety. In addition, to enhance customer satisfaction, tighter controls must be implemented on the fulfillment processes in terms of tracking and rerouting.

Technology can be a powerful external driver. Much has been written about how the internet transformed the supply chain. It transformed from being linear as shown in Fig. 1.2, where information could only be accessed sequentially between functions, to a nonlinear system shown in Fig. 1.3, where any function could access information from any other.

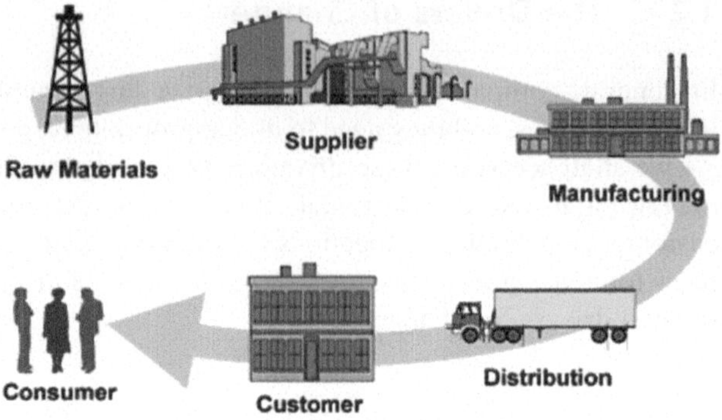

Fig. 1.2 A linear supply chain (Source: web search)

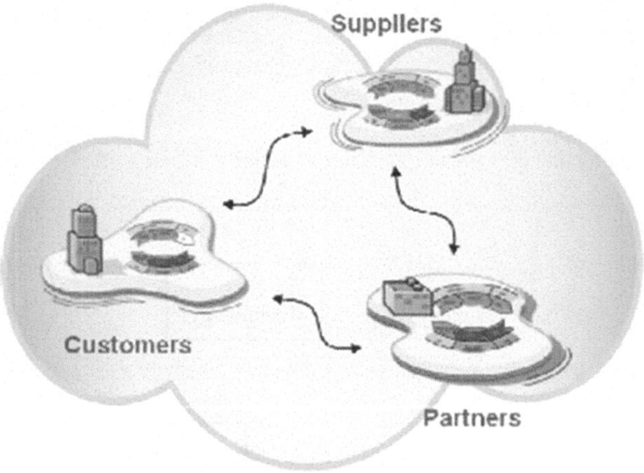

Fig. 1.3 Nonlinear network (Source: web search)

This provided opportunities for disintermediation in the value chain, and enabled companies like Amazon.com and Dell to set up online retail stores bypassing traditional brick-and-mortar retailers and distributors. Nonlinear information access also contributed to overall efficiency, as the supply chain became more transparent.

Cloud technology provides a common platform where companies can be interconnected on-demand through information exchange (refer to Chap. 6).

Biotechnology, in the form of smart drug delivery system and targeted treatment solutions are in the process of shaking up the pharmaceutical industry. While drug delivery methods may use slow-dissolving tablets and nanotechnology to deliver medicine to specific tissues at specific doses and times, targeted treatment solutions are aimed at specific patient subpopulations. In future, companies will manufacture a complete mix of medical nanochips, and biopharmaceuticals. These products have the potential of generating higher revenues but will require restructuring of the pharmaceutical supply chain.

The supply chain may transform in two possible ways. It may splinter into supply chains for mass producers of generics and traditional drugs, and the supply chain for the new innovative drugs that would be very asset intensive and a lot riskier than the traditional supply chains. The supply chain can also transform by creating a new intermediary, the so called "channel master". The channel master would control the design, R&D, and the markets of new drugs and get independent manufacturers to produce them (both traditional and innovative drugs) and deliver them to retailers, just as Cisco does in the telecommunication industry.

Emerging economies have recently become important external drivers for change. Companies have sourced material and components from low cost countries for many years. Over the years, some of these countries have emerged with large markets that have high GDP growth rates and significant purchasing power.

GE expects 60% of its revenue growth to come from the emerging markets over the next 10 years. Clearly they cannot be ignored, but they also represent significant departures from established business practices. These include a weak logistics infrastructure with congested transportation network, informal transactional procedures that mostly depend on payments in cash, deliveries made in small quantities to a large number of retailers and intermediaries, and a lack of reliable data for coordination and performance appraisal (refer to Chap. 7).

When Dell took their printer business to China (Farhoomand 2006), they experienced a different market than what they were used to, and had to rethink their direct-to-customer (DTC) model. The driver for change was the Chinese customer who did not possess computer savvy, and was uncomfortable in using the web. A second, more interesting, issue was their suspicion of online payment systems. Another example is WalMart's supply chain where efficiency is derived from a tight control of formal procedures. Obviously, such a supply chain would not perform well in China or India because of the informal business practices prevalent in those countries. Therefore, the question is whether companies like Dell and WalMart should create independent supply chains for an emerging country that is based on production and delivery of functional products and/or sequential information access? Dell decided to restructure their supply chain with more brick-and-mortar intermediaries, which they had cut out in their DTC approach. WalMart is considering to enter India as a joint venture with a local partner: the partner will be in control of retailing and the physical supply chain infrastructure, while WalMart becomes responsible for the back-office supply chain. Both WalMart and Dell will each have to manage two different supply chains.

There can be other solutions, of course. They could build a single robust supply chain, or a single resilient supply chain. A robust supply chain creates redundancies by building-in country specific features in addition to the features that are common among all markets. Thus China-specific features would only be used in China. Redundancy may also exist in resource use, as in using multiple suppliers for a single component. A resilient supply chain, on the other hand, engineers a new capability at the meta-level that can adjust the supply chain processes, so that they can be adapted to exigencies in the system as they arise. Examples are Li and Fung adjusting its supply plan rapidly to meet customer demand during a currency crisis in Indonesia, and Dell modifying its pricing strategy just in time to satisfy customers during supply shortage following the 1999 earthquake in Taiwan (Tang 2006).

The likelihood of severe disruptions in a supply chain can be a driver for change, as it may require protection or real time relief. Waffle House Restaurants have structured their supply chain to effectively respond to hurricanes (Ergun et al. 2010). They have integrated prepositioning of inventory with procurement strategies, and contract negotiation. Waffle House Restaurants have built strong relationships with suppliers by renting refrigerator trucks to help transport supplies. Waffle House Restaurants must also take into account the perishable nature of inventory in hurricane supply plans and inventory management.

The uncertainties associated with natural disasters create complexities in the supply chain, and they need to possess the capabilities of disaster prevention and fortification, preparedness, relief, and recovery. Both proactive and real-time response decisions should be enabled in the supply chain. Redundancies such as prepositioned inventories, fortified homes and businesses need to be balanced with the cost of logistics and procuring supplies at short notice. As the disaster may strike at only one of potential sites, the supply chain must be structured for transshipment of supplies from other sites where disaster did not strike. As human lives are at stake, the supply chain also needs to be endowed with a rapid infrastructure-repair capability (humanitarian supply chains are discussed in Chap. 8).

Environmental risk from partners' is a powerful driver for restructuring the supply chain. There is growing recognition that issues of environmental pollution should be coordinated with the supply chain processes. Many European countries have enforced legislation that hold manufactures responsible for used-products and waste (Robeson et al. 1992; Fleischmann et al. 1997). Therefore, issues such as recycling, waste disposal, and control of industrial pollution must be addressed in an integral fashion (refer to Chap. 9).

Dow Chemical, like other chemical companies, faced a high risk of ecological disaster during transportation of chorine using rail cars. This is so because the derailed cars can puncture upon impact causing leakage of chlorine, thereby, endangering the neighboring communities (Darnall et al. 2008; Sheu et al. 2005). Dow lacked the internal expertise to address the problem on its own. It partnered with a transportation equipment supplier to design rail cars that were two times thicker than was required. The new rail car soon became the industry standard. Similarly, in the automotive industry, some manufacturers have developed environmentally friendly material through partnerships with their paint and chemical suppliers (Geffen 1997; Geffen and Rothenberg 2000). Both examples above imply major restructuring of the supply chain because of ecological concerns – design of a new rail car, and formation of partnership with suppliers.

The market forces of supply and demand do not work well in the healthcare industry because the consumer (patient) does not often know what medical services he/she may require, and does not know how to assess the quality of services provided. The supplier (providers) determines the services the patient needs.

Therefore, the information asymmetry constrains the patient from shopping around for price and quality. In addition, new technology is causing healthcare to mutate in different ways by broadening the span of the service delivery network, enabling patient-pulled services, segmenting the service markets, and altering ownership structure.

The impact of the external drivers on the contemporary supply chain is discussed in depth in Chaps. 2, 3, 4, 5 and 6. The evolution of supply chains driven by emerging paradigms such as humanitarian relief, emerging economies, sustainability, and healthcare delivery are elaborated in Chaps. 7, 8, 9 and 10.

1.2.2 Internal Drivers

Opportunities for improving the effectiveness of operations act as the internal drivers for change. They include asset utilization, productivity improvement, cost minimization, waste reduction, and work place safety, as shown in Fig. 1.4. For example, it may be possible to improve product performance and cut manufacturing cost by substituting one component with another. To implement it, however, one may have to modify some manufacturing processes, modify procurement, work with new suppliers, modify marketing plans for the product, and train workers.

Fig. 1.4 Internal drivers

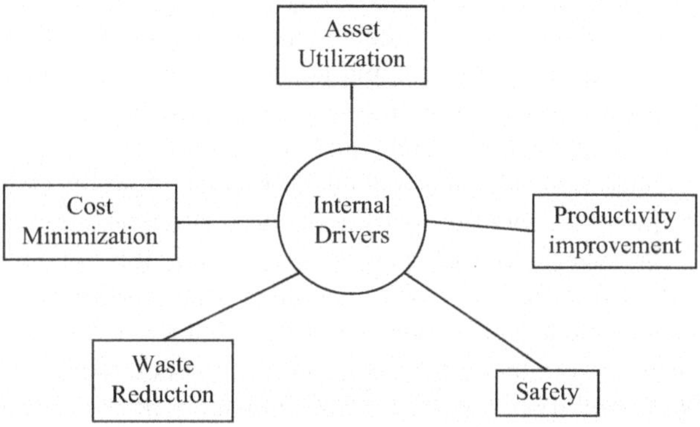

Asset utilization is defined as the ratio of actual output to maximum possible output. Some plants produce products whose demand is seasonal. During periods of off-peak demand, the plant may run at reduced capacity, causing asset utilization during those periods to drop to below 100%.

In the above example asset utilization can be increased by smoothening demand, or by setting plant capacity to equal demand during off-peak periods and outsourcing the excess demand during peak periods. Another common reason for asset underutilization is lengthy repair time when equipment breaks down. Moreover, there could be scheduling conflicts in the system whereby one machine is overloaded while another starves for jobs. A bottleneck in the system, similarly, causes all machines following the bottleneck to be underutilized. Setting up a machine for a product changeover, if the changeover time is lengthy, causes machine downtime, increasing underutilization. Clearly, increasing flexibility that permits an asset to be used in multiple ways increases the degrees of freedom in scheduling the asset and, hence, its utilization.

There are multiple reasons for the productivity of a plant to be lower than expected. It includes task complexity, untrained workers, and error-prone equipment. Complexity of task involving multiple tool changes slows its completion. It follows that the task should be broken down, if possible, into a number of simple tasks that the operator can complete with greater confidence. In this context, worker training cannot be overemphasized.

To control production cost, procurement, processing, energy, and human resource (compared to outsourcing) costs at each step of the supply chain must be minimized. To cut procurement cost, suppliers must be evaluated based on on-time delivery, low unit cost, and a low defective rate (refer to Chap. 4). Buy back contracts reduce the buyer's risk in unused components. To control production cost, it is imperative not to produce defective parts. The root causes of defects in products must be identified and the associated problems resolved. Consumption of power, water, and other utilities must be controlled, and recruiting, training, and retention of workers must be a high priority.

To ensure the greatest possible protection for employees in the workplace, cooperative efforts of both employers and employees will help in establishing and maintaining a safe and healthy work environment. In general, employers are responsible for performing a "hazard assessment" of the workplace for identifying and controlling physical and health hazards, identifying and providing appropriate personal protection equipment for employees, training employees in the use and care of the personal safety, and periodically reviewing, updating and evaluating the effectiveness of the personal protection program (refer to Chap. 9). In general, employees should properly wear protection aids, and attend training sessions.

Waste affects every part of supply chain; it is not only environmentally damaging but also expensive. Both suppliers and customers play a part in the creation, management and disposal of waste. While waste can be reduced through better design of products, processes, and equipment, the knowledge base for designing for sustainability is very thin. Some companies have shown that significant reductions in waste are practical and profitable, but many others fail to factor in the costs of waste, and do not understand how to improve production processes. Businesses can take the lead by demanding good practice from their suppliers. In order to promote waste reduction, simple methodologies should be developed to allow businesses to analyze the lifetime implications of the materials, products and services they produce.

Redesign of a manufacturing or service process provides opportunities for waste reduction, productivity improvement, and safety enhancement. Consider the example of a company that made a product used in the manufacturing of hybrid car batteries. Production occurred in a 9-step manual batch process. The process caused potential exposures of nanoparticles and titanium tetrachloride gas to employees at each step. The redesign of the production process changed the way the final nano-product was manufactured. New capital equipment was purchased that eliminated seven manual handling operations through reengineering, and eliminated use of titanium tetrachloride. Manual handing was reduced to only at the beginning and end of the production process and, therefore, the need to comply with air quality permits associated with chlorine was eliminated. Operators who were required to wear acid/dust respirators could now wear simple air-purifying respiratory protection at only the first and last process steps (AIHA 2013).

As manufacturing processes are interlinked, altering one process may require changes in many others in the supply chain. This was the case with Maria Yee, a furniture maker in California, who in 2008 decided to switch to water-based finishes that eliminated toxic emissions during manufacturing (Caroll 2009). The problem was that Maria Yee was using a US base solvent supplier Valspar, and the furniture making plant was set up in China. Valspar agreed to replace nitrocellulose lacquer solvent with water-based finishes and supply them to Yee's Chinese factories at what she considered competitive prices. In addition, it committed to helping Maria Yee switch to water-based finishes on all of its furniture. In return, Maria Yee agreed to let Valspar use its Hunan plant to test mass production methods for water-based finishes and to participate in Valspar's customer education seminars on green manufacturing in China.

Another example comes from a parcel delivery company using a hub and spoke supply chain. The company has been using a sorting process for incoming parcels received at local depots, to see which were local and which needed to be sent to the regional hub. At the hub another sorting process separated out parcels to be trucked to other hubs. Parcels at their correct hubs were sorted into truck loads for each local depot. At local depots another sorting process matched up parcels with van routes. Too many expensive manual sorting processes were taking place, and the sorting was inefficient at each stage. The modified process had elements such as a redesign of parcel labels to incorporate human-friendly icons, and a novel feature to improve automatic scanning, and a large investment in buildings and IT to create automated hubs. Labels with sorting code identifying hub and final destination were used for automated sorting. Parcels were sorted directly into containers to allow fast deployment. Vans, designed for quick access to the right parcels, were used. High capacity trucks were used for a fast load/offload process.

Radio Frequency Identification (RFID) is a technology used to help identify, authenticate, track, and trace objects and processes. It is also used to gather and store information about them and their environment. RFID tags can be attached to objects and they can be scanned remotely to collect status information about the object. As RFID helps create transparency, its implications to supply chain design is far reaching. RFID implementation can drastically reduce costs associated with the downstream flow of physical goods and the upstream flow of demand information (refer to Chap. 6).

Consider a power outage that that can be repaired by replacing a component. The operator calls the maintenance center where the engineer must identify the nearest store where the component may be available. If the part is not in inventory, the store must locate it in the common inventory that it may share with other store locations. Finally, the store operator must complete documentation for transfer of the part.

This can be a very expensive and time consuming process. A redesigned process integrating RFID technology would have automatic "access" to the store network for the "automatic picking" of a "specific product" in order to fix a power outage that just occurred.

RFID has revolutionized supply chains such as that of Wal-Mart, which implemented the use of RFID chips in its supply chain in order to increase efficiency. As inventory must be handled by both Wal-Mart and its suppliers, Wal-Mart requires its suppliers to use RFID technology as well. Because of the costs involved, suppliers are reluctant to implement this technology. Wal-Mart's objective is to have product-level RFID tagging for all of its US distribution centers. WalMart requires its suppliers to share the cost and benefit of using this technology. According to Barut et al. (2006), Wal-Mart realized a 16% reduction in out-of-stocks since it introduced RFID technology, and products were replenished three times as fast as items that only used bar code technology.

1.3 Capabilities

Capabilities include customer-management, procurement, production, and fulfillment, as shown in Fig. 1.5. With respect to the emerging paradigms, new capabilities must be emphasized: fund management and social cost in both healthcare and humanitarian relief; rapid fulfillment in humanitarian relief; infrastructure performance, and social norms in emerging economies; and corporate social responsibility (CSR) in sustainability.

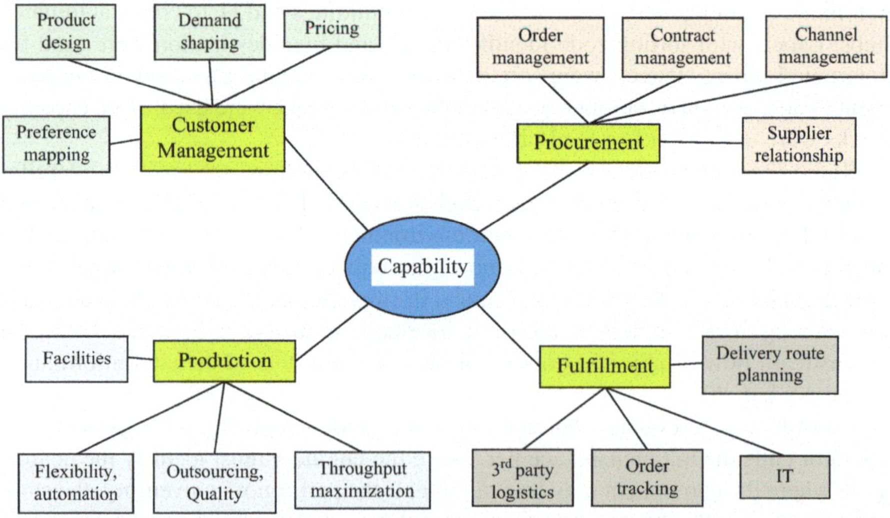

Fig. 1.5 Supply chain capabilities

Restructuring generally involves assessing what the customers want, and then satisfying them by making adjustments in the capabilities of the supply chain. Therefore, at the front end of any transformation would be customer management.

1.3.1 Customer Management

Customer preferences for product attributes, response time, and price may vary widely. Also, in today's "flat" world, a large number of customers may be located in foreign countries with widely varying preferences. Even if customer preferences in any two countries are identical, they may have to be served by different supply chains because the country-specific infrastructure may not be the same. Thus "getting to know the customer" should figure at the top of any customer management effort. One may use traditional approaches such as market surveys and focus groups, but the web based data-capture approaches where one may glean their preferences from the history of purchases is also being used at an increasing rate. Buyers express their value perceptions in many ways – through what they purchase and what they do not, through what they say is important in the buying process, and the steps they actually take to buy.

The strategic decision would be to establish the range of preferences to satisfy, given that it would be prohibitively expensive to satisfy all. There are two related issues: loyal vs. new customers, and domestic vs. global customers. Should one primarily target the loyal customers by attempting to maximize the percentage of a customer's total purchase satisfied? If the number of loyal customers is large, this can be an effective strategy. If not, there would be tradeoffs to consider. The real challenge is to capture a large percentage of purchases of new customers. In terms of restructuring, the choice would be between a supply chain dedicated to loyal customers and one that is flexible enough to serve new customers as well. The location of customers – domestic or global must also be considered. Consider a domestic market with high margin but a GDP growth of 3% and a foreign market with low margin and a GDP growth of 10%. The decision as to which market to target is not simple. If it is also known that the preference in the domestic market is for complex high-end products and in foreign market it is for simple products, the decision becomes simple – target high volumes of simple products for the foreign market. As mentioned earlier, the supply chain for the foreign country will be affected by infrastructure considerations (refer to Chap. 7 on emerging economies).

There could also be a tradeoff within the preference matrix. While most customers who prefer functional products do not care for quick delivery (Fisher 1997), value can be created by breaking this assumption in some cases. Consider home delivery of food, which by all means is a functional product. Yet, as we are witnessing, the internet technology has created a significant value for rapid response food delivery, especially in large cities such as New York, Boston, and Chicago. Clearly, the supply chain for food delivery is very different from the supply chain for mail-order goods.

Product design is a customer management function, as it determines how the external drivers such as product features and quality are addressed in the supply chain. The design function aims at reconciling what the customers want with what the manufacturing and procurement functions can deliver. Therefore, the built-in designed features will be a subset of all desired features, and the actual performance level may be lower than what the customers desire. As the chosen features in a product determine a unique supply chain (refer to Chap. 3 on products and processes), changes in external drivers may spawn a modified supply chain through a new product design.

The design of supply chain can be greatly simplified, if the volatility in demand could be minimized by *shaping the demand*. If there is capacity shortage (excess), demand can be decreased (increased) by increasing (decreasing) price. Sales revenue will depend on the shape of demand curve. Generally, revenue increases and then decreases as price increases from a small value, as revenue is a concave function in price. Hence, revenue is maximized at a specific value of price (and the corresponding demand). If the revenue-maximizing demand is less than the available capacity, it will not be optimal to use up the whole capacity by cutting price further. If the revenue-maximizing demand exceeds the capacity, capacity will be fully utilized but revenue will be less than its unconstrained maximum value.

Peak-time pricing practiced by the utility companies, scheduling of flights by airlines, and appointments at health clinics, are some of other ways of shaping the demand curve, where demand is shifted to the time period when capacity is available. It is difficult to imagine the complexity of an airline's supply chain if it did not use scheduled flights. In that case the airline with its aircraft would wait for customers to arrive, and only if a sufficient number of customers wanting to travel to Boston (say) materialized, the flight to Boston would take off!

1.3.2 Procurement

The procurement capability such as order processing, contract management, and supplier relationship management, must be designed in response to the external and internal drivers.

An *order processing* system captures order data from customer relationship management (CRM), directly from customers, and from other procurement channels, shown in Fig. 1.6. It stores the data in a central database and sends order information to the accounting and shipping departments. It provides tracking of data on orders and inventory for every step of processing. It interacts with other procedures such as pick up, packaging, and delivery to the shipping carrier. For example, an apparel designer with global operations may use an order processing system to capture online orders, and transmit them to a manufacturer in a different country. The manufacturer includes the order details when shipping the product to retail stores. Alternatively, the item may be shipped directly to the customer using shipping information from the system.

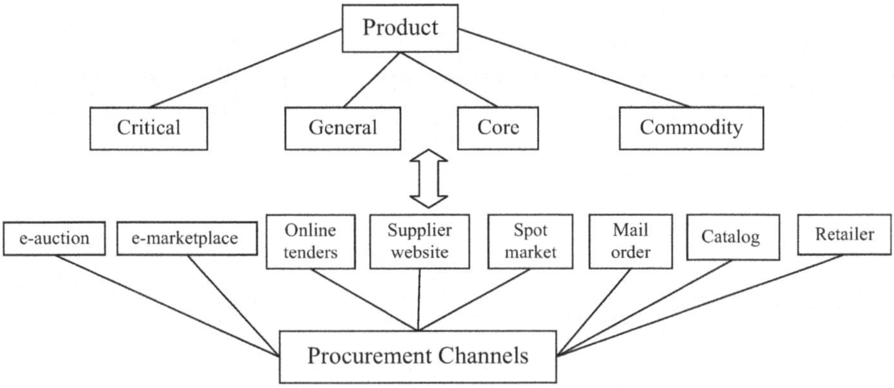

Fig. 1.6 Procurement channels

A supply *contract* is a formalization of the terms of agreement between buyers and sellers. It specifies the expected performance and, in some cases, the corrective and/or punitive actions for under-performance. Supply contracts typically contain clauses concerning decision rights, pricing strategies, transportation, and buyback/return policies. In addition, financial and organizational regulations can be included. Contracts that are based on measurable attributes such as quantity, quality, time, and price, are easy to implement while contracts based on unobservable values such as partner's cost and effort are not.

Companies can purchase components and material from multiple *procurement channels*, both online and offline (Fig. 1.6). Examples include direct purchase from retailers, catalog purchase, mail-order purchase, spot market, supplier's websites, e-marketplaces such as Amazon.com, online tenders, and electronic auctions. There are strengths and weaknesses with each channel. In general, e-procurement has a better fit with commodity products where touch and feel is not important. For complex products such as machine tools it would be preferable to test the product before purchase and, hence, they are not ideal for online purchase. Therefore, with a large product variety, the company may need to work with a large portfolio of channels, each channel earmarked for a set of products. The channel- specific sets of products must be created based on the characteristics of the channel and the characteristics of products: critical, general, core, and commodity. For most companies, building and maintaining a channel portfolio would be a new capability.

Reduction of procurement cost is necessary but not sufficient for competitiveness. One needs to ensure that suppliers are doing their best in terms of innovation in products and processes, quality, on time delivery and, of course, price (refer to Chap. 4 on suppliers). As Toyota has shown, building close relationship with suppliers minimizes uncertainty in most transactions. The idea is to sense and eliminate problems in the supply line, before they happen. The way relationships are structured is a key to setting expectations. Good relationships depend on good communications, and a sense of fairness. As an example, Toyota has found a way to integrate different functions to form a supplier-partnering hierarchy (Liker and

Choi 2004). These include conducting joint improvement activities, sharing relevant information, developing suppliers' technical capabilities, coordinating suppliers' processes, getting suppliers to cooperate with one another, and understanding suppliers' problems and work environments (see Chap. 4). Another example is Ford's hub for designing a world car that included qualified suppliers on the hub with complete transparency.

1.3.3 Production

Production capability includes process flexibility, outsourcing, quality assurance, and work scheduling.

Creating *flexibility* in production processes is a way of coping with the dynamic nature of the business environment today, which requires a wide variation in the associated production processes. Many approaches have been proposed to build flexibility. These include altering the product mix to satisfy need for variety, facilitating smooth changeover from one product to another, altering the volume of production to cope with demand uncertainty, facilitating sourcing of different sets of materials to support product design changes, enabling design modifications to accommodate products with different characteristics, and altering the process plan and rerouting products in the shop floor to overcome bottlenecks and breakdowns.

Production and service processes can be automated by using control systems to replace the human brain and vision, and robotic arms to replace human arms. Vending machine for soft drinks is a good example of a very rudimentary form of *automation*. ATM goes one step further as it also communicates with the customer's account in a bank. In an automated manufacturing shop, jobs on pallets are moved on carts using programmable guidance system embedded on the factory floor. Tools are loaded on machines by robotic arms, and operations such as cutting and drilling are completed by appropriately moving the jobs or the tools. Clearly, the capability to automate is heavily based on the use of advanced IT for designing control systems.

The *outsourcing* capability enables value creation for the company outside the company. That is, the principal gets other companies (agents) to complete certain projects on behalf of the principal, for a certain fee. One may outsource for two reasons: capacity shortage or capability shortage. Outsourcing capability includes identifying suppliers, assessing supplier capability, negotiating contract, transferring product specifications, contract monitoring, and termination.

Printronix, a market leader in designing and manufacturing a broad range of printers, outsources manufacturing of ribbons and other parts to China (Greco 2007). Shipments from China and other Asian countries are brought first to Singapore for printer assembly and then to California. If there is a problem with the product, additional time is required to repair the items on the spot, as returning them to China or Singapore for repairs is prohibitively expensive and time consuming. As the proprietary ribbon represents a core competency, the company chose to move it to Mexico to be closer to its headquarters in California because of the quality issues and the long supply chain from China, mentioned above.

Quality management capability seeks to improve the strength of the attributes of a product so that they meet or exceed customer expectations. This can be achieved by integrating all improvement-related functions and processes throughout the company. Total Quality Management (TQM) looks at the overall quality measures used by a company including quality in design and development, quality control and maintenance, quality improvement, and quality assurance. TQM takes into account all quality measures undertaken at all levels. The capabilities needed for TQM include process monitoring, screening, and redesigning systems to drive the percentage of defectives to approach zero; the so called quality at source. Statistical techniques such as SPC (statistical process control) are used for process monitoring and screening.

A major problem in a production system is that processes can be delayed due to bottlenecks or inappropriate *scheduling* of work and equipment. To increase throughput of a facility (output of the facility per unit time) the following steps are usually considered: reduction of setup time, reduction of processing time (per part), minimization of move time (between work stations), minimization of in-process wait time and its variability, and reducing congestion by increasing machine capacity (Johnson 2003).

To reduce setup time one must reduce the number of setups used and/or the time needed per setup. The drive for reducing time per setup has been led primarily by the Japanese auto companies. Shingo (1985) outlines how this can be achieved through using equipment with short setup times, improving setup procedures, and dedicating workstations to families of parts that use common fixtures. Large batch sizes reduce the number of setups required to complete production. Processing time per operation can be reduced by a better design of processes for the part, using advanced technology, and reducing scrap and rework. Scrap can be reduced by using poka-yoke (fail-safe) devices, and single-piece flow that can quickly detect defective parts and eliminate it. The time required per move can be reduced by increasing the speed of the material handling equipment through the installation of conveyors or other automated systems, or by reducing the move distance required. Move distance can be reduced by reorganizing the layout of departments by forming manufacturing cells. Waiting time can be reduced by controlling transfer batch sizes, processing time variability, arrival variability, resource utilization, and/or the number of queues.

1.3.4 Fulfillment

The fulfillment capability includes delivery route planning, real time tracking of orders in transit, use of communication technology, and management of logistics outsourcing. The fulfillment process starts with order placement, followed by logistics planning for shipment, delivery and pick-up arrangements, and the actual pickup of the freight. After the freight is picked up, it is tracked in transit until it is delivered to the customer. Logistics planning involves preparation of estimates for truck, van, and trailer-usage, and determination of the pickup and delivery locations during the shipment. Shipment cost is estimated based on the number of packages and the total volume of the freight. The Shipment Notice provides a summary of the shipment which includes the shipment number, the order number, the number of packages, the pickup and delivery schedules and whether the shipment is a full truck load. The progress in delivery of orders is tracked and compared with the scheduled pickup and delivery of delivery trucks.

Customers expect goods and services to be *delivered quickly* or when promised. This requires the ability to accurately predict and execute on-time deliveries to the consumer's home or business location. The route plan determines the optimal assignment of shipments to vehicles, and finds the best sequence of pickups and deliveries at customer sites for each vehicle. Given a heterogeneous vehicle fleet and trucks with limited capacities, the route planning problem can become complex. As customers want to know the status of their delivery and the revised time of arrival if the order cannot be delivered as promised, route optimization needs to be dynamic. This must incorporate the consequences of contingencies such as traffic jams and vehicle breakdowns. Clearly, real-time data for arrival and departure times (of transit vehicles) must be used for modifying the route plan. As such, live data-streams from GPS and cellular data services can be very useful in dynamic scheduling (refer to Chap. 5).

Mobile and wireless devices can act as data collection devices for *tracking* information such as delivery status, consumer information, and scan of delivery packages. This information can be used for re-optimizing schedules, re-routing vehicles in real-time, and improving driving and service time estimates. These devices are also important for receiving new delivery instructions, or receiving repair or installation instructions.

Increased complexity of the logistics function has led many companies to outsource logistics activities such as order-delivery to *3rd party logistics* (3PL) providers. There are multiple reasons for wanting to outsource logistics that include, the need to focus on core activities, better transportation solutions, cost savings, customized services, reducing inventory, penetrating markets, gaining the use of sophisticated technology, and avoiding locking capital in warehouses, trucks, containers, and other assets. Another factor influencing the adoption of 3PL services is globalization. As firms expand their markets beyond national boundaries, the need for services such as ensuring compliance with international trade rules, increases. Clearly, companies need to decide which fulfillment functions they would outsource and which they would keep in-house. As a general rule, activities that enhance the competitive advantage of the company should be retained in-house so as to protect the intellectual property (IP).

1.4 Building and Transforming Capabilities

As shown in Table 1.3, the supply chain capabilities can be organized into two groups: those that are geared to managing customers (customer facing), and those that enhance the efficiency of the supply chain (internal). Nokia was more

Table 1.3 Customer facing and internal capabilities

Customer facing capabilities	Internal capabilities
Assessing customer preferences, rapid response, satisfying a portfolio of loyal and new customers, targeting domestic as well as international customers, product design, demand shaping and allocation	Order processing, contract management, managing multiple procurement channels, supplier development, altering product mix, facilitating changeover, rerouting, automation, production outsourcing, quality management, SPC, work scheduling, bottlenecks, setup time reduction, dynamic route planning, tracking orders, logistics outsourcing, recycling

concerned about the customer facing capabilities whereas Anna's Linens focused on improving its internal capabilities. Which capabilities a firm chooses to strengthen will obviously depend on the external and internal drivers it chooses to respond to. While the conventional wisdom is to build core competency on either differentiation – product variety, rapid response etc. – or cost reduction (Porter 1985), today's hypercompetitive world requires differentiation at low cost as a recipe for success.

Consider Dell Computers. With PCs becoming more like a commodity product, Dell decided to compete in the printer market. It clearly needed to build a new capability of designing and manufacturing printers. In addition, it built the capability of high productivity printer cartridges to differentiate itself from HP. HP relied on customer buying cartridges more frequently by offering them low productivity cartridges. As discussed earlier, Dell had to reengineer its distribution chain in China by creating brick and mortar stores to work with their established DTC supply chain. Dell was forced to adopt a portfolio of supply chains, as the emerging markets of China and India became too large to ignore. As discussed earlier, the strength of emerging economy as a market driver has also caused Wal-Mart to reevaluate its business practice in relation to India.

Flextronics International has excelled as a contract manufacturer for electronics that included cell phones. As it worked on contracted volumes, it was sheltered from market uncertainties and had ample opportunity to stand out in its internal capabilities. In early 2000s, it decided to compete as an ODM (original design and manufacture), that required it to assess customers' preferences, design appropriate features in the cell phone so as to satisfy the customer, manufacture it, and sell it to a service provider through an OEM (Huckman and Pisano 2006). In doing so, it had to develop the customer facing capabilities of market-analysis and development of product architecture. In addition, it was now exposed to the market side risk in price and sales volume. The transition was successful, as the company took care to build the additional capabilities.

Zara created a unique set of interlocking capabilities to support its strategy of quick response. These capabilities are shown in blue boxes in Fig. 1.7. They defined quick response as satisfying customer's needs quickly with an existing or a new product. This permitted reduction in inventories of existing products, but required them to keep introducing a stream of fashion-sensitive new products rapidly. They

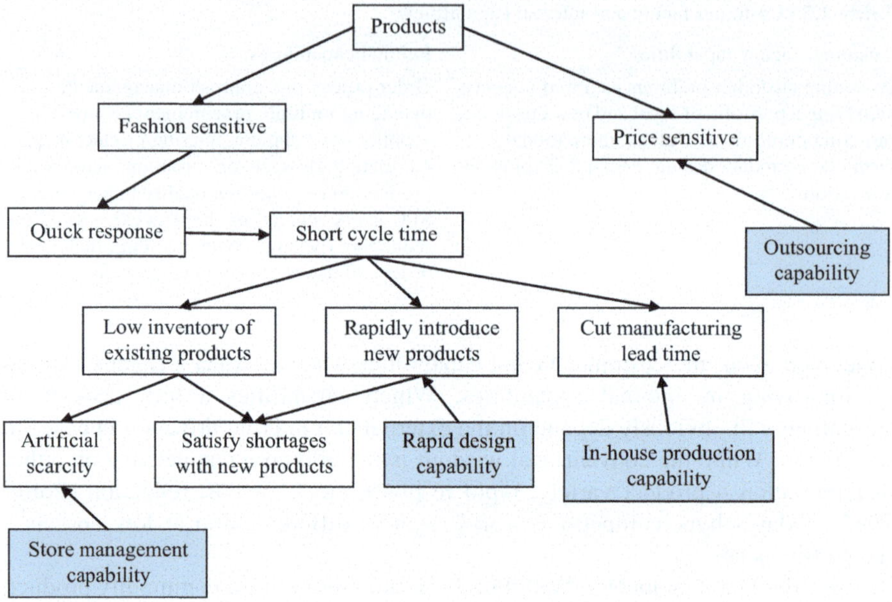

Fig. 1.7 Zara's approach

sharpened their product design capability for speed in keeping with fashion trends. They organized production in two segments: in-house for fashion sensitive products, and outsourcing for price sensitive products. They integrated backward into manufacturing for better control of throughput, and not necessarily for cost reduction. Two production lots were used: a small lot for sensing demand and, a larger lot if the demand picked up after the first production run. If not, the product was discontinued.

Creation of functional capability may require a company to reconfigure organizational designs and relationships, and allocation of people, capital and equipment. Definition of basic tasks and roles of supply chain processes may need major changes. For example, instead of a service station providing the same service to a large number of customers, it may require multi-tasking for a providing a variety of services. To incorporate changes one may require new manufacturing processes, regrouping of machines and tasks, and redeployments of overheads. Restructuring may also lead to new working arrangements with partners. For example, if customization necessitates postponement of procurement, both the logistics partners and the suppliers will be impacted. To the extent the knowledge base embedded in the existing set of capabilities can be leveraged for the new capability, it would be a success.

A new capability may necessitate redeployment of employees, implying training or new hires, and new responsibilities. It may cause friction within the organization if the new hires represent a new culture in terms of technology adeptness, entrepreneurial initiatives, informality, and work habits. It may also lead to changes in organization such as replacement of pyramid structure with flat or network structure. New technology may imply shifts in the supply chain processes. Examples include use of flexible equipment and processes, reengineering for e-procurement, and RFID based supply coordination. New capability may also imply changes in the content as well as the direction of flow of information. For example, enterprise resource planning (ERP) systems mandate increased integration among key activities, including order entry, manufacturing and distribution. Current owners of information would have to adjust to a new reality. External linkages may require determination of business channels that will be supported.

According to a McKinsey survey (McKinsey 2010), the top organizational challenges in building new capability are organizational resistance to change (36%), lack of resources (35%), clear vision or objectives (34%), inconsistent application of processes (30%), insufficient funding (22%), lack of credible metrics (22%), identification of who is accountable for execution (19%), inability to gain attention and buy-in from line managers (17%), lack of senior-management support (16%), and ineffective training approaches (10%).

Building a "totally new" capability may require changes in basic values and objectives of the company. For example for a cost-driven company, building a capability driven by quick response and short product life cycle may expose the company to the culture of rapid changes, and high risk taking, something it may not be prepared for.

Bakhru and Grant (2007) propose a 4-stage capability development process shown in Fig. 1.8: defining the strategic intent, building the capability architecture in terms of systemic and functional capabilities, creating routines or processes that capture functional capabilities, and codifying and standardizing for implementation.

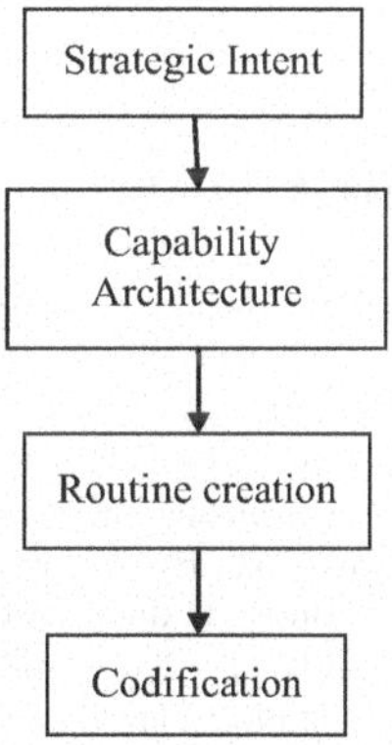

Fig. 1.8 Capability building process

The authors identify the four stages using Self Trade (Bakhru and Grant 2007) as a case study. Strategically, Selftrade's intent was becoming a leading on-line broker through a comprehensive range of products that included share dealing, savings products, and a fund supermarket.

Selftrade's systemic capability was aimed at seamless integration of the customer interface at the front end of the business with order processing at the back office that required regulatory compliance for operating in the securities market. Functional capability comprised recruitment of clients for trading (marketing, providing information, and training the clients adequately), operations (cash functions, stock functions, reconciliation functions, dealing, and call centers), settlement, and compliance.

Creation of routines focused on areas such as opening an account, closing an account, withdrawing money, and making a complaint. Initially the routines were manual. Routines were developed around relatively simple administrative tasks. They identified between 20 and 25 key processes. They then created documentation in consultation with the compliance department.

Tasks such as cash-handling and customer registration were codified in the first instance. Such technology-based processes were characterized by modular structures. Codification permitted standardization which was a prerequisite for automation. Efficiency was especially important for online brokerage. For well-structured routines such as employee appraisals, codification was relatively easy. Over time, Selftrade intended to codify functions related to customer services.

Weerakkody and Currie (2003) discuss the case of a company that wanted to implement a new capability for managing customer complaints through standardization. Ten key business processes were identified as central to this relationship. Several IT systems were developed to support these processes, many of which needed to integrate with the existing system. Of these 60% of the new processes were supported by new systems being developed (without interfacing with the current system), 30% required integration of the new and current systems, and 10% relied on modifying the existing system to match the new processes.

References

AIHA (2013). Case study 11: Chemical substitution; Process containment. American Industrial Hygiene Association. http://www.aiha.org/votp_new/study/chemicalsubstitution.html

Bakhru, A., & Grant, R. (2007). Creating organizational capability in new businesses: Building sets of complementary capabilities. http://www.gsom.spbu.ru/files/upload/niim/seminar/creating_organizational_capability_in_new_businesses-_building_sets_of_complementary_capabilities.pdf

Barut, M., Brown, R., Freund, N., May, J., & Reinhart, E. (2006). RFID and corporate responsibility: Hidden costs in RFID implementation. *Business and Society Review, 111*, 287–303.

Caroll, G. (2009). *Maria Yee Inc.: Making 'Green' furniture in China, Case S1–110*. Stanford: Stanford Graduate School of Business, Stanford University.

Chayet, S., Kouvelis, P., & Yu, D. (2011). Product variety and capacity investments in congested production systems. *Manufacturing and Service Operations Management, 13*, 390–403.

Darnall, N., Jolley, G., & Handfield, R. (2008). Environmental management systems and green supply chain management: Complements for sustainability. *Business Strategy and the Environment, 18*, 30–45.

Ergun, O., Heier, J., Keskinocak, P., & Swann, J. (2010). Waffle house restaurants hurricane response: A case study. *International Journal of Production Economics, 126*, 111–120.

Farhoomand, A. (2006). *"Dell: Overcoming Roadblocks to Growth", Case # HKU575*. Hong Kong: Asia Case Research Center, University of Hong Kong.

Fisher, M. (1997). What is the right supply chain for your products. *Harvard Business Review, 75*, 105–116.

Fleischmann, M., Bloemhof-Ruwaard, J., Dekker, R., Van der Laan, E., Van Nunnen, J., & Van Wassenhove, L. (1997). Quantitative models for reverse logistics: A review. *European Journal of Operational Research, 103*, 1–17.

Geffen, C. (1997). *Innovative environmental technologies in automotive painting: The role of suppliers*. Doctoral Thesis, Massachusetts Institute of Technology, Cambridge, MA.

Geffen, C., & Rothenberg, S. (2000). Suppliers and environmental innovation: The automotive paint process. *International Journal of Operations & Production Management, 20*, 166–186.

Greco, J. (2007). Outsourcing offshore to China – A case study. http://www.articlesbase.com/business-articles/outsourcing-offshore-to-china-a-case-study-189080.html

Huckman, R., & Pisano, G. (2006). *"Flextronics International", case study 9-604-063*. Boston: Harvard Business School.

IBM (2006). Supply chain transformation is becoming a critical element for driving business results. IBM Global Services, IBM Corporation. https://www-935.ibm.com/services/us/gbs/bus/pdf/bcw00617-scm_executive_summit_2006_insights.pdf

Johnson, D. (2003). A framework for reducing manufacturing throughput time. *Journal of Manufacturing Systems, 22*, 283–298.

Liker, J., & Choi, T. (2004). Building deep supplier relationships. *Harvard Business Review, 82*, 104–113.

McKinsey (2010). *Building organizational capabilities* (pp. 1–7). McKinsey Quarterly. https://www.mckinseyquarterly.com/Building_organizational_capabilities_McKinsey_Global_Survey_results_2540

Porter, M. (1985). *Competitive advantage*. New York: Free Press.

Robeson, J., Copacino, W., & Howe, R. (1992). *The logistics handbook*. New York: Macmillan, Inc.

RPE (2010). Supply chain transformation. Retail Process Engineering. http://www.rpesolutions.com/WP_Supply_Chain.pdf

Sheu, J., Chou, Y., & Hu, C. (2005). An integrated logistics operational model for green supply chain management. *Transportation Research Part E, 41*, 287–313.

Shingo, S. (1985). A revolution in manufacturing: The SMED system. Productivity Press. http://books.google.com/books?id=ooXVVIfqEQwC&printsec=frontcover&source=gbs_ge_summary_r&cad=0#v=onepage&q&f=false

Tang, C. (2006). Robust strategies for mitigating supply chain disruptions. *International Journal of Logistics Research and Applications, 9*, 33–45.

Weerakkody, V., & Currie, W. (2003). Integrating business process reengineering with information systems development: Issues & implications. In van der Aalst et al. (Ed.), *Business process management*, Lecture notes in computer science (Vol. 2678, pp. 302–320). Berlin: Springer-Verlag. http://link.springer.com/chapter/10.1007%2F3-540-44895-0_21#page-1

Part II

The Contemporary Supply Chain: Building Blocks and Reconfiguration

Three dominant themes are discernible in transforming the supply chain, irrespective of their drivers: modification of operational control systems, leveraging of emerging technologies, and redesign of the supply chain architecture.

Managing the Customer

2

2.1 Introduction

In a competitive marketplace customers serve as an important source of revenue, which may be lost if the value created for customers is inadequate. Maximizing the fit between what is being offered and what the customers want is therefore imperative. Customer management must include the objectives of creating value for the customer while shaping their preferences, and responding to demand volatility through innovations in technology and processes.

Customers prefer high quality products at low cost. Most customers like a variety of products to choose from, and value convenience in shopping such as home delivery. The uncertainty in customer choices related to quality, variety, and convenience causes the demand to be volatile. Demand volatility may decrease supplier and retailer surpluses, as it forces them to hold high inventory. As a result, companies invest in flexible processes and, at the same time, provide incentives for customers to alter their purchasing behavior. Customer management therefore must include the objectives of creating value for the customer while shaping their preferences, and responding to demand volatility through innovations in technology and processes.

We are interested in customer management in the context of a supply chain. As shown in Fig. 2.1, this can be achieved in multiple ways: adjusting capacity to match demand volatility, influencing demand so that it matches available capacity, targeting inducements to customer segments, consolidating demand for maximizing capacity utilization, creatively using agile and efficient capacities, decoupling upstream production from the vagaries of downstream demand, and improving communication with the customer through customer relationship management.

As conceptualized above, customer management is not just a marketing responsibility of the supplier; it requires coordination of multiple partners in the supply chain with evolving expertise. While demand shaping is more of a marketing role, order decoupling is more of a supply role, and revenue management may involve

A.K. Chakravarty, *Supply Chain Transformation*, Springer Texts in Business and Economics, DOI 10.1007/978-3-642-41911-9_2,

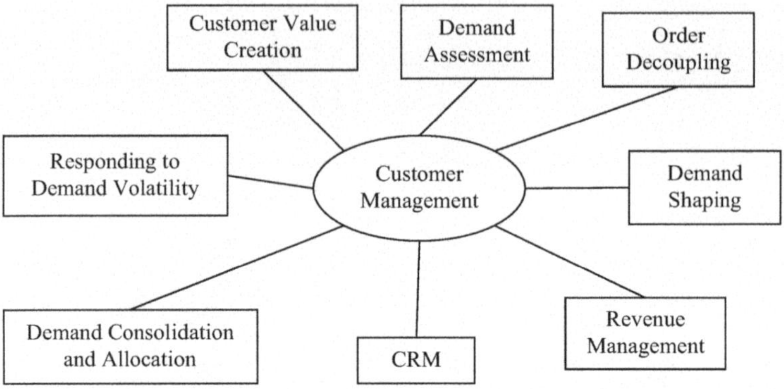

Fig. 2.1 Customer management

third parties. In what follows, we describe the elements of customer management (shown in Fig. 2.1).

2.2 Customer Value Creation

The value created by a company in a supply chain equals the surplus accrued to all individual stakeholders – customers and suppliers – in the company's supply chain. Surplus for each stakeholder is defined as its revenue minus cost. As the revenue of one company in a supply chain can be the cost of another, value creation can be intertwined. Therefore, if value creation were a zero sum game, the stakeholders would operate as adversaries. However, if the total value-pie were to increase, the stakeholders would be mutually supportive. This can happen only if the supplier is interested in increasing customer's value and vice versa. Conversely, to increase the total value, the supply chain would need to increase benefits to customers, lower the costs of its suppliers, use its resources more effectively, and develop relationships with suppliers and customers in innovative ways.

Consider a computer manufacturer that makes two types of computers. Assume three customers with willingness to pay (WTP) of $1,500, $1,800, and $1,900. As the total demand of these three customers exceeds supply of two computers, the manufacturer will increase price until the customer with the lowest WTP is driven out of the market. At that point the price of the computer will be set to $1,500. At that price point the value created for the remaining two customers will be $300 and $400, respectively. Assuming the manufacturer procures parts from suppliers at $1,050 for each computer, the manufacturer obtains a value of $900 {=2 (1,500 – 1,050)}. Assuming the suppliers' cost of producing the parts equals $500, the suppliers' value equals $1,100 {=2(1,050 – 500)}. The total value created in the supply chain is $2,700 (= 1,900 + 1,800 – 2*500). Thus, value is created for all three parties: $700 for customers, $900 for the manufacturer, and $1,100 for the suppliers. Obviously, if the manufacturer could negotiate a lower price for the parts, his value would increase and the suppliers' value would decrease.

The real world problem is much more complex for several reasons: customers purchasing from other competing manufacturers, a large number of customers with uncertain buying habits, raw material price uncertainty, and the uncertain manufacturing lead times. Therefore, the challenge each partner faces is to maximize its share of value, while increasing the total supply chain value.

Broadly speaking, the approaches for creating customer value include: designing products that customers like, satisfying customer's total needs through a consortium of suppliers, offering ad-on services to facilitate use of the product, and providing advisory services so as to minimize customer's cost in using the product. These are shown in Fig. 2.2.

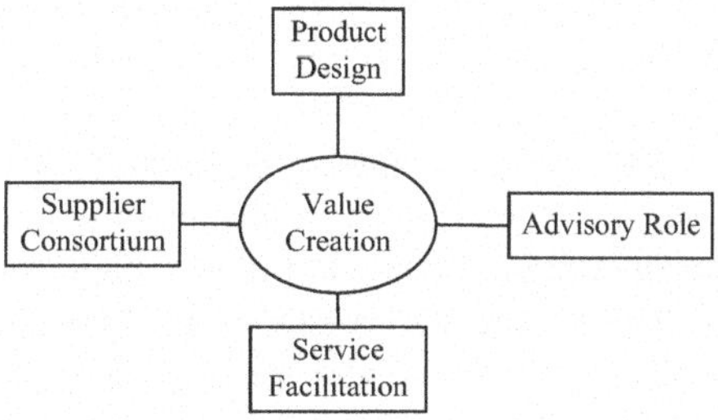

Fig. 2.2 Customer value creation

In the digital age, with the full power of digital technology at its disposal, companies can create customer value by manipulating time, space, and matter, more creatively. Digital products require little physical space, can be "transported" instantaneously, and are not comprised of physical substances. As Pine and Korn (2011) explain, FedEx lets customers access the physical whereabouts of any package instantaneously; Starbucks enhances the value of its business premises through Wi-Fi access to virtual places; and Lego enables consumers create models with bricks in the virtual space which can then be recreated in the real space using real Lego bricks.

2.2.1 Product Design

We may define the value of a product or service as the *ratio* of the perceived value (by customers) and its cost of ownership (Christopher 1996). The total cost of ownership includes the purchase cost, usage cost (maintenance, training, system integration, modifications, etc.), and disposal cost. Obviously, a reduction in the cost of ownership can be achieved in many ways. First would be the product-design that should retain overall fit with customer's preferences even if they change somewhat (robust design); it eliminates the cost of having to realign it frequently. The product should be simple to use, eliminating the cost of learning and training of users; and it should be inexpensive to maintain, reducing the cost of utilities, updates, and modifications. For example, the cost of implementing the ERP system can be huge – several times the purchase price of the software – as it must be integrated carefully with the customer's current system. Dell Computer spent millions of dollars on an ERP system only to scrap it because the system was too rigid. ERP failures include Boeing, Dow Chemicals, Applied Materials, and Kellogg's (Chen 2001). Note that a judicious choice of manufacturing processes and material-use, while designing the product, increases efficiency and can substantially reduce cost.

Perceived benefits of a product or service to a customer can be very different from what the manufacturer may claim to be. Customers buy products and services to satisfy different needs. For example, a marine biologist who buys a watch that works under water has different expectations than that of a person who uses a watch for time keeping. Both watches tell the time, but they also offer other benefits that differ widely. Perception of benefits therefore is largely dependent upon whether the product satisfies the immediate needs of the buyer as well as whether it possesses other built-in features that the customer may need some day. If the variability in customer preference is high, one approach for managing them would be to group them in segments based on similarity of preferences, and offer product features targeted to specific customer segments.

The manager of a truck manufacturing company was shown a screw that cost five cents to make and sold for 30 cents. He asked "What happens if that screw breaks?" The workers explained it was a vital part of the engine and that without it the truck wouldn't work. The manager then proposed to raise the price of that screw to $100. The employees were appalled, but the manager explained that the manufacturing price should not determine the product price; the perceived benefit of the clients should.

Involving the customer in the product design process is known as co-creation. Customers can participate at just any stage of the value chain including production and distribution. Therefore, managers must look for novel approaches for involving customers as a way to successfully monitoring and managing them. The Fiat of Italy implemented co-creation by inviting potential customers to visit its Web site and select from an array of features (Kambil et al. 1999). Fiat was able to capture valuable insights into consumer preferences, and test different design concepts at low cost. More recently, Toyota's website allowed users to fully customize car design from a few basic models and it has seen success with its Scion line doing so (Forbes 2012).

Apple, Amazon.com, Cisco Systems, eBay and other electronic commerce innovators often partner with their customers to *co-create* value and capture greater value than what either party could have done independently. Gateway satisfies the customer need to sell or dispose-off obsolete computers by allowing customers to upgrade their computers easily. Dell adds value through customization by selling computers that are built to order. Computer manufacturers like HP are meeting the need for convenience in the integration process by opening web stores that facilitate purchase of related accessory-products that customers can use with their computers.

2.2.2 Supplier Consortium

It is possible that a company may not possess capabilities to satisfy the set of all needs of its customers. If so, it can form alliances with other companies with complementary skills, and capture customer's order. Nagarajan and Sosic (2009) provide multiple examples of supplier coalitions from different industries. They cite Stallkamp (2001) who discusses consortium formation between Delphi (auto part manufacturer), Lear (interior trim manufacturer), and the suppliers of wiring, carpets, and molded plastic, to satisfy the total cockpit needs of one of the Big Three auto manufacturers. Similarly, the alliance between Symbol Technologies (scanner manufacturer) and Paxar/Monarch, a leader in bar code labeling enabled them to supply a bundle that included a bar code labeler and scanner and thus win valuable customer orders.

In the travel industry, customers routinely need airline tickets, hotel rooms, and rental cars. An intermediary such as Expedia.com creates bundles of these services to match the demand of different customers. In Fig. 2.3a Expedia operates as a consolidator for three types of services, and obtains volume discounts for customers for specific service bundles. Demand consolidation reduces variability that the vendor needs to handle. A variety of bundles is possible: bundling several services, bundling by service type, and bundling by regions of the country. These bundles are high margin products for companies like Expedia, and real-time bundling based on previous click data allows online travel sites to increase conversion rates. In Fig. 2.3b the customer negotiates for services individually with each vendor.

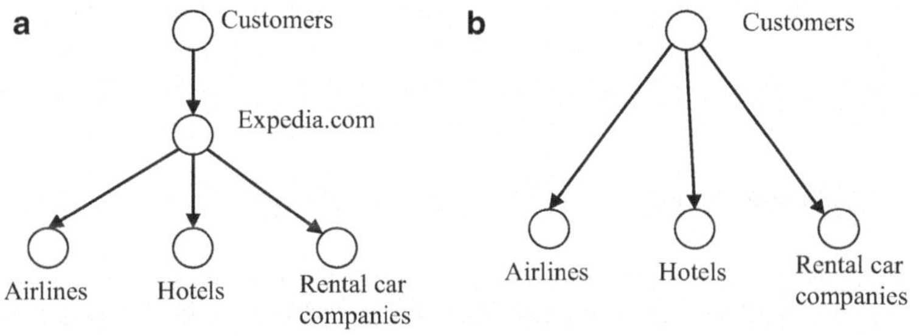

Fig. 2.3 (**a**) Supplier consortium. (**b**) Independent suppliers

The Ontario Motor Coach Association represents several alliances between bus operators, hoteliers, and retail outlets in Canada, and it sells service packages to tour operators, who in turn sell tour packages to customers (Nagarajan and Sosic 2009).

Consolidators or system integrators play an important role in facilitating supplier alliance. They add value in the supply chain in multiple ways: reduce the number of transactions, win customer orders, decrease the cost of distribution, and increase search efficiency.

Consider 10 manufacturers each supplying one item to 100 retailers. If each manufacturer sells directly to the retailers, there would be 1,000 transactions.

The complexity of distribution plan can be reduced by adding wholesalers as consolidators between manufacturers and retailers. With a single consolidator, the number of transactions is reduced from 1,000 to just 110 – ten transactions between the manufacturers and the consolidators, and 100 transactions between the consolidator and the retailers. Reducing the number of transactions enhances efficiency by eliminating duplicate efforts in such processes as ordering, processing, and shipping. An added benefit is that they help reduce cost, as transactions can be standardized in terms of lot size, frequency of delivery, payment, and communication. Seller and buyer no longer have to bargain over every transaction. With routine transactions, the costs associated with transactions decrease.

Consolidators provide assortment services (goods and services). Typically a manufacturer produces large quantities of similar products, while consumers want small quantities of many different products. In order to smooth the flow of goods and services, consolidators perform such functions as sorting, aggregation, allocation, and creation of assortments. In sorting, consolidators organize supplies of different items from different suppliers into similar groups, as exemplified by graded agricultural products. Aggregation means that consolidators bring together items from a number of different sources to create a larger supply for their customers. Finally, they select products from different sorted groups (allocation) to build up an assortment of products to give their customers a wider selection.

The use of consolidators also aids the search processes of both buyers and sellers. Customers must search for certain products and services, while producers search to determine their customers' needs. The degree of uncertainty in both search processes can be reduced by using consolidators. For example, consumers are more likely to find what they are looking for when they shop at retail institutions, organized by the lines of trade such as grocery, hardware, and clothing stores.

2.2.3 Advisory Role

A customer centric organization lays a high priority on working with customers in an advisory role to define the needs of the customer. The notion of customer-centric can vary. In an emerging economy with a high GDP growth rate it may imply just filling large customer orders on hand, while in a mature economy that is sensitive to the unique needs of customers, companies must also be cognizant of the cost the customer may incur in using the product or service. Companies operating in such mature economies are realizing that value cannot be created simply by focusing on the product, however unique or well-engineered it might be. Unfortunately, most companies today are narrowly focused at transactions that are routine and mundane. As a result they develop little insight into what motivates the customers to buy their products and services, or what their future needs would be (Borg 2003). It also means that suppliers must understand the product in a contextual sense: what the product is used for, and how that use can be satisfied better. This can happen only if companies work closely with their customers in solving their short and long term issues. The key is in striking the right balance between the value of the planned level of customer satisfaction and the cost of "delivering" that level of satisfaction. In the digital space the latest trend is the Lean Startup approach for creating products – adapted from Toyota's lean/agile methodology (Ries 2011). This employs a build-test-learn approach whereby the company gets a minimum viable product in the market and learns how to improve it based on customer interactions, in real time.

Customer-centric companies develop a holistic view of each customer's life-cycle needs (perceived benefits as well as the cost of providing it) as shown in Table 2.1.

Table 2.1 Customer centric versus sales centric

Characteristics	Sales-centric	Customer-centric
Customer orientation	Transactions	Work with customer to solve both short and long term issues
Solutions orientation	Multiple use of products off the shelf	Bundles that combine product, service, and knowledge
Customer interface	Focus on product economics	Focus on customer economics
Value creation	Product features, new applications	Customizing for total solution
Business processes	One size fits all	Tailored business processes
Organization	Profit center	Empowerment for customer satisfaction
Process focus	New product development	Customer relationship
Metrics	Market share	Customer share

Frank et al. (2007) discuss the case of Breeze2 meter from Bayer that earned the ease-of-use commendation from the Arthritis Foundation. The Breeze meter helps make monitoring easier for people with diabetes. The Breeze2 meter uses a small sample size of one micro liter, and produces results in just 5 s. It uses a unique disc that eliminates the need to load, unload and handle tiny test strips, which often can be a challenge for people with arthritis who test their blood regularly. There is also no coding required with the Breeze system, which means fewer steps in the testing process, and it also helps ensure accurate readings. As another example, Eaton Corporation, a diversified power management company, teamed up with Caterpillar to integrate its switches with Caterpillar's generators so as to provide its customers with a complete electrical power backup system (Borg 2003).

Companies in many industries have migrated from selling "off-the-shelf" products to customizing solutions. They do so through product bundles that can deliver tailored solutions to client-specific problems. Customer-centric companies even work with competitors in alliances to fulfill customer needs. For example, GE, instead of manufacturing its own jet engines, now sells solutions for reducing life-cycle costs to airlines; it also services engines of its competitors as it does its own (Booz Allen 2004).

Leading banks have increased penetration of the mass market banking segment by developing innovative "advice bundle" and technology-enabled tools. Citibank employs a financial planning tool called Citipro that generates customized Citibank solutions for clients, based on their financial information such as account statements, tax records, and insurance (Borg 2003). Citibank gained an early advantage in securing a larger share of customer orders. Suppliers may purchase financial options such as the option to sell an asset at a pre-determined price (put option) to hedge against the risk from uncertainty. A structured certificate of deposit (CD) that partially protects the investor's principal is an example of a bank sharing market risk with the customer.

Amazon.com has led the way in deploying artificial-intelligence to discern patterns in customers' book reading habits. As new books emerge, the company matches attributes of individual books with customers' revealed reading preferences, and alerts customers of the new book, if a match exists. This also acts as an effective way of neutralizing the negative effects of demand variability on Amazon, as the company develops advanced intelligence on what the customer will be likely to order, and will have adequate time to get ready. Netflix and other companies have recently followed suit, utilizing big data mining for explicit and implicit customer signals to recommend them relevant products online.

Helping cut customer's cost is a proven way of winning customer order. This can take multiple forms including implementing vendor managed inventory (VMI) to simplify customer's procurement process, sharing risk with customers through buy-back contracts, and redesigning products to avoid customers needing to modify it for their use. Vendor managed inventory, where the supplier is responsible for replenishing customer's inventory, can significantly reduce the customer's cost of warehouse management. The customer needs to make inventory status information transparent to the supplier.

North Coast Electric (NCE 2012), the largest distributor of electrical materials in the Pacific Northwest and Alaska, implemented VMI to manage approximately 25% of its inventory. Currently, 13 of their manufacturing suppliers also use a wide variety of VMI systems in managing inventory replenishment. In implementing VMI, NCE had hoped to achieve two main goals: higher inventory turns and improved fill rates. As expected, NCE is realizing substantial business benefits in both these areas. Today, they have increased inventory turns by 20%, which puts them considerably above the industry average. This is even more impressive as they have also achieved 40% increase in sales, while the inventory to support those sales only increased by 18%.

Most products are manufactured with standardized specifications. As mentioned earlier, a customer with unique needs may end up modifying the product at own cost. The Baltimore based Zentech Manufacturing did it differently for Sleep Solutions, a national sleep diagnostic provider located in Glen Burnie Md (Zentech 2012). To be consistent with the FDA approved existing hardware of the customer, Zentech reassigned an existing product from a tier-1 manufacturer. Zentech also embedded software to utilize current component parts to ensure improved device functionality.

2.2.4 Service Facilitation

Facilitation services are a subset of complementary features that enhance the probability of selling the core service. Examples include "drive-through" windows at banks and fast food restaurants, home delivery of pizza and groceries, the ticket-master, and the iPhone's apps store. Some facilitating services may be so important that they become independent of the core service. For example, some airline flight reservation systems earn a significant portion of the airline's revenue, although they might have been developed originally as facilitating services. Sabre Holdings that started as a reservation system for a major airline, is now a key technology company on its own right.

British Airways provides another good example (Kolesar et al. 1998). In addition to the normal services provided by an airline, British Airways provides "travel clinics" for full medical and immunization advice. In addition, passengers can secure hotel and car rental arrangements. On the day of departure, British Airways passengers can take advantage of British Airways Express Car Parking service. When a traveler drives to Heathrow Airport, a British Airways agent will take his car and park it; when the passenger returns, his car will be waiting. From pre travel planning to the travel itself, British Airways has identified other services that its customers require. In fact, ancillary products are the fastest growing product lines within many travel companies and encompass everything from airport transportation to travel insurance.

Supplying packaging material is a commodity business, but a company that specialized in this found a way to create new values for its customers. The problem faced by the customer was one of leakage, with a host of negative implications such as increased returns, wastage, and end customer dissatisfaction. The cause of this problem was a change in the chemistry of the product. The supplier's solution was a modest change in the packaging materials and the filling process. The solution required the customer to make a few minor process changes, but she managed to hold the price line and launch her new product on time. The supplier, understanding how the customer used packaging to protect her product, was able to reengineer his offering to achieve the required level of protection with a simpler packaging solution.

The one-stop shopping option can be a powerful incentive for a customer, as it minimizes the inconvenience factor. Consumers increasingly prefer to concentrate a substantial part of their purchases with a single retailer. According to a study conducted by the UK Competition Commission cited in von Schlippenbach and Wey (2011), the ability to shop one-stop determines the choice of store to a large extent. The report states that customers spent 85.3% of their overall expenditures on purchases at major supermarket chains. While a single item shopper engaged in frequent shopping and bought only one of the goods per shopping trip, the one-stop shopper bundled her purchases in a single shopping trip, and economized on shopping costs. The buying decision of one-stop shoppers, therefore, depends on overall expenses rather than on individual product prices. One-stop shopping behavior results in higher wholesale prices than if suppliers operated separately.

Examples include energy markets, which traditionally have been monopolies but are now open to varying degrees of competition. Consumers are often able to buy their gas and electricity from a single supplier. Similarly, consumers nowadays can often source telecommunications, cable television and internet services from a single supplier.

The reason companies fail in their efforts to shift from a product focus to a customer focus is the organizational complexity that may result. Mi Adidas tried to customize shoes by building a layer of specific one-off customer-focused processes on top of old product-focused processes, which created a havoc in its production system. Companies may organize their basic products with stable demand through efficient and least expensive processes. Less predictable and complex elements can be diverted to customized processes. This technique isolates complexity within the system and minimizes its costs.

2.3 Assessing Demand

Because of the uncertainty in customer demand and supplier deliveries, manufacturers and retailers find themselves holding huge inventories at high cost. Proficiency in demand assessment is absolutely critical in producing the right amount of product at the right time, developing the right relationships with suppliers, and forging a mutually beneficial relationship with customers. As shown in Fig. 2.4 demand sensing is the process of assessing the total demand from multiple sources including the point-of-sale, harnessing the voice of customer, using statistical forecasting, and collecting market intelligence. The assessed demand must then be translated to obtain demands of individual components comprising the system.

Fig. 2.4 Assessing demand

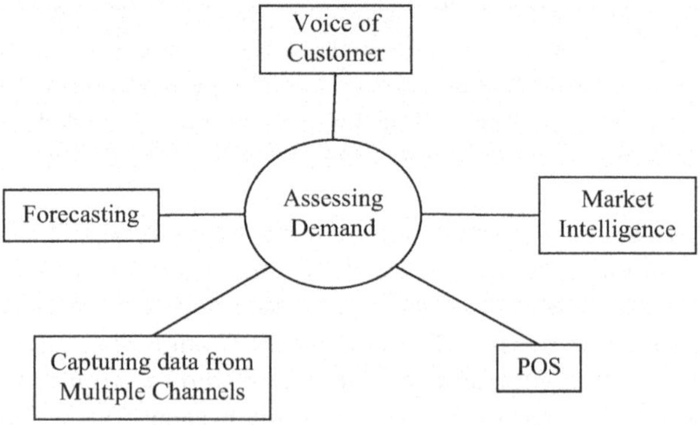

Unfortunately, most companies still rely on past-usage as a proxy for future demand. Such approaches suffer from deficient end-user demand data as it may hide the causes leading up to this data, ignorance of customer buying patterns and market potential, and inability to access actual demand at the point-of-sale. Clearly, the demand assessment approach needs to be broadened.

An AMR study of 300 North American companies found that 56% of the companies took more than 2 weeks to sense true channel-demand (Cecere 2006), and only 26% could see channel demand in days or hours. In this context, the database of demand signals by attribute type can be a rich source for demand modeling approaches (Cecere 2009). Essentially, a demand model creates effective links between the sources of market data and customer buying behavior. Such links are not static and, therefore, to obtain a better sense of demand one must track the underlying attributes. Demand assessment must also include procedures to tie forecasts to shifts in demand insights.

Demand response capability must ensure customer satisfaction and increase supplier's profit. Companies must first convert independent demand of products to determine the dependent demand for location-specific products. Failing to convert demand to each location may lead to over or under stocking cost (see Tesco's problem below). It may also incur unnecessary shipping expenses. Dependent demand focuses on demand at each location, demand in the factory, demand for specific parts for the suppliers, and the demand for logistics within the entire network.

The competitive driver for Tesco, a large UK-based grocery chain is product availability (EIU 2009). But determining product availability is a challenge, as Tesco has operations in 14 countries with six different store formats, and more than 50,000 products on its shelves globally. It is not enough to know how many chicken legs are sold annually; managers must know how many were received and sold by day at each store location. Clearly, increasing inventory across the board is not a solution, as there would be shortages and excesses. In addition, grocery items such as fruit and dairy products that are highly perishable can lead to waste and loss of revenue.

Thus, to ensure effective demand response, companies need to build capabilities for the following: ensuring transparency between sales and operations (S&OP) and the daily consumption, translating the dependent demand to material requirements, and keeping major component suppliers informed about production requirements.

2.4 Coping with Demand Volatility

Companies can take a number of actions to cope with demand volatility. These include, as shown in Fig. 2.5, holding inventory, building excess capacity, using flexible technology, and managing risk with financial securities.

2.4.1 Inventory and Capacity

Inadequate inventory causes delays to customers, resulting in negative utility. Therefore, a popular way to cope with demand volatility is to hold inventory, which can be expensive. Inventory acts as a hedge against demand volatility as suppliers can dip into it in times of high demand. The used items are made up and inventory is built up to its original level during periods of low demand, thereby

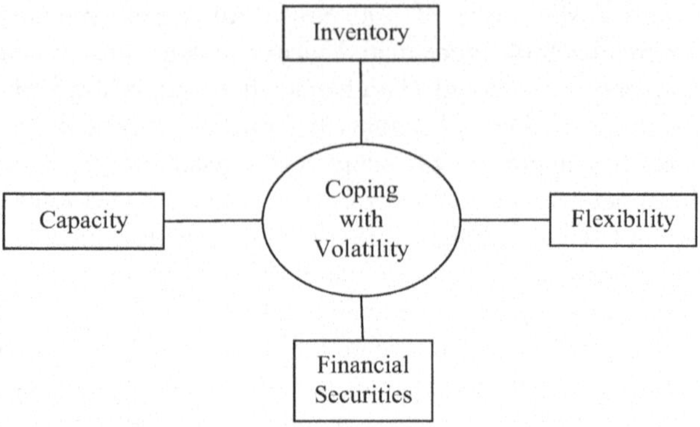

Fig. 2.5 Coping with demand volatility

increasing product availability. Inventory can become unavoidable in many scenarios even without demand volatility – uncertain production and transportation lead times, and uncoordinated flows between successive stages in a supply chain. Therefore, insufficient inventory can cause lost sales and, in some cases, financial penalties (Silver et al. 1998). It follows that reducing inventories in an uncontrolled way can adversely impact customer service. Hence, setting the correct customer service target (hence, inventory level) is a critical dimension of supply chain competition. However, investment tied up in inventory acquisition and warehousing can be a significant part of total cost. In addition, inventory may "hide" problems that should be solved by other means such as setup time reduction and close relationship with vendors. In certain settings such as car rentals, companies may share their individual inventory of cars with others as a coping mechanism.

Conceptually, excess (or buffer) capacity acts to cushion the upswings of demand just as inventory does. However unlike inventory, excess capacity is not consumed. Although building excess capacity is more expensive than holding inventory, its reusability can make it attractive. An alternative is to use contingent capacity that can be adjusted up or down depending on need. Use of additional production shifts, overtime work, and outsourcing are some of the options for providing contingent capacity. Holt et al. (1955) discuss the tradeoff in coping with fluctuations in demand using outsourcing, changing productivity, and inventory. Indeed, this idea of trading off capacity with inventory is the basis of aggregate production planning.

2.4.2 Aggregate Plan

The aggregate production plan is a proactive plan that effectively utilizes the organization's resources to satisfy demand as forecast. It determines not only the planned output levels but also the appropriate mix of resource inputs to be used. Inputs include inventory levels, back orders, subcontracting, and the available workforce. The outputs are a master production schedule, raw material usage, shop floor loading, and cash flow. The aggregate plan is then disaggregated to determine the schedules for components. There are three strategies that can be followed in any aggregate plan: level strategy, chase strategy, and a mixed strategy. Level strategy aims at maintaining a constant production rate with a constant workforce size, accumulating inventory to cushion the periods with higher than normal demand (as forecast). The chase strategy, on the other hand, plans to produce exactly the amount forecast. Therefore, production rate must change with forecast, necessitating frequent hiring and firing of workers. Mixed strategy, as the name suggests, combines the above two strategies. Thus, instead of altering production rate every week as in chase strategy, the mixed strategy may allow production rate changes once a month or even less frequently. It is more flexible than the level strategy, which may have a constant production rate for the entire year.

Unlike manufacturing, services cannot be inventoried or transported. Therefore, industries such as banking and hospitality depend on their ability to adjust capacity and/or build excess capacity as the mechanisms for coping with volatility. As in the case of inventory sharing, capacity may be shared between companies. Airlines with infrequent flights in and out of a particular airport may share gates, ramps, baggage-handling equipment, and ground personnel with others.

2.4.3 Flexible Technology

Flexible manufacturing technology can be used for coping with demand volatility. Flexibility is obtained through general-purpose machines and equipment, capable of performing multiple tasks with negligible setup time. It provides process flexibility that enables rapid switching of models to cope with product proliferation. Such a system must be supported by advanced IT for controlling the movement of parts, and loading of right tools at right time in the tool banks. A manual flexible system can be obtained with cross-trained workers. Flexibility must also be built in the use of raw materials and, therefore, vendors supplying raw material and components to the firm must be proficient in rapidly adjusting to variations. The firm may also purchase an option for placing a second order in a period, if the actual demand is high.

2.4.4 Financial Hedging

Derivative instruments for risk management have been used widely in the financial service industry. Companies are now beginning to use options to hedge against volatility in consumer demand. These resemble the call and put option contracts in the financial market. However, unlike financial options that are based on the assumption of a perfect market, pricing of supply contracts is established through private negotiations between two firms over prices and quantities (Cheng et al. 2003).

The call option in a supply chain works as follows. At the beginning of the period, the retailer places an order of quantity Q, paying a price W for each unit. She also purchases q (call) option contracts from the supplier (at a price c per contract). She now has the right to purchase up to q additional units at the end of the period after demand is realized, at a cost w per unit (exercise price of the option). Under this arrangement, the supplier is committed to produce $Q + q$ units, but can salvage any unexercised options at the end of the period. Thus, the call option transfers some risk from the buyer to the supplier. In scenarios where production lead time is high the buyer must place her order early, so that some uncertainty still remains when the call option is exercised. In this situation, a combination of call and put options can be employed (Erkoc and Wu 2005) that incorporates a put option when the call option is exercised; the buyer has the right to return excess orders to the supplier after all demand uncertainty is resolved.

Firms are also experimenting with insurance plans such as weather-related derivatives for protection against demand volatility (Sumitomo Group 2002). The idea of using derivatives, a financial product, was based on the realization that if part of such risks could be traded in the markets to hedge a wide variety of risks, the markets would be capable of absorbing even such large potential losses. In 2002, Mitsui Sumitomo Insurance issued a weather-derivative contract to a soft drink wholesaler based on the number of hours of sunshine, which is correlated to the demand for soft drink. If the number of sunshine hours fell below a certain threshold, Mitsui Sumitomo Insurance would pay the company a pre-determined amount to cover the losses. Mitsui Sumitomo Insurance has also developed, small-payout weather-derivative products in fixed forms for smaller businesses, and partnered with some 50 regional banks to sell them. Payouts from weather derivatives, however, are based on an agreed, pre-determined amount, which means the products can be designed more flexibly than insurance policies. In addition to temperature, other weather conditions sometimes specified in weather derivatives include rainfall, snowfall, and wind.

2.5 Demand Consolidation and Allocation

Aggregation of demand from two or more sources can lead to better utilization of capacity (Gerchak and He 2002). As shown in Fig. 2.6, demand consolidation can take multiple forms that include use of distribution centers, product bundling, demand aggregation (over a wide "window"), combining customer deliveries as in home-delivery, and online selling that combines customer orders.

The intuition is that without aggregation, if demand in one location is higher than expected it may run out of inventory even if there are extra units available in the other location where demand is lower than expected. Thus, if the inventories for the two locations are consolidated, high demand at one location can be offset by low demand at the other location. If demand at both locations is high, there may be shortages even with consolidation. Clearly, consolidating demands that are negatively correlated, leads to greater savings.

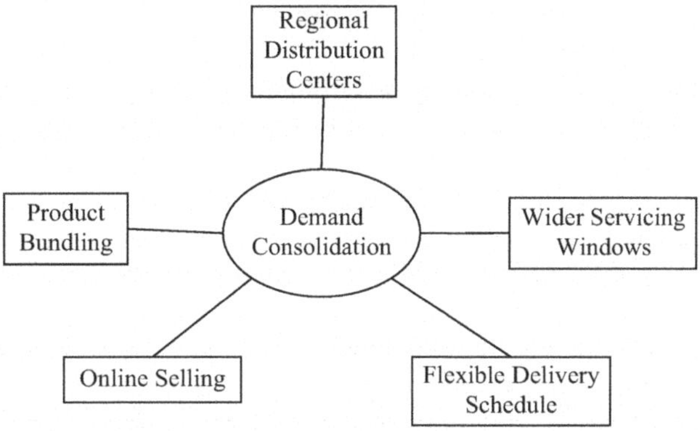

Fig. 2.6 Demand consolidation

Consolidation can occur in three different ways: by location, by time periods, and by products. Consolidation by location has resulted in companies setting up regional distribution centers, and serving multiple cities from the same distribution center. Staples has used a centralized distribution network (Gourley 1997). Disney Stores uses a dedicated Central DC (CDC) in Memphis to supply more than 100,000 types of products to 360 stores (Jedd 1996). Benetton uses one CDC in Ponzano, Italy, to serve over 6,000 stores in 83 countries (Dapiran 1992).

Aggregating demand across several periods allows cost-effective smoothing of production and consolidation of orders. For example, use of wider service windows permits aggregation of customer demand, reducing *variability* of orders. Cisco Systems often uses a 21-day delivery window despite the fact that many orders can be filled in 10–15 days. It achieves almost perfect order fulfillment with more time to fill each order. As delivery-window is not a competitive differentiator for Cisco Systems, increasing it from 10 to 21 days does not hurt them. Another example is flexible delivery commitments that enable reliable and cost effective demand fulfillment. The online retailer Amazon.com consolidates orders to optimize picking, packing and shipping operations by quoting a flexible service window of 5–9 days. Dell consolidates parcel packages for each of its hubs and uses zone-skipping freight rates in conjunction, to reduce logistics cost.

Consolidation by products is possible if they are complements. In the aviation industry it has led to the formation of horizontal alliances and code sharing. Code sharing allows an airline to sell tickets on a partner airline on routes where the partner provides connecting services. There are many examples of code sharing alliances that include United and Lufthansa, JAL and British Air, and JAL and United. Retailers in food and software industries frequently offer bundled products (Stremersch and Tellis 2002; Chakravarty et al. 2013). They may bundle two products with relative ease without altering the identity of individual products, and selling them together as a "pack". McDonald sells both drinks and burgers together at a discount. Amazon.com offers to sell bundled books at a discount price.

2.5.1 Demand Allocation

Demand allocation, in a sense, is opposite of demand consolidation. Consider demand allocation to two "buckets" – an amount Q produced in advance with a view to exploiting the scale economy, and an amount $D - Q$ produced in real time after observing the actual demand D. This is an example of postponement of purchase for the volatile part of demand. Clearly, the Q units can be allocated to a supplier that produces in mass volume at a low unit cost (it aggregates orders from several customers). However, as such suppliers are usually located in a low cost area (probably overseas) they require long delivery lead times. The amount to be produced in real time must obviously be allocated to an agile supplier, located in a close vicinity. As the real time order would vary from period to period, the agile supplier would charge a premium for responding to the variations. For example, if the demand for a product varies between 1,500 and 1,800. It would be risk free to produce the first 1,500 units in advance, which is the stable part of demand. Therefore, if the observed demand is 1,700, only 200 units need be produced in real time. It follows that the agile supplier must put together capacity in a reactive mode to satisfy the demand surge of 200. Demand allocation strategy is shown in Fig. 2.7 below (refer to Chap. 4 on supplier management for further discussion).

Fig. 2.7 Demand allocation

	Stable demand	Demand surge
Agile supplier		Reactive capacity
Low cost supplier	Known capacity	

Order splitting is another form of demand allocation and it is used when very little or no advance information on demand is available. In phase 1, retailer purchases a small lot to be sold in his store to observe how customers respond; the store essentially acts as a laboratory for demand assessment. If the product sells well, in phase 2, the retailer places a larger second order. If not, the sale of the product is discontinued. Clearly, this policy can succeed only if the supplier is able to deliver the second order quickly in phase 2. Because of this, the supplier is likely to charge a premium for the second order. Zara, as discussed in Chap. 1, uses a 2-period split order to sense demand, and it acts as its own supplier to produce the second order. This way it can introduce new products in rapid succession.

2.6 Order Decoupling

Using the principle of order decoupling (order penetration point) is one way of responding to demand volatility in real time. For example, in a build-to-order (BTO) environment the supplier builds the product from scratch (component manufacturing and assembly) according to customer's needs, thus providing a fully customized product. There is however a finite lead time for the completion of the manufacturing and assembly processes, and the customer must be prepared to wait. In contrast, in a make-to-stock (MTS) environment the customer has no choice but to buy what is available in the finished goods inventory, but her wait time is almost zero.

From the supplier's perspective, BTO is risky if his supply chain is not decoupled from demand. The supplier is exposed to the risk of having to switch processes back-and-forth to satisfy customer-specifications but, on the positive side, he need not hold much inventory in the supply chain. The supplier would obviously charge a premium for customization. In MTS, on the other hand, supplier can produce at a steady rate and he is cushioned from demand volatility, but he may end up holding substantial inventory throughout the supply chain. MTS implies that manufacturing is totally decoupled from market demand.

Assemble-to-order (ATO) can be thought of as a hybrid, where the supplier manufactures the components in advance and assembles them to create different products according to customer choices. Thus, component manufacturing is decoupled from demand as it is cushioned from volatility and only the assembly process is exposed to demand risk. What the customer buys is not entirely customized, but they are not generic either. Thus, the supplier's tradeoff is between holding inventory and exposing his processes to the risk of demand volatility, while the customer must weigh between partial customization and waiting time.

From a strategic perspective, moving forward (downstream) the point at which order decoupling occurs enables the supplier to compete based on quick-delivery. Moving it backwards enables him to compete based on customization. A forward move can be achieved through completion of more of the upstream production stages in advance of order realization. Completing component manufacturing before the order becomes known is one example. A backward move, on the other hand, implies increased customization and requires a reorganization of manufacturing processes such as use of flexible manufacturing, and elimination of bottlenecks (refer to Chap. 6 for further discussion).

2.7 Demand Shaping

Demand shaping is the process of shifting demand, using appropriate incentives, to appropriate time slots or channels where capacities exist. These tactics help increase or decrease demand to match it with existing capacity (Jinhong and Shugan 2001). The major means of shaping demand, as shown in Fig. 2.8, include the use of product promotions, offer of rain-checks, scheduling of airline flights at fixed time slots, servicing customers by appointment, and offering financial incentives to customers for delaying servicing (pay-to-delay).

Cecere (2009) summarizes these tactics in a supply chain, as shown in Table 2.2.

Fig. 2.8 Demand shaping

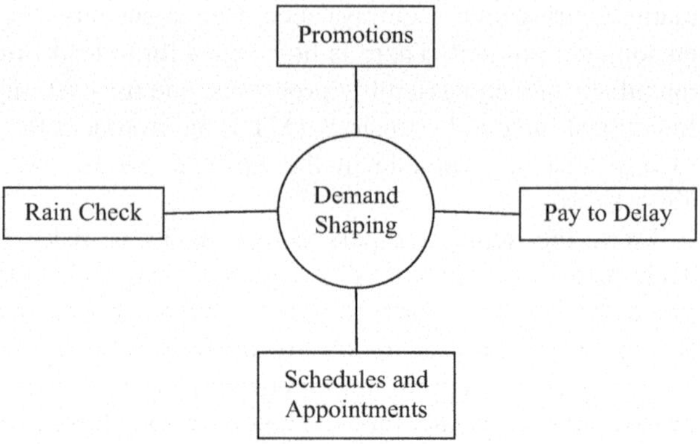

Table 2.2 Shaping demand (AMR research, Cecere 2009)

Direct Initiatives	Brand Building
New product launch Pricing	Service and social responsibility programs to build brand presence
Sales incentives	Use of partners
Marketing	
Trade promotions	

2.7.1 Promotions

Price markdowns help sell off excess inventory (overstock.com). However, if markdowns become predictable, customers wait until the period-end, thereby negating the very notion of markdowns; customers would rarely buy at regular prices, decreasing the company's revenues. Promotions through advertisements or other marketing techniques create additional demand to fill unused capacity or to get rid of excess inventory (Bursa 2009). For seasonal products promotions may be used during the off season so that the supplier may retain a constant rate of production throughout the year. During bad economic times in the 70s Campbell Soup used promotion to increase demand in the month of January (McKenney and Clark 2008). While sales increased, Campbell's cost soared because they needed to produce and accumulate a large inventory to be able to sell a huge quantity in January. Because of the bad economy the promotion had to be continued for several years and the customers (wholesalers) came to expect it every year. They increased their profits substantially by forward-purchasing whole year's supply in January and selling them at the regular price for the rest of the year. Campbell could not withdraw promotions for fear of losing customers. To cut their losses Campbell found a way to make adjustment to their promotion – they offered discount prices only on amount *sold* by the wholesalers in January (not on the amount they bought), eliminating the incentive for forward purchase.

Dell uses promotions creatively to get rid of excess components. They promote certain computer configurations that use more of the excess components in their warehouses. They compute the discount percentage by trading it off against the revenue from the sale of the excess components.

2.7.2 Schedules and Appointments

Customers usually must wait for service as evident from the waiting lines at bank-teller windows, hair salons, and check-out counters. One way of minimizing customer wait is to schedule the services and have customers fit in, as in scheduled airline flights and appointments at a doctor's clinic. In principle, it shifts demand to the time when capacity becomes available. The underlying assumption is that the negative utility of delay arising from demand-shifting is the same for all customers. Therefore, customers with high negative utility of waiting may be lost if competitors can accommodate them. Thus, the tradeoff for service providers is between losing some customers and having to provide additional capacity (such as additional service providers and/or equipment) during the high-demand periods. Note that the concept of appointments is impractical in retail and manufacturing industries. Utilities such as telecom companies practice peak-load pricing to smooth high demand for service during regular work hours with lean demand in the evenings. Customers with high negative utility of waiting now have an option of using the service during peak hours by paying more.

2.7.3 Pay to Delay

Another way of influencing demand is to pay customers for waiting. This payment could take the form of a price discount that could be proportional to the length of delay. The practice of rain-check is a variant of this policy, where the retailer promises to provide the discounted item (not on the shelf) at a later time. In general, the retailer gains from not having to carry a high inventory or capacity, though the effective per unit revenue is reduced. The customer, on the other hand, must trade off the savings from lower unit price with the negative utility of waiting. Online firms with no inventory can compete by discounting the unit price for delayed fulfillment.

The negative utility of waiting has been studied by McGuire and Kable (2012) in an experimental setting. They establish a relationship between the willingness to wait and the reward expected for doing so, and discovered that higher the wait time, higher the reward expected for waiting. Most logistics companies including UPS, FedEx, and DHL, have published rates that charge a lower rate for slower service. For example, a 3-day ground delivery by UPS is significantly less expensive than overnight delivery. To provide a faster service, the company would need to expand its service capacity.

2.8 Revenue Management

Talluri and van Ryzin (2004) outline two primary strategies for revenue management: capacity control, and price control. Capacity control is based on allocating capacity while keeping the price fixed. Price control strategies include dynamic pricing as in auctions, and price discrimination in different customer segments. As shown in Fig. 2.9, revenue management may also include overbooking and capacity reservation (early commitment) by customers.

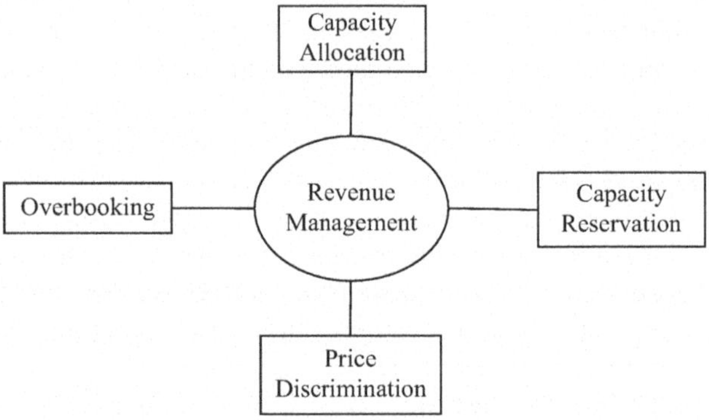

Fig. 2.9 Revenue management

2.8.1 Capacity Allocation

Capacity based revenue management attempts to maximize revenue with given prices (Mcgill and Van Ryzin 1999). An airline can dynamically increase or decrease the number of seats it sells in a fare class. It typically uses fare classes such as Y, M, B, and Q to which customers fit in. The number of seats assigned to a segment (fare-class) is a function of the type of fare-class and the demand pattern. Consider an airline that needs to allocate 100 unsold seats between two classes of customers who are willing to pay price p_1 and p_2 ($p_1 < p_2$) for the same seat. Assume that the average demand for class 1 seats is 80 and for class 2 it is 40, and there is a nonzero probability that demand in any class would exceed 100. If a customer requests a class 1 seat, with 100 seats yet to be sold, should the airline sell the seat or wait for a potential class 2 customer to materialize? If the airline *knew* that it could only sell 99 or fewer seats to class 2 customers, the decision would be simple – sell the seat to the current customer. Thus, if the class 2 demand is 100 or more (which would happen a fraction of time), revenue from the seat in question would be p_2. The fraction of time when class 2 demand exceeds or equals 100 can be expressed as the probability, $P(D_2 \geq 100)$. Therefore, the expected revenue from the policy of targeting class 2 customers (not selling a specific seat to a class 1 customer) equals $p_2P(D_2 \geq 100)$. Note that this revenue will be less than p_2, and can even be less than p_1. In case of the later, that is, $p_2P(D_2 \geq 100) < p_1$, the supplier will be better off selling the seat to the current class 1 customer. The above condition can be expressed as $P(D_2 \geq 100) < \frac{p_1}{p_2}$ or $100 > f(p_1/p_2) = y$, where $f(\cdot)$

is a decreasing function in the ratio p_1/p_2, and can be determined from the distribution function of demand for class 2. In general, denoting the inventory of unsold seats as n, we must have $n > y$, for a seat in the lower price segment to be sold. It follows that for given segment-prices, the seat inventory threshold can be determined uniquely. Thus a simple decision rule of the type, sell if $n > y = f(p_1/p_2)$, is sufficient for real time implementation. Clearly, the expected revenue from denying a specific unsold seat to a class 1 customer ($= p_2P(D_2 \geq n)$) would increase if n is low. It also follows that if n decreases, p_1 must increase for the condition $n > y = f(p_1/p_2)$ to hold. That is, the price of a seat in class 1 increases if the inventory of unsold seats decreases – a common practice in inventory pricing. For a uniform distribution with $0 \leq D_2 \leq 150$, we may express the decision rule as $n > 150(1\text{-}\frac{p_1}{p_2})$. It follows that if $p_1 = \$300$ and $p_2 = \$500$, the airline would protect 60 seats for the high-paying customers. Note also that if $n = 50$, the airline will increase the discounted price to $333.34.

Leakage between demand segments can be a sticky problem in allocating seats to the segments. To avoid it, companies create fences between classes. In the airline industry the fence may take the form of a high penalty for making changes, advance-purchase, or Saturday-night-stay.

2.8.2 Overbooking

Another way of capacity control is the use of over-booking to minimize the risk of no-shows and hence maximize seat utilization. In effect, it creates a hedge against no-shows, but incurs the risk of having to buyback sold seats through auction at a high price (when most customers show up). To illustrate, consider a Boeing 757 jet flying from Chicago to Boston. The aircraft has 180 seats. Based on the past travel pattern, it is observed that an average of 5% do not turn up at the time of boarding the flight. If the airline books all seats for this leg, it will fly with only 171 occupied seats on average. However, there is a small chance that on some occasion all 180 passengers would show up. Therefore, if they book 182 passengers instead of 180, they are likely to end up with only 173 passengers. In the odd event of exactly 182 passengers reporting, the airline would need to *auction* for two passengers to relinquish their reserved seats. The expected revenue would be the revenue from 182 passengers minus the *expected* cost of compensating the two passengers through auction. As the probability of all 182 passengers turning up is very low, the revenue from selling the overbooked two seats may more than compensate for the expected cost.

Capacity reservation (early commitment) is another way of moving demand to "capacity categories", and it acts like a financial option. Customers may reserve capacity ex ante for a lower unit price. They use the reserved capacity first and, if needed, the unreserved capacity at a higher price. They forfeit all unused reserved capacity. Discount pricing for early commitment provides an opportunity to sell a part of the capacity with certainty.

2.8.3 Price Control

Determining prices based on how customers value products and services, at any point in time, is known as dynamic pricing (Bitran and Caldentey 2003). It has become a feasible strategy in online transactions, as sellers and buyers can make rapid adjustments to prices at which products are bought and sold. Research on dynamic pricing started in 1960s, initiated by American Airlines. They soon discovered that the problem was too complex to be solved efficiently, and divided it into three smaller problems: overbooking, discount allocation, and traffic management. American Airlines estimates the benefit from these models at $1.4 billion.

Dynamic pricing methods can be categorized into two groups: posted price and price discovery. Posted prices are determined by the seller and hence not flexible. However, they could be dynamic in the sense that the seller could change posted prices periodically depending on demand, supply availability, and the time of sale. In price discovery mechanisms, prices are determined through a bidding process such as auction. Customers on sites such as eBay have the option of "buy it now" based on posted prices or make a bid to purchase in an auction.

2.8.4 Price Discrimination

It is now possible to charge different prices to different consumers for the same product. For example, sellers may offer special deals targeted to individual customer segments, and buyers quickly scan to compare prices and services. Price discrimination relies on being able to create customer segmentation based on willingness to pay, and the need for convenience and comfort. For example, negotiation may result in different prices for urban and rural customers buying the same service. Similarly, special orders such as made-to-order services can be created for customers who are willing to pay more. A well-known example is the patented drug that fetches a higher price in US than in Canada. Other examples are group pricing, and differential telecom pricing for businesses and households.

2.8.5 Capacity Shortage

Capacity shortage is a major issue in a robust and growing economy. The way it usually impacts decisions is that capacity is committed to early customers on a first-come-first-serve (FCFS) basis, and customers who delay their decisions end up with denied service. The fact that some customers in this later group would be willing to pay a higher price, leads to lost opportunity, as money is left on the table. Obviously if all customers arrive simultaneously, the supplier could cherry pick high-paying customers to maximize her profit. A simple approach would be to increase price for all so that demand is reduced to match available capacity. However, with volatility in demand, the decision becomes complex.

2.9 Customer Relationship Management (CRM)

The concept of CRM is built around the customer (instead of product or process), in the belief that the development of a close relationship with the customer will help determine, fulfill and even predict the needs of the customer (Beckett and Camarata 1998). By learning customer preferences and focusing on long-term relationships, firms can provide products and services that better fit customers' needs. They can also do this in a way that ensures loyalty and builds sufficient trust for customers to share strategic information about their preferences and needs. As shown in Fig. 2.10, the major emphasis is on acquiring and retaining customers, and extending relationships to future.

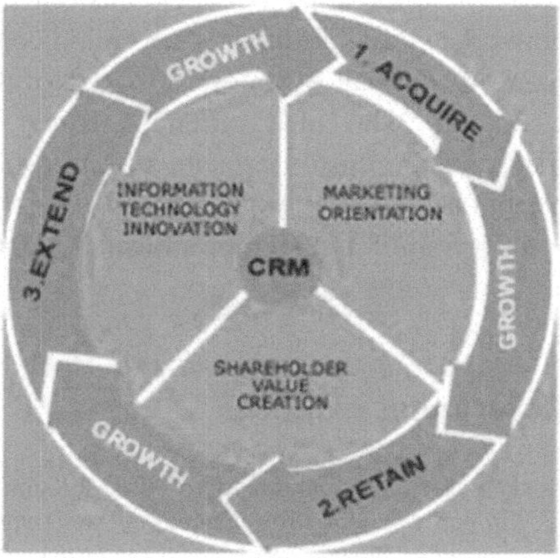

Fig. 2.10 Customer relationship management (Source: www.shringen.com)

Winer (2004) outlines the basic elements of CRM as a database of customer activity, procedures for analyzing the database and targeting customers, building relationships with targeted customers, and building metrics for measuring the success of the CRM program. The database contains information on purchase-transactions, customer contacts, descriptive information for data analysis purposes, and customer response to marketing initiatives.

A plethora of statistical methods have been devised to group together customers with similar behavioral patterns which are then used to develop different product offerings or direct marketing campaigns. Data analyses can also reveal which products are often purchased together. In addition, Internet clickstream analysis of customer purchases can be used to better understand and predict customer behavior. The goal is to increase yield, measured as the percentage of browsing customers who actually make one or more purchases.

Segmentation analysis (based on preferences, demography etc.) on purchasing behavior can help determine which customers would be selected first. The wireless provider PageNet raised monthly rates for unprofitable subscribers with the intent of driving them away. Southwest Airlines's e-mail-based Click 'n Save program has 2.7 million subscribers. Every Tuesday, the airline sends out e-mails to this database of loyal users containing special fare offers. Relationship building programs include loyalty programs, customized services, rewards programs, and community building.

Customer service programs are normally of two types. Reactive service is where the customer needs a problem resolved and contacts the company to solve it. This could be a product failure, clarification about a bill, or product return. Proactive service, on the other hand, establishes a dialogue with customers prior to receiving complaints. The intent is to resolve potential problems and preempt customer complaints.

Loyalty programs provide rewards to customers for repeat purchase. Companies build networks of customers for exchanging product-related information and to create relationships between customers and the company or its brand. These networks and relationships are called communities. The goal is to build an environment which makes it more difficult for the customer to leave the "family" of other customers who also purchase from the company. Examples of CRM-based measures are customer acquisition cost, conversion rate from browsers to buyers, retention rate, same-customer sales rate, loyalty measures, and customer-share (proportion of a customer's business satisfied).

References

Beckett, E., & Camarata, M. (1998). Integrating internal and external customer relationships through relationship management: A strategic response to a changing global environment. *Journal of Business Research, 41*, 71–81.

Bitran, G., & Caldentey, R. (2003). An overview of pricing models for revenue management. *Manufacturing and Service Operations Management, 5*, 203–229.

Booz Allen (2004). The Customer-Centric Organization: From pushing products to winning customers. http://www.booz.com/media/file/The_Customer_Centric_Organization.pdf

Borg, S. (2003). Enabling a Customer-Focused Organization. Center for Digital Strategies at Tuck School of Business and Cisco Systems, Inc. http://digitalstrategies.tuck.dartmouth.edu/cds-uploads/publications/pdf/Round_Overview_EnablingCustomerFocus.pdf

Bursa, K. (2009). How to effectively manage demand with demand sensing and shaping using point of sales data. *Journal of Business Forecasting, 27*, 26–28.

Cecere, L. (2006). *DDSN vs. SCM: Look to the sky for the differences part 1* (p. 6). Boston, USA: AMR Research.

Cecere, L (2009). *Want to be demand driven? 15 tactics to consider: Industry value chain strategies* (pp. 1–6). Boston, USA: AMR Research Inc.

Chakravarty, A., Mild, A., & Taudes, A. (2013). Bundling decisions in a supply chain. *European Journal of Operational Research*. http://dx.doi.org/10.1016/j.ejor.2013.06.026, doi:10.1016/j.ejor.2013.06.026#doilink

Chen, I. (2001). Planning for ERP systems: Analysis and future trend. *Business Process Management Journal, 7*, 374–386.

Cheng, F., Ettl, M., Lin, G., Schwarz, M., & Yao, D. (2003). *Flexible supply contracts via options* (Working Paper). Yorktown Heights: IBM T.J. Watson Research Center.

Christopher, M. (1996). From brand values to customer value. *Journal of Marketing Practice: Applied Marketing Science, 2*, 55–66.

Dapiran, P. (1992). Benetton – Global logistics in action. *Asia-Pacific International Journal of Business Logistics, 5*, 7–11.

EIU. (2009). *The Demand-Driven Supply Chain: A Holistic Approach*. London: Economist Intelligence Unit, The Economist.

Erkoc, M., & Wu, D. (2005). Managing high-tech capacity expansion via reservation contracts. *Production and Operations Management, 14*, 232–251.

Forbes (2012). Bringing up Scion: Toyota overhauls brand and products. http://www.forbes.com/sites/dalebuss/2012/05/20/bringing-up-scion-toyota-matures-the-brand-and-products/

Frank, J., Blok, T., Carter, J., Mullen, L., Tideman, A., & Parkes, J. (2007). Performance and ease of use of the Breeze®2 Blood Glucose Monitoring System (Clinical Trial Report). Bayer Health Care. http://www.bayercontour.com/resources/pdf/propdf/Article_Breeze2_Blood_Glucose_Monitoring_System

Gerchak, Y., & He, Q. (2002). *On the relation between the benefits of risk pooling and the variability of demand* (Working Paper). Tel Aviv: Tel Aviv University.

Gourley, C. (1997). Supplies and demand: Staples puts pen to paper and centralizes its DC. *Distribution, 6*, 60–62.

Holt, C., Modigliani, F., & Simon, H. (1955). A linear decision rule for production and employment scheduling. *Management Science, 2*, 1–30.

Jedd, M. (1996). Walt Disney's logistical magic. *Distribution, 6*, 64–66.

Jinhong, X., & Shugan, S. (2001). Electronic tickets, smart cards, and online prepayments: When and how to advance sell. *Marketing Science, 20*, 219–243.

Kambil, A., Friesen, B., & Sundaram, A. (1999). Co-creation: A new source of value. *Outlook*, Number 2, pp. 38–43, http://kambil.com/accenture/cocreation2.pdf

Kolesar, P., Vanryzin, G., & Cutler, W. (1998). Creating customer value: Through industrialized intimacy. *Strategy+Business, 12*, 2–12.

Mcgill, J., & Van Ryzin, G. (1999). Revenue management: Research overview and prospects. *Transportation Science, 33*, 233–256.

McGuire, J., & Kable, J. (2012). Decision makers calibrate behavioral persistence on the basis of time-interval experience. *Cognition, 124*, 216–226.

McKenney, J., & Clark, T. (2008). Campbell Soup Company: Selling channel innovation to customers. HBS Case Study, 9-608-141.

Nagarajan, M., & Sosic, G. (2009). Coalition stability in assembly models. *Operations Research, 57*, 131–145.

NCE (2012). http://www.datalliance.com/North_Coast_user_story.html

Pine, J., & Korn, K. (2011). Creating customer value on the digital frontier. HBR blog network. http://blogs.hbr.org/cs/2011/09/creating_customer_value_on_the.html

Ries, E. (2011, October). Creating the lean startup. *Inc. Magazine*.

Silver, E., Pyke, D., & Peterson, R. (1998). *Inventory management and production planning and scheduling* (3rd ed.). New York: Wiley.

Stallkamp, T. (2001). Fixing a broken economic model: a case for supplier alliances (Management Briefing Seminars). Traverse city: MSX international. http://ptgmedia.pearsoncmg.com/images/9780137145638/samplepages/0137145632.pdf

Stremersch, S., & Tellis, G. (2002). Strategic bundling of products and prices: A new synthesis for marketing. *Journal of Marketing, 66*, 55–72.

Sumitomo Group (2002). Hedging risk with weather derivatives (Special Report). http://www.sumitomo.gr.jp/english/discoveries/special/90_05.html

Talluri, K., & van Ryzin, G. (2004). *Theory and practice of revenue management*. New York: Kluwer.

Von Schlippenbach, V., & Wey, C. (2011). *One-stop shopping behavior, buyer power, and upstream merger incentives* (DICE discussion paper, No. 27). https://www.econstor.eu/dspace/bitstream/10419/48602/1/663926440.pdf

Winer, R. (2004). *Customer relationship management: A framework, research directions, and the future* (Working Paper). Berkley: University of California.

Zentech (2012). Zentech Manufacturing Inc. manufacturing today. http://www.manufacturing-today.com/index.php/sections/profiles1/527-zentech-manufacturing-inc

Products and Processes

3

3.1 Introduction

In the competitive markets of today, positioning a company simultaneously in the domains of product, process, and supply chain is an imperative. However, the tradeoffs among the dimensions can be quite complex, as they are interdependent. The building blocks of a framework for such tradeoffs can be the product-architecture that maps the functions of a product to physical components and determines the configuration of product-variants. The framework must include tools for assessing attributes such as component commonality, product quality, technology, manufacturing processes, and postponement, as shown in Fig. 3.1. Thus, to obtain the optimal tradeoff, one may tweak the product architecture appropriately in view of the customer preferences.

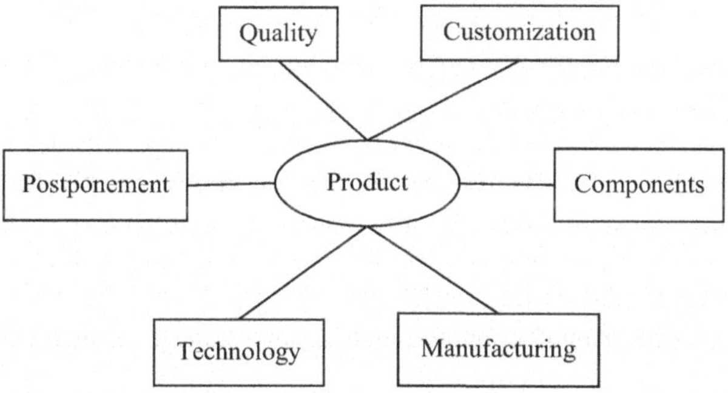

Fig. 3.1 Product management

A product can be conceptualized as a bundle of attribute-values. Attributes such as customization may possess multiple values and therefore there can be a large number of bundles possible. Each such bundle is a version of the product and, the set of versions of a product offered by a manufacturing company is defined as product variety. While attributes such as quality and customization are stipulated by

A.K. Chakravarty, *Supply Chain Transformation*, Springer Texts in Business and Economics, DOI 10.1007/978-3-642-41911-9_3,

the customer, the attributes of technology, manufacturing processes, components, logistics, and postponement are controlled by the manufacturer along with the distributor and retailer.

Product quality depends on the choice of material, standardization, and design for manufacturability (DFM), among others. Product cost, in addition to the above factors, is affected by procurement-efficiency, sharing of components, outsourcing, and the economies of scale that the company can achieve.

Products that are relatively inexpensive and do not change much exhibit stable demands. They are also known as functional products. At the other extreme are products that evolve rapidly due to new technology and/or new composition, positioned for volatile demand, and are expensive. They are known as innovative products, and companies need to be very responsive to customer needs to remain competitive in such products. As Fisher (1997) points out, the structure and operations of supply chains for innovative products need to be markedly different from those of functional products in terms of: inventory positioning and replenishment, capacity configuration, logistics contracts, information sharing, and collaboration.

Unlike manufacturers, retailers do not create product variety per se, but they can control customer access by deciding which variants of a product they would carry in stores. A retailer's objective is to maximize sales while minimizing its cost of sales: inventory, warehousing, transportation, displays, pilferage, and markdowns. Consumers have idiosyncratic preferences for different styles of products such as CD's, books, shoes, toys, clothing, and videos. Therefore, a retailer that carries a broad range of product versions is more likely to make a sale. That notwithstanding, a retailer that sells shoes would increase variety – by adding a new style of shoes – only if the costs of additional inventory and needed display space are low relative to the probability that customers would buy the new-style shoe. In addition, a large retailer must consider the capability of the supplier, the ordering process, and the delivery lead time – the characteristics of its supply chain.

The shoe manufacturer, in turn, must configure its upstream supply chain by selecting appropriate parts, processes, and vendors. The problem becomes intricate because a certain raw material can be supplied by more than one vendor, the product may require several machines or processes for manufacturing, and multiple shipping options may be available to deliver the product to customers. The fact that the company must plan the supply chain in the context of its existing supply chain for other products – using some of its current vendors and logistics providers – may induce additional complexity.

The product related decisions are summarized in Fig. 3.2.

3.2 Mapping Functions to Components

Concept development decisions determine product specifications and the product configuration. A vector of attributes (such as speed, price, reliability, capacity) is an effective way of describing the product (Krishnan and Ulrich 2001). The choice of this vector is driven by customer preferences as well as manufacturing and supply

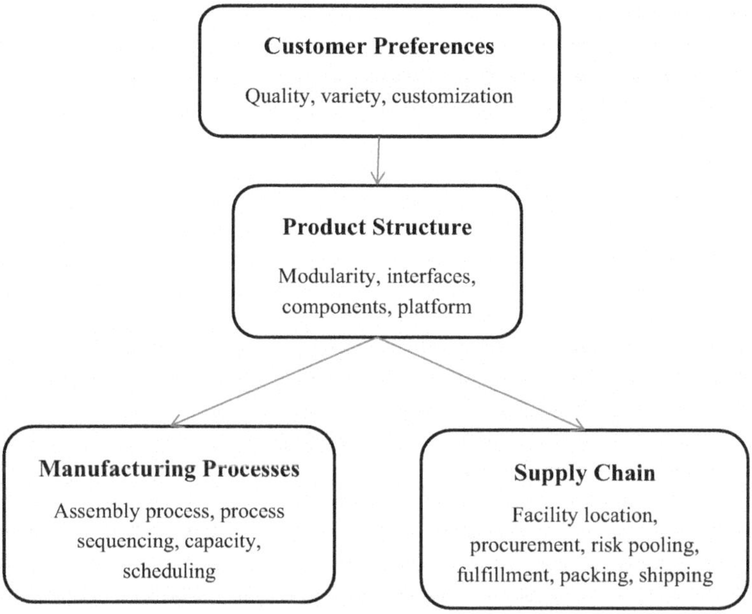

Fig. 3.2 Product structuring

chain concerns, as shown in Fig. 3.2. The first step would be developing a product concept by setting target values for the attributes of the product and its variants that the company wishes to offer. The second step is the determination of variants. Steps 1 and 2 must proceed concurrently, as target setting must be done in conjunction with the determination of product variants. The attribute values preferred by customers typically have a wide range, and target setting balances this preference-range against the economies of standardization in design and production. Target setting is followed by determination of components of the product that satisfy the targeted attribute values. Multiple variants of the product may be created to accommodate customer preferences, and any two variants of the product may use several common components. The final step is to establish the component-specifications, and the physical shape of the product.

The mapping from the functional elements to the components of a product can be one-to-one (only one function is performed by a component), many-to-one (many functions are performed by a component), or one-to-many (one function is performed by many components). Complexity of a function increases if it is performed by more than one component. For example, the function of car air conditioning is to cool the inside of the car. This requires a group of components – the compressor, the radiator, and fans – to work together even if they are constrained by interfaces with other components (Muffatto and Roveda 2002).

Ulrich (1995) uses two examples – a traditional trailer for cargo, and a special trailer – to explain how product functions are mapped to physical components. In

Table 3.1 Modular product

	Components				
Functions	Box	Hitch	Bed	Springs	Wheels
Protect cargo from weather	X				
Connect to vehicle		X			
Support cargo load			X		
Absorb "shocks" to trailer				X	
Move loads on road					X

the first case, the cargo is placed in a box and the box is secured to the base of the covered trailer; the trailer is hitched to a truck. The relationship between the functions and components are shown in Table 3.1.

Note that each of the five components supports only one specific function, in effect creating a set of five modules.

The second trailer has a more integral architecture. Upper and lower halves come together to form a cylindrical shape trailer. The strip of metal, between two slots in the upper half of the trailer, acts as a leaf spring (Ulrich 1995). The cargo is hung by straps from the two springs in the upper half. The axle is attached to a similar spring in the lower half. The nose piece is the component containing the trailer hitch. Covers are attached over the slots. The function map for this trailer is shown in Table 3.2.

Table 3.2 Integral product

	Components					
Functions	Upper half	Lower half	Nose piece	Cargo hanging springs	Slot covers	Wheels
Protect cargo from weather	X	X			X	
Connect to vehicle			X			
Support cargo load	X	X	X	X		
Minimize air drag			X			
Absorb "shocks" to trailer	X	X				
Move loads on road						X

Each of the functions, except "Connect to vehicle", and "Move loads on road", requires multiple components for it to be completed. Hence, the trailer has an integral architecture. It is clear from Table 3.2 that "upper half", "lower half", and "nose piece" are shared components, as each supports multiple functions. Also, note that "wheel" is the only component shared by the two trailer designs. Component sharing can thus be defined at the functional level (single product) or at product level (multiple products).

The functions, described above, may not all be of interest to the customer. For example, if the customer is only interested in having his/her cargo delivered, the functions related to air drag and shock absorbers would be superfluous, as long as the cargo is delivered undamaged and on time. On the other hand, if the customer is the one who is interested in purchasing the trailer, functions relating to protecting the cargo and the trailer from damage would be very relevant.

These pattern, in an increasing order of architectural complexity, can be described as function-driven, component-driven, and combined-complexity (Muffatto and Roveda 2002). Complexity of a function increases when it is performed by more than one component. In the car air conditioning example, mentioned earlier, the cooling function is performed by a group of components which must work together. Hence, the cooling function is quite complex. Component driven complexity arises if one component performs many functions. Modifying one component function would imply re-designing the whole product. For example, the body of a car performs the product's structural function, as well as the aesthetic, aerodynamic and weight distribution functions. It is therefore difficult to change one of these aspects without a major re-design of the body. Combined complexity is the combination of the previous two. It arises from a many-to-many mapping between the function (aesthetics, power, air-conditioning, structural strength, weight distribution) and the parts (body, chassis, engine), as they become mutually interdependent.

As a product strategy, therefore, component sharing is of critical importance (Fisher et al 1999). In designing cars, for example, components such as steering-wheel and engine are typically shared by multiple models of a car. The objective is to maximize the variety of the consumer product while minimizing component variety; it implies maximization of component sharing. Clearly, the production volume of a component increases in the number of products or functions that share this component, enabling scale economy (including learning) to set in. On the other hand, a defect in a shared component can cause a bigger financial damage to the company than a similar defect in a unique (not shared) component. The financial risk of a shared component increases further, if the product has an integral architecture, as the component provides support to multiple functions. In general, to enable component sharing, the components must be standardized: AA batteries that are used in multiple products such as the TV remote-control, and wireless mouse. It follows that if a component performs more than one function, its procurement must be controlled carefully. Thus, while the supplier's price remains the same, the buyer's effective cost of using the component increases because of the risk of bigger damage.

3.3 Modular Structure and Variety

A module of a product is a cluster of functions and components that are operationally self-contained. Such clusters can be created in multiple ways: quality function deployment (QFD), design structure matrix (DSM), statistical clustering, and artificial neural network (ANN). In a modular structure, the components are connected tightly within (structural relationships), and relatively loosely to elements in other modules. For a module to function well, the interface between any two interacting module components must be well-defined (Ulrich and Eppinger 1995). For examples, the protocols for transmitting power from the engine of a car to its wheels, through the gears and the shaft, must be clearly laid out.

3.3.1 Product Variants

Variants of the product can be created by adding, removing, or replacing some of the modules from the original product (conceptually similar to replacing components). For mass customization, it must be possible to switch to manufacture different product variants quickly and with ease. For that to happen, the components and interfaces used in modules must be standardized. They can then be produced in mass volume, creating a scale economy advantage. Thus, firms can upgrade the functions of a product by manipulating a few modules, while other components remain unchanged. However, if the design-performance is the sole objective, an integral product would be preferred, as it is likely to outperform the modular product.

To plan for variety, a company must first ascertain the number of variants of a product it would like to introduce, and the numbers of options per module to support this level of variety. A product typically has a fixed number of *module-types* such as brake system, fuel system, and transmissions, in case of a car. Each module-type may have several *module-options* depending on the choice of components, part geometry, and manufacturing processes. If a product comprises three module-types (A, B, and C) with the number of module-options in each being N_A, N_B and N_C respectively, the total number of possible product variants would be $(N_A)(N_B)(N_C)$.

The modules are produced in a proactive mode because they can be preassembled. So manufacturing or component-assembly is not a bottleneck for mass customization with modular products. If products can be designed so that the customers can configure the modules themselves (such as configuring a home music system), the supply chain can be greatly simplified.

ITT Automotive has applied the modularity concept in designing the antilock braking system (ABS) module (Pisano 1994). Components such as pump, valve block, and motor together form a module-option for the ABS; Fig. 3.3 shows four such options (M_1 to M_4). The ABS options M_j are created by choosing from an assortment of each of these components. For example, the ABS M_2 includes the pump p_1, valve v_2, and motor t_2; $M_2 = (p_1, v_2, t_2)$. Each option fitted to the end product creates a product variant. For example different car models can be fitted with different ABS configurations, and so can trucks and SUVs. In Fig. 3.3, two car variants are shown with ABS options M_1 and M_3, and the two truck variants have ABS options M_2 and M_4.

ABS Module		ABS Options j				Product Variants			
						Product 1 (car) $i = 1$		Product 2 (truck) $i = 2$	
Component	Option	M_1 $j = 1$	M_2 $j = 2$	M_3 $j = 3$	M_4 $j = 4$	$j = 1$	$j = 3$	$j = 2$	$j = 4$
Pump	p_1	1	1			car, M_1	car, M_3	truck, M_2	truck, M_4
	p_2			1					
	p_3				1				
Valve	v_1	1		1					
	v_2		1						
	v_3				1				
Motor	t_1	1			1				
	t_2		1						
	t_3			1					

Fig. 3.3 Components, modules and product-variants

In Fig. 3.3, the implied assumption is that M_2 and M_4 are not compatible with product 1 (car), and M_1 and M_3 are not compatible with product 2 (truck).

3.3.2 Market Share and Profit

In the context of Fig. 3.3, we denote $I(i \in I)$ is the set of product types (such as car, truck, and SUV), and $J(j \in J)$ is the set of ABS module-options, where a module-option is a choice of components (one from each component type). Observe that products – car, truck, and SUV – can all be fitted with an ABS module, but the module must be configured to be specific to the product type. Thus, the variant of a product will be denoted by the (i,j) pair, assuming module-option j is compatible with product i. Denote r_{ij} as the rating of product i containing module-option j, and $a_{ij} = \exp(r_{ij})$ is the attraction factor of this product variant to customers; $a_{ij} = 0$ if i and j are incompatible. As in Lilien et al. (1992), we define the market share of (i,j) as,

$$m_{ij} = a_{ij} \Big/ \left(A + \sum_{n \in I} \sum_{m \in J} a_{nm} \right)$$

where with A is the attraction factor of the competitors' products.

The demand of the product variant (i,j), with M as the total market size, is therefore written as,

$$D_{ij} = m_{ij}M = Ma_{ij} \Bigg/ \left(A + \sum_{n \in I} \sum_{m \in J} a_{nm} \right)$$

The profit margin of variant (i,j) is written as $u_{ij} = P_{ij} - c_j$, where P_{ij} is the unit sales price and c_j is the unit cost of the jth module-option. The OEM's profit is written as,

$$\pi = \sum_{i \in I} \sum_{j \in J} P_{ij} Q_{ij} - \sum_{j \in J} c_j q_j$$

where Q_{ij} is the quantity of product variant (i,j) sold, and q_j is the quantity of module-option j produced, so that $q_j \geq \sum_{i \in I} Q_{ij}$ and $Q_{ij} \leq D_{ij}$.

3.3.3 Module Choice Decisions

The OEM needs to decide what options of the ABS module to acquire, in what quantities, and what product variants ((i,j) pairs) to offer to the customers. Chakravarty and Balakrishnan (2001) show that it is optimal to cherry-pick the options of ABS modules from an ordered-list of module-options. Denoting the set of module-options that the OEM purchases as $S(S \subset J)$, the OEM's profit is expressed as,

$$\pi(S) = \sum_{i \in I} \sum_{j \in S} P_{ij} Q_{ij} - \sum_{j \in S} c_j q_j$$

Subject to $q_j \geq \sum_{i \in I} Q_{ij}$, $j \in S$; and $Q_{ij} \leq D_{ij}$.

For a given S, known demand D_{ij}, and $P_{ij} \geq c_j$, it must be the case that $Q_{ij} = D_{ij}$. In addition, with a single-period profit maximization as the objective, we would never have q_j exceeding $\sum_{i \in I} Q_{ij}$. Therefore, the above two constraints will be satisfied as equations, and the objective function will be rewritten as

$$\pi(S) = \sum_{i \in I} \sum_{j \in S} P_{ij} D_{ij} - \sum_{j \in S} c_j \sum_{i \in I} Q_{ij} = \sum_{i \in I} \sum_{j \in S} (P_{ij} - c_j) D_{ij}$$

$$= \frac{\sum_{i \in I} \sum_{j \in S} Ma_{ij} (P_{ij} - c_j)}{A + \sum_{i \in I} \sum_{j \in S} a_{ij}}$$

Letting, $x_j = \sum_{i \in I} Ma_{ij} (P_{ij} \text{-} c_j)$ and $y_j = \sum_{i \in I} a_{ij}$, we have

$$\pi(S) = \sum_{j \in S} x_j \Bigg/ \left(A + \sum_{j \in S} y_j \right)$$

Further simplification with $X_S = \sum_{j \in S} x_j$ and $Y_S = \sum_{j \in S} y_j$ leads to

$$\pi(S) = X_S/(A + Y_S)$$

The optimization problem can now be stated as

Maximize $\pi(S) = X_S/(A + Y_S)$

Subject to $S \subset J$

Chakravarty and Balakrishnan (2001) establish that the optimal S satisfies the property that $j \in S$ and $u \notin S$ implies $x_j/y_j > x_u/y_u$. This property allows us to rank order the module-options ($j \in J$) in decreasing order of x_j/y_j. Items from the top end of this list are added to S, so long as $\pi(S)$ keeps increasing.

The quantity of module-option $j \in S$ purchased by the OEM are determined as,

$$q_j = M \sum_{i \in I} a_{ij} \Bigg/ \left(A + \sum_{i \in I} \sum_{k \in S} a_{ik} \right), \quad j \in S$$

Note that if the OEM builds the modules instead of purchasing them, he would also incur a fixed development cost for each module-option, independent of the quantity built. Chakravarty and Balakrishnan (2001) establish a procedure for modifying the ranking scheme of module-options that includes the development costs.

3.3.4 Organizing Module Development in the Supply Chain

With standardized modules, it is clear that suppliers need not be involved in the design of the product, and they need not be co-located. Suppliers make independent decisions and, in fact, they will be at liberty to sell the modules to other customers. As the supplier is not constrained by the decisions other suppliers make, coordination is much simplified. Routine maintenance is also simplified, as the defective modules can be replaced as a whole. Thus, the supply chain can benefit from global sourcing of standardized modules.

The product development team can be divided into module-specific teams to work autonomously and concurrently, and they can even be outsourced. With simplified coordination between the teams, the product development process can be quickened. However, in extreme cases, modularization may lead to fragmentation of the supply chain.

The cost in a supply chain may increase in product variety, for several reasons. Production cost increases as assembly processes must be switched over from one variant to another. Inventory holding cost of the end product increases as safety stock must be held individually for each product variant. Depending on the product, production cost may dominate the marketing cost, and vice versa. For example, variation in body styles of an automobile is production-cost dominant because of the huge tooling investments associated with creating that variety. On the other hand, an increased variation in trim colors of an automobile is marketing-cost dominant because the impact of additional colors on production costs is minimal, while the impact on inventory and stock-out costs can be substantial (Randall and Ulrich 2001).

Companies may choose to reduce production costs through scale economy, which would require it to pool production volume from different markets into one facility (MacCormack et al. 1994). The pooling increases the distance from regional markets, and hence replenishment lead times, making the firm more vulnerable to higher demand uncertainty. It follows that to reduce the marketing cost, companies would locate production close to customers in the target market.

In the realm of new product development, Verdouw et al. (2010) discuss how changes in product structure can influence the dynamics in supply chains. They observe effects on supply chain network related to components, tier-structure, bargaining power of suppliers, and formation of strategic alliances. As noted earlier, modularization motivates outsourcing. This, in turn, results in transfer of an increased number of components to suppliers.

With new product development where modules may not be standardized, outsourcing often motivates collaboration of suppliers, as it can help them in developing innovative new modules and products. Modularization may lead to consolidation of first-tier suppliers into large suppliers, enabling them to manage the development and production of large sets of components as modules (Takeishi and Fujimoto 2001). Because of their larger role as integrators, these suppliers become powerful and can increase their bargaining power. It can also lead to the formation of strategic alliances between the OEMs and suppliers. Diez (2000) points out that developing new modularized products leads to improved information sharing, especially tacit knowledge sharing through physical co-location and face-to-face communication. Suppliers make more relationship-specific investments to facilitate the design of common modules.

3.4 Platform Strategy

We discussed modification in an automobile design for replacing the brake system with an anti-lock braking system (ABS) module. A typical vehicle, however, comprises other modules – the steering system, body frame, transmission, and the engine – with multiple options for each module. Therefore, in designing a product with a large number of module types, the company must resolve multiple tradeoffs. These can be categorized by positioning product variants in the customer-preference space, matching the module options with the characteristics of a product-variant, and leveraging the scale economies in production/procurement processes and knowledge accumulation. While customer preferences and scale economy influence, respectively, the revenue and cost; matching the module-options with product-variants is essentially a search and evaluation problem.

A platform is a structured “library” of the knowledge base on designing and developing a set of common components, processes, and markets, from which a stream of variants can be created. It is typically organized for satisfying the objectives of revenue, cost, and search effort. The concept of a platform has evolved significantly from being confined to common module-options and design reuse, to a broader perspective to include all shared assets: modules, suppliers, customer segments, and production processes (Robertson and Ulrich 1998).

3.4.1 The Cost Objective

The platform approach helps reduce the cost and time of product development, cost of manufacturing and assembly, and the cost of procurement. For example, the platform ensures that the common parts and assembly processes developed for one product-variant do not have to be re-developed and tested for other variants. Second, companies achieve economies of scale from large-volume production because common components are used in multiple product variants. Third, companies can cut investment in manufacturing-equipment, tooling, and the time and effort needed in developing shared processes and assets. Fourth, companies can simplify the supply chain – decreasing the number of parts and processes; and decreasing the costs of procurement, transportation, and warehousing.

In addition to asset-sharing that produces economy of scale benefits, the platform assets display strong interdependence with other assets on the platform. The platform products are characterized by low variety and high commonality, while the non-platform products reveal high variety and little commonality. With a platform, companies can easily increase the breadth of their product portfolio and create differentiation by adding non-platform assets as add-ons. This combination of platform and non-platform assets enables companies to serve multiple market segments.

Williams et al. (2007) discuss the example of a manufacturing enterprise wishing to offer a line of customizable cantilever beams. The company wishes to be able to provide customers the choice of beams of any desired length under any loading condition. The company decides to explore three options: beams with different cross-sections, beams with different material, and beams that are cut into standardized pieces for offering customized lengths. These decisions must be made in a market with demand volatility (need to hold inventory in reserve); to provide the lowest average cost, and to ensure the lowest average of the maximum beam deflection (a quality metric). The company must determine how to structure the platform in terms of beams with multiple customizable specifications, comprising different materials, varying production volumes, and the tradeoff between commonality and product performance.

Piezunka (2011) offers the Sony Walkman as an example of a product platform. Sony included customer segments as assets, and created three product platforms: to build miniature Walkman, to create a new market for hi-fi enthusiasts, and to build models for price sensitive customers. Each of these platforms includes a shared set of modules – motor, tape, and drive. Sony added attributes such as radio, remote-control, and music-search to differentiate the products.

3.4.2 The Search Objective

Consider n types of modules with m options per module. There would be nm possible module options, and m^n possible product variants. Clearly, the search space for creating the optimal product family would be very large if n or m is large. To better organize this search, the company may maintain a knowledge base of a selective few modules, components, product variants, production and assembly processes, and customer-segments. The structured form of such a library is called the platform as mentioned earlier.

Consider a product family comprising three products A, B, C. As shown in Fig. 3.4, product A comprises p_1 unique modules, and it shares r_1 and r_2 modules with products C and B, respectively. Products B and C comprise p_2 and p_3 unique modules, respectively, and share r_3 modules. Finally, r_4 modules are shared by all the three products. We use the notation p_i, r_i to denote the identity as well as the number of modules.

Fig. 3.4 Product family

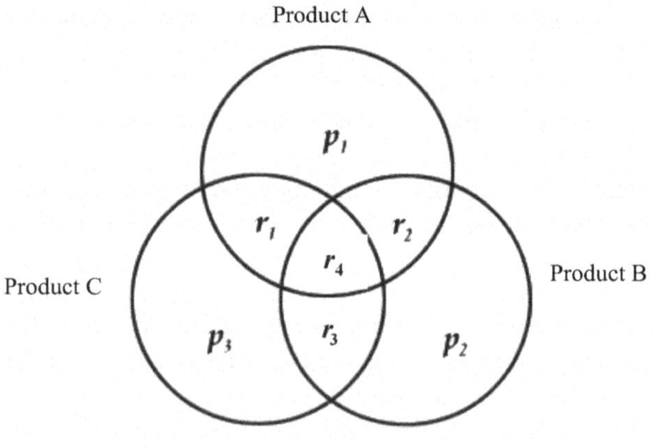

Without the structure of a platform, all the modules of a product must be identified and then assembled to create that product. This must be done independently of other two products. For example, to build product A, the number of modules that the designer would need to search and evaluate would equal $p_1 + r_1 + r_2 + r_4$. Similarly, the search space for products B and C would be $p_2 + r_2 + r_3 + r_4$ and $p_3 + r_3 + r_1 + r_4$, respectively. Therefore, the total search space for the three products would be expressed as $S = p_1 + p_2 + p_3 + 2r_1 + 2r_2 + 2r_3 + 3r_4$.

The simplest platform that can be built would possess the knowledge base for building r_4 modules given that this set is common to all three products, so that no search is involved for using this knowledge base. This implies that every product will have the module-options in r_4 built into them ex ante, reducing the search space to $S = p_1 + p_2 + p_3 + 2r_1 + 2r_2 + 2r_3$; $p_1 + p_2 + p_3$ searches are performed outside the platform and $2(r_1 + r_2 + r_3)$ searches are performed inside the platform.

Next, consider enlarging the platform to include the module options in the sets r_1, r_2, r_3, and r_4. The module options r_4 will be built in all products; r_1 earmarked for products A and C; r_2 for A and B, and r_3 for C and B. The search space outside the platform will equal $p_1 + p_2 + p_3$, as before. Note that to build product A, the company needs to search and eliminate r_3 modules. Therefore the search space for the three products inside the platform will equal $r_1 + r_2 + r_3$. Hence the total search space will equal $S = p_1 + p_2 + p_3 + r_1 + r_2 + r_3$. If the company decides to reduce variety by cutting out product C, it can build a platform comprising r_2 and r_4, and the search space will equal $S = p_1 + p_2 + r_2 + r_3$.

Thus, one of the trade-off in the selection of the number and characteristics of platforms relates to the size of the design space. The library of functions, module-options, and communication components form the design space that can be explored at the appropriate level of abstraction. The downside is that they restrict the design space with a structured search space on the platform that prevents adoption of other design options that could lower costs and produce a better success-rate (Carloni et al. 2000).

3.4.3 Economics of Platform Design

Creating a platform involves *ex ante* design and development of a set of assets (components, modules, production equipment etc.) so that they may be used by more than one product variants centered on the platform. The design decision, therefore, pertains to the identification of assets and the product variants that should be based on the platform. The interrelated issues are the fit between the platform assets and planned product variants, and the costs of assets on and off the platform. The market share, in turn, would depend on how well the attributes of the product variants are liked by customers (Gupta and Krishnan 1998).

Consider a scenario with a set of design alternatives (product variants), $I(i \in I)$; a set of assets considered for the platform design, $T(t \in T)$; and a set of customer desired product attributes, $B(b \in B)$. Next, for a set of known product variants (i.e. given I) and known customer desired product attribute (i.e. given B), define three types of constants:

$r_{bi} = 1$, if product variant i incorporates customer attribute b; 0 otherwise. Therefore, the set of attributes packed into the j th product variant is expressed as $I_j = \{b | r_{bj} = 1\}$.

$u_{tb} = 1$, if asset t is needed for customer desires attribute b; 0 otherwise. The set of assets needed to engineer the bth customer desired product attribute is expressed as $T_b = \{t | u_{tb} = 1\}$. It follows that $\bigcup_{b \in B} T_b = T$.

Assets × Product Attributes

Assets	b=1	b=2	b=3	b=4
t=1	1		1	
t=2		1		
t=3	1		1	1
t=4	1			1
t=5		1		1

Product Attributes × Product Variants

Product Attributes	i=1	i=2	i=3	i=4	i=5
b=1			1	1	
b=2	1	1			1
b=3		1		1	1
b=4	1	1	1		

Fig. 3.5 Platform design consideration

α_{ti} = n, if n units of the asset t is needed for product-variant i; 0 otherwise. Clearly, $\alpha_{ti} = \sum_{b \in B} u_{tb} r_{bi}$.

Examples of u_{tb} and r_{bi} are shown in Fig. 3.5. Note that while the product attribute #1 requires three assets, attribute #3 requires only two assets. Similarly, product variant #2 is defined by three product attributes, whereas, variant #2 is defined by only two attributes.

It follows from Fig. 3.5 that $\alpha_{32} = u_{32}r_{22} + u_{33}r_{32} + u_{34}r_{42} = 2$ etc.

Denote w_b as the relative weight (ranking) customers associate with attribute b. Therefore, the attraction factor for product variant $i = \sum_{b \in B} w_b r_{bi} = W_i$. In Fig. 3.5, assuming attribute rankings of 100, 350, 65, and 225, respectively, the attraction factors of the five product variants can be verified to be 575, 640, 325, 165, and 415, respectively.

The market share of product variant i can now be expressed as,

$$m_i = W_i \Big/ \left(\Gamma + \sum_{j \in I} W_j \right)$$

where Γ is the attraction factor of competitors' products.

Assuming, a market size of M, demand for variant i is written as,

$$D_i = MW_i \Big/ \left(\Gamma + \sum_{j \in I} W_j \right)$$

Next, assume Q_i and P_i to be the production volume and unit price, respectively, of the product-variant i. Therefore, the quantity of asset t needed $= \sum_{i \in I} \alpha_{ti} Q_i$, and the variable cost of manufacturing the asset $t = c_t \sum_{i \in I} \alpha_{ti} Q_i$. We assume $C^d{}_t$ to be the one-time development cost of asset t; $C^d{}_t = 0$, if $\sum_{i \in I} \alpha_{ti} = 0$.

Letting $\beta_t = 1$, if $\sum_{i \in I} \alpha_{ti} > 0$, and observing that $Q_i = D_i$, the expression for profit is written as

$$\pi(T, I, B) = \sum_{i \in I} P_i Q_i - \sum_{t \in T} \sum_{i \in I} c_t \alpha_{ti} Q_i - \sum_{t \in T} C^d{}_t \beta_t$$

Substituting $Q_i = D_i = MW_i \Big/ \left(\Gamma + \sum_{j \in I} W_j \right)$ we have,

$$\pi(T, I, B) = \frac{M \sum_{i \in I} W_i \left(P_i - \sum_{t \in T} c_t \alpha_{ti} \right)}{\Gamma + \sum_{i \in I} W_i} - \sum_{t \in T} C^d{}_t \beta_t$$

Note that W_i is a function of B.

For given I and B, we express the optimization problem as

$$\underset{T}{Max} \{\pi(T, I, B)\} = \underset{T}{Max} \left\{ \frac{M \sum_{i \in I} W_i \left(P_i - \sum_{t \in T} c_t \alpha_{ti} \right)}{\Gamma + \sum_{i \in I} W_i} - \sum_{t \in T} C^d{}_t \beta_t \right\}$$

Denoting $T^*(I, B) = Argmax_T \{\pi(T, I, B)\}$, the maximum value of $\pi(T,I,B)$ for given I and B is $\pi(T^*,I,B)$. Next, denoting $I^*(T^*, B) = Argmax_I \{\pi(T^*, I, B)\}$ the maximum value of $\pi(T,I,B)$ for a given B is $\pi(T^*,I^*,B)$. Clearly, $\pi(T^*,I^*,B) \geq \pi(T^*, I,B) \geq \pi(T,I,B)$.

Market Uncertainty

A major source of market uncertainty is the customer preferences. The configuration of B as well as the attribute rankings w_b may not be known with certainty. As we have seen in the context of Fig. 3.5, both B and w_b impact the value of the attraction factor W_i. Therefore, W_i will be a random variable, and so would be D_i. The mean value of D_i will be $\underset{W_i}{E} \left\{ MW_i \Big/ \left(\Gamma + \sum_{j \in I} W_j \right) \right\}$.

It follows that with a production quantity of Q_i, it is possible to have $Q_i < D_i$ as well as $Q_i > D_i$.

The platform profit will now be expressed as

$$\pi(T) = \sum_{i \in I} P_i \underset{D_i}{E}\{Min(Q_i, D_i)\} - h(Q_i - D_i)^+ - s(D_i - Q_i)^+ - \sum_{t \in T}\sum_{i \in I} c_t \alpha_{ti} Q_i - \sum_{t \in T} C^d{}_t \beta_t$$

h is the unit carrying cost of inventory, and *s* is the unit shortage cost.

The profit maximization problem can be expressed as

$$\underset{T, Q_i}{Max}\{\pi(T)\} = \underset{T, Q_i}{Max}\left[\begin{array}{l}\sum_{i \in I} P_i \underset{D_i}{E}\{Min(Q_i, D_i)\} - h\sum_{i \in I}(Q_i - D_i)^+ - s\sum_{i \in I}(D_i - Q_i)^+ \\ -\sum_{t \in T}\sum_{i \in I} c_t \alpha_{ti} Q_i - \sum_{t \in T} C^d{}_t \beta_t\end{array}\right]$$

3.4.4 Other Platforms

Though product platforms have been discussed the most in the literature and used by companies, one should note that a complete platform must also include processes, customers segments, brands, and operations. Some of these platforms are described below.

3.4.4.1 Process Platform

It refers to a set of processes that are shared by multiple tasks (Vuuren and Halman 2001), and determines the range of products that can be produced on the platform (refer to Sect. 3.9). For example, in producing different models of a car, the production processes developed for the platform become the determinants of the possible sizes and weights of modules – dashboards, seats, and doors – that can be assembled to the car. These processes also determine the assembly sequence. Therefore, modifying the process platform can be very expensive. The cost can be reduced by using technology that is not rigid, such as flexible equipment, programmable automation, robots, guided vehicles, and computerized scheduling. Such systems can be quickly reprogrammed and retooled for performing different tasks, as needed. Sony's flexible assembly with a programmable part-orientation system that permits model change with ease is a good example of a flexible process platform (Sanderson and Uzumeri 1995).

3.4.4.2 Customer Platform

A customer segment that a firm chooses in entering a new market is a good example of a customer platform. Such a customer segment serves to determine the company's product and service offerings, and serves as a basis for expansion into related products and markets. Companies need to create the knowledge base of customer preferences and relationships, and leverage commonalities to reduce the cost of products and services, while increasing customization.

3.4.4.3 Brand Platform

A brand platform provides the foundation for specific product brands such as the iphone and Organics, or Philips and BMW (Vuuren and Halman 2001). From this brand platform sub-brands can be created, reflecting the same image and worth: Philishave, Hugo Boss perfumes, and Organic shampoo. With a small set of brand platforms and a relatively large set of sub-brands, a firm can leverage its brand equity across a diverse set of offerings.

3.4.4.4 Global Platform

A global platform enables standardized offering of global products. For example, designing software for a global market can be very challenging, as the goal is to support different countries with applications without modifying the source code. A global plan details the aspects of the product that can be standardized as well as those that must be adapted to be country-specific customer preferences. Customization can involve physical changes in the product; and adaptation in pricing, service, positioning, message, and channel.

3.4.4.5 Industry platform

The owner of an industry platform provides the core components and defines the system architecture. Companies, called complementors, often add components that are consistent with the system architecture and add value to the entire system. They build "on top" of the industry-platform, using services and functionalities of the platform. The industry platform serves as a point of coordination for all industry-participants; it motivates innovation because a large number of companies can either build-on or add to the system. This flexibility leads to greater variety, as the number of possible product configurations can increase rapidly. Platform-based industries also offer high efficiency.

Good examples of industry platforms are the Microsoft Windows and Apple's iphone. The Windows operating system is the central component, and all other products running on Windows must interface with it strongly: many applications rely on program-libraries that enable compatibility with Windows. The platform provides services that the complementary firms make use of in developing complementary applications; it leads to an increase in product variety. As application programmers rely on the toolbox provided by the Windows, they incur lower fixed costs; and this drives down the costs of the entire system.

3.4.4.6 Multiple Platforms

A related issue is whether the company should offer a single platform for all of its products or multiple platforms (Suh et al. 2007; Suh 1990). With multiple platforms, product performance may improve, but the cost of production would also increase. In general, if both product variety and product changes are high, the firm should opt to develop multiple platforms. For example, two platforms producing two different cars share several parts of instrument panels such as mounting screws and small brackets with little loss of distinctiveness. As additional components are shared (gauges, environmental controls, audio systems etc.), the two instrument panels lose distinctiveness.

As established in Section 3.4.3, the demand of the product variant i on the kth platform can be expressed as,

$$D_{ik} = MW_{ik} \Big/ \left(\Gamma + \sum_{k \in K} \sum_{j \in I} W_{jk} \right)$$

Therefore, the platform optimization problem for given I and B will be expressed as,

$$\underset{T_k, Q_{ik}}{Max} \{\pi(T_k, Q_{ik}, I, B)\}$$

$$= \underset{T_k, Q_{ik}}{Max} \left[\begin{array}{l} \sum_{i \in I} P_{ik} \underset{D_{ik}}{E} \{Min(Q_{ik}, D_{ik})\} - h \sum_{i \in I} (Q_{ik} - D_{ik})^+ - s \sum_{i \in I} (D_{ik} - Q_{ik})^+ \\ - \sum_{t \in T_k} \sum_{i \in I} c_t \alpha_{tik} Q_{ik} - \sum_{t \in T_k} C^d{}_t \beta_t \end{array} \right]$$

Profit is maximized in T_k and Q_{ik}, simultaneously for all $k \in K$.

3.5 Postponement

Postponement is a strategy that delays customization of a product until the latest possible point downstream in the supply chain, often by restructuring the supply chain. This implies that the product is partially produced *ex ante*, and customized *ex post*. Postponement decreases the finished goods inventory, without compromising responsiveness to customers.

However, to implement postponement, companies need to restructure the supply chain – redesign the product and realign the manufacturing and delivery processes. The supply chain must be capable of supplying the basic product to the customization facilities, and it must be able to deliver the finished customized goods to customers quickly. As we discuss next, the modularization of products and processes makes it easier, in a relative sense, to restructure a supply chain (Fine and Whitney 1996; Fine et al. 2005).

The HP's narrative on postponement is well-known to many. They used the notion of "core plus peripherals", which is a variation of modular design. HP built a core printer with a large number of common modules that did not vary with customers. They also created peripheral components, each with specific functionality, that could be assembled to the core at the distribution centers or at customer sites. This allowed HP to postpone customization of printers for its European customers, who needed multiple language support. Lucent Technology used the same principle for successfully bidding on a Saudi contract. By building the core in advance of the order, they could satisfy a stringent delivery time constraint in the Saudi contract (Hoyt et al. 2007; Chakravarty 2010).

As the specially designed common components in the core product, which can be standardized, may increase the cost of materials, companies must carefully assess whether the benefits of standardization outweigh the added costs. Using standardized components decreases the size of the needed buffer inventory. As buffer inventory increases in the uncertainty, standardized components in the core helps reduce cost in a volatile market. For example, products with shorter life cycles increase uncertainty, increasing the benefits of standardization.

The supply chain must be restructured to implement postponement. Based on the extent of customization needed, the development team must redesign the product into a modular form. Manufacturing and assembly processes must be restructured to produce the modified components and the modules. Distribution logistics must coordinate the shipment of the core product to distribution centers with product customization at those centers (refer to Chap. 6 for further discussion).

In another division of HP (disk-drive), the existing processes for testing made it difficult to accommodate customer orders because of frequent revisions. This process required a customer-specific printed circuit board to be inserted prior to the test-process (Feitzinger and Lee 1996). HP restructured the test process by splitting it into two sub processes: standard tests for all products, and the customized tests specific to individual products. HP tested its disk drives through the standardized test process first, and held the generic units in storage. When an order arrived, it added the circuit board that the customer required, performed the customized tests, and shipped the product.

In the garment industry, postponement involved re-sequencing of processes. Instead of first dyeing the yarn into different colors and then knitting it into finished garments, Benetton changed the order of the dyeing and knitting sub processes. The company dyed the colorless knitted sweaters when it received an order. By rearranging the sub processes, Benetton effectively postponed the point of product differentiation and thereby saved millions of dollars in charges for obsolete inventory.

3.6 Customer Economics

The design of a product not only determines its manufacturing and logistics cost, it also impacts the cost the customer may incur in using it. First, there is a cost to the customer in switching from the product or service it currently uses, to the new product or service. Thus, in a broader sense, the customer's cost would be reduced, if the supplier solved customer's total problem and deemphasized "making a sale".

Installation of products and systems is a critical issue in any switchover. The supplier may need to modify its products and systems so that installation by the customer is smooth and without a major downtime. Typical installations require many special components from multiple suppliers, and these components, if missing or delayed, hold up installations for weeks at times. Therefore, the suppliers' delivery schedules need to be coordinated so as to have the right components delivered to the sites at the right times and in right quantities. Companies create special functions called "design for installation" (DFI), similar to design for manufacturing (DFM) to minimize the cost of incorporating a new product in the current system.

To ensure that customer receives other components of the system that the company does not make (as discussed in Chap. 2), the company can create a consortium or partnerships with other component manufacturers or service providers. Cisco Systems, manufacturer of Internet routers, created a partnership with KPMG, a management consulting company, to provide service in network management and other software to customers who need those services to use Cisco's routers. Lucent Technologies (now Alcatel-Lucent), on the other hand, acquired a fast growing consulting company called International Network Services in a bid to move beyond the manufacture of complex communication equipment and expand its ability to help its corporate clients install and use them. The Cisco/KPMG deal is a typical example of an extended enterprise, for which issues such as seamless interfacing and transparency become critical in designing an appropriate supply chain.

3.7 Introducing New Products

Introduction of new products often causes disruptions in the supply chain. Such disruptions may impact the inventories of components or they may require restructuring the architecture of the supply chain.

3.7.1 Product Rollover

To introduce a new or modified product, companies need to decide what it should do with the current product. They can discontinue the product instantly, they can phase it out gradually over time, or they can continue to produce and sell the current product in different geographical regions. If they adopt the second option, they would need to manage the leftover inventory of the current product when rolling over to the new product. If it is decided to discontinue production of the current product, the finished product can be disposed of in different ways: markdowns, product-dismantling for use as spare parts, and recycling of material. The leftover inventory can also be sold at discount stores, outlet malls, and online stores such as Overstock.com. If the current product is assigned to a new market (usually overseas) the company would need to put together a distinct supply chain, to incorporate the new market.

A pervasive problem is the inventory of *components*, whether the product is phased out or discontinued instantly. In industries such as aeronautics, the investment in such components can be substantial. Therefore, making use of this inventory can be a powerful economic argument for phasing out (not instant discontinuation) the product. It can be a complex issue, as the new and current products may have components in common. While production of the common components must continue, components which are relevant only to the current product become obsolete, and must be discontinued. Disruptions in the supply chain can be significant, requiring a restructuring.

In the third option, if the new market is overseas, the company would need to decide where to locate component production and assembly. If the supply chain straddles multiple countries, the company would need new expertise in freight forwarding, warehousing, and local supply chain practices. We discuss these further in Chap. 7.

A related option is to sell the current product through a new channel such as mail order or the online channel.

The company will need to manage three groups of suppliers if the phase-out option is chosen. Suppliers of components that are not used in the new product must draw down their capacity or look for new customers. As these suppliers are negatively impacted, the company can explore opportunities to transition them to supplying some of the components of the new product. Suppliers of common components must increase capacity immediately, and then reduce it gradually as the current product phases out, and reach a steady state when only the new product is in production. Finally, new suppliers must be sourced for components unique to the new product. As new specialized capacity may be needed, the company would need to start working with potential suppliers much earlier than when the new product is introduced.

The company would also need to manage its customers. Discontinuation of the current product may not sit well with some customers. Instead of losing them to competitors, the company may want to explore whether the customers' business processes can be modified sufficiently for them to use the new product (with some customization). For the new customers of the new product, the company would need to work on pricing, contracts, and returns.

If the new product is not too different from the current product, the company may be able to use its existing warehouses, transportation carriers, and fulfillment-contracts. However, if the new product is bulky, heavy, perishable, or fragile (in relation to the current product) the fulfillment chain would need to be totally redesigned. For example, more secure packaging may be necessary for protection against damages, and a faster transportation mode may be required to minimize damage from perishability. If the new product generates demand from new customer segments, the unique fulfillment needs of those segments would also need addressing.

The new product may result in an increased capacity requirement, and if additional in-house capacity is not available in the same period, it would lead to an increase in subcontracting and/or backorders. A botched product rollover can have significant implications, as unforeseen delays in rolling over to the new product may create a vacuum where some customers receive no deliveries of products, current or new.

Another aspect of product rollover is responding to the variant-specific demands. Manufacturers and key suppliers must be updated quickly with accurate information on demand and consumer preferences for different variants. For example, products may first be introduced into those markets where there are is a strong preference for a product-variant, but if this is not communicated to suppliers and as the demand for different variants changes over the lifecycle, suppliers would not be able to keep pace. In a different scenario, if some product variants are not introduced until later due to supply constraints, but the supply constraint for a variant is not communicated to all suppliers, suppliers will be subjected to very low demand early and a huge demand surge later.

Hewlett-Packard Laboratories has developed the Inventory Optimization for Product Rollover (IOP), a tool for the enterprise that allows product planners to analyze various trade-offs and scenarios (HP 2002). IOP uses business and manufacturing information such as component and assembly stock on hand, finished product inventories, component and product costs, and product demand forecasts. The objective is to provide recommendations about quantities of products to sell, quantities of products or components to build, quantities of additional raw materials or components to purchase, and quantities of specific parts to scrap or sell at discount.

While IOP is a useful tool from the perspective of a single enterprise, it is not sufficient for the supply chain restructuring needs. Such tools can be useful, however, as an input for the supply chain wide decisions.

For a successful product rollover, while companies need to assess the uncertainties associated with manufacturing and delivery, as well as with the product's market potential, it must be done in the context of the supply chain. They need to figure out the timing (cut-in of the new product and cut-out of the current product), component inventory utilization, supplier relationships, new capacity needs, production/assembly cost, pricing, fulfillment plan, and customer relationships, in the context of both the new and current products (Balakrishnan and Chakravarty 1996).

3.7.2 Complex Products

To reduce the development time and cost, Boeing decided to develop and produce the Dreamliner (787) by modularizing its supply chain design (Tang and Zimmerman 2009). Boeing's primary motivation was to share the development risk of the new aircraft with its suppliers. The supply chain was structured in tiers, with significant decision rights vested in the tier-1 suppliers. With n tier-1 suppliers, such a supply chain can be sliced into n segments, each segment being the responsibility of a tier-1 supplier. The tier-1 suppliers became Boeing's strategic partners, and Boeing was able to build close relationships with them. This was in sharp contrast to the supply chain for the 737 aircraft, which required Boeing to procure and assemble many parts and subsystems, produced by thousands of suppliers. Specific restructuring required for the 787 aircraft, in relation to the 737 aircraft, are compared in Table 3.3.

As shown in Table 3.3, tier-1 suppliers were responsible for delivering entire sections of the aircraft such as fuselage and wings to Boeing, which allowed Boeing to assemble these complete sections in a few days at its plant in USA. The tier-1

Table 3.3 Restructuring supply chain for a complex new product (Source: Tang and Zimmerman 2009)

Design elements	737 supply chain	787 supply chain
Sourcing strategy	Moderate outsourcing	Outsourcing increase by over 40%
Supplier relationship	Transaction based	Strategic relationship with tier-1 suppliers
Supplier responsibilities	Suppliers responsible for specific components	Suppliers responsible for entire sections of the aircraft
Number of suppliers	Boeing managed thousands of suppliers	Boeing needed to coordinate with only 50 tier-1 suppliers
Supply contracts	Fixed price contract	Risk sharing contracts
Assembly operations	Boeing needed about 30 days to assemble all components and sub-assemblies	Boeing needed 3 days to assemble complete sections provided by tier-1 suppliers

suppliers in the 787's supply chain, shown in Fig. 3.6, served as "integrators" who assemble different parts and subsystems produced by tier-2 suppliers.

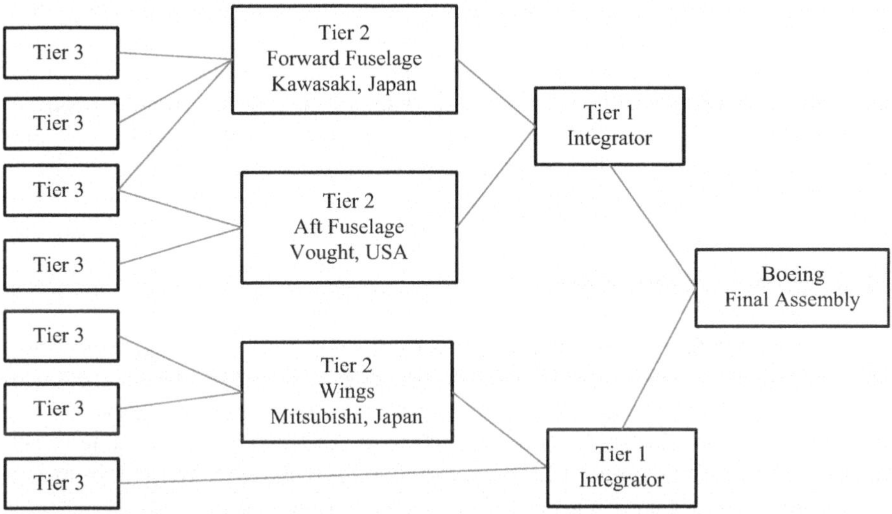

Fig. 3.6 Boeing 787 supply chain

By outsourcing a large chunk of the development and production activities to its suppliers, Boeing managed to shorten the development time considerably, and it permitted them to develop different segments concurrently. Boeing implemented a web-based tool called Exostar to facilitate collaboration among suppliers and itself (Tang and Zimmerman 2009). Exostar was intended to provide supply chain visibility, improve control and integration of critical business processes, and reduce development time and cost.

3.8 Process Structure

The major supply chain processes were outlined in Chap. 1, and are shown in Table 3.4, as customer management, procurement, production, and fulfillment.

Table 3.4 Supply chain processes

Customer management	Procurement	Production	Fulfillment
Product design	Order management	Facilities	Delivery route planning
Demand shaping	Contract management	Flexibility, automation	Tracking of orders in transit
Preference mapping	Channel management	Outsourcing quality assurance	Information technology
Pricing	Supplier relationship	Throughput maximization	Third party logistics

Each of these processes comprises several sub processes. For example, the fulfillment process can be broken down into vehicle routing, shipment tracking, creating information flow, and transportation. How these processes function can have significant impact on the supply chain structure. For example, if contract manufacturing is used in the production process and drop-shipping in the fulfillment process, the supply chain would need significant restructuring to enable the drop shippers to pick up the products from the contract manufacturers at the right time and deliver them directly to customers. Conversely, if the supply chain is restructured to include a foreign supplier, the procurement process including quality control and ordering processes would need major restructuring.

3.8.1 Process Definition

Processes are the fundamental building blocks of a company's operations, and both the understanding and improvement in processes determine their excellence. Processes transform inputs (actions, methods, and operations) into outputs. They are the steps by which value is added to satisfy the expectations of customers. The processes interact with other processes in the supply chain, as outputs from one process form the inputs to another. Each process is therefore a part of a larger process so that an organization can be seen as a network of interconnected processes. In addition to the inputs and outputs, a process is defined by controls (a set of rules that determine how the process is to be performed), and the needed resources. Hence, we may describe a process as, *process* = *f*(*inputs*, *outputs*, *controls*, *resources*).

The inputs of a process are the entities that are transformed by the process into a product or service. Inputs can be tangible such as physical components, or intangible such as verbal requests. The outputs of a process (products or services) should conform to the stipulated specifications as designed. Process controls may be imposed externally such as that by a regulatory authority, or internally based on the process-chemistry or the quality imperative. While customer specifications, environmental requirements, and copyright laws are externally imposed, quality checks and temperature controls are designed by the organization. Processes need resources to convert the inputs into outputs. Resources may be tangible (people, computer, money), or intangible (skills).

There are two types of business processes: operational and strategic. Operational processes involve routine transactions and resource allocation, while strategic processes help set direction for the company to be competitive from a long term perspective. Redesign of strategic processes involve reshaping of the generic process capabilities, enabling broader interdisciplinary and boundary spanning processes, developing core competency, incorporating new technologies, and streamlining the process of collaboration and coordination.

Each of the above processes can be further categorized as physical or informational. Physical processes create or modify products, as in manufacturing. Informational processes involve flow of information, as in making a decision, preparing a marketing promotion, and monitoring a project. Most processes are driven by some combination of physical and information flow characteristics. For example, information is often added to create value in a product, while it transitions from the upstream supply chain to downstream. Logistics tasks combine the movement of physical products with locational information to track movements of packages.

3.8.2 Integrating Knowledge

Knowledge must be integrated by firms to create process capability. Explicit knowledge, such as drilling a hole or assembling components, poses few problems of integration because it can be communicated easily. Information technology facilitates codification, storage, retrieval, transmission, and assimilattion of the explicit process knowledge. Tacit (implicit) process knowledge, on the other hand, is difficult to codify. There are two approaches that companies usually adopt for managing the tacit knowledge. One is providing *direction* for completing tasks, and the second is creation of *routines*.

McDonald's has created an operating manual which provides direction for almost all the processes of the restaurant such as cooking, nutrition, hygiene, engineering, marketing, procurement, HR, accounting and finance. In creating a process *routine*, a set of activities is said to be routinized if it can be implemented through fixed responses to defined stimuli. For example, a surgical team in an operating room coordinates the specialist knowledge of each member of the team through a working arrangement; it relies on commonly understood roles and interactions established through years of training and constant repetitions, and supported by a series of explicit and implicit signals.

Increasing the span of knowledge in a process increases the potential for establishing and sustaining advantages. The greater the scope of knowledge being integrated in a supply chain process, the greater the difficulty faced by competitors in replicating that capability. Toyota's lean production system combines cost efficiency, quality, flexibility, and innovation, each of which require a different type of integration. While sequential interdependence may lead to efficiency, complex patterns of reciprocal interdependence must be integrated to obtain flexibility. For example, the suppliers of high fashion clothing need to manage a very complexknowledge integration, in combining a low costs objective with differentiation and quick response.

Maintaining superior performance may require new process capabilities. This can be realized by either extending existing process capabilities to encompass new knowledge, or reconfiguring existing knowledge into new processes. If new knowledge is explicit, integrating it does not pose much difficulty. In designing the 777 aircraft, Boeing provided access to its knowledge of electronics and new materials through an advanced CAD system that could be shared by specialists from different suppliers with different backgrounds (Grant 1996). In contrast, General Motors' adoption of lean manufacturing was slow, as it attempted to incorporate the tacit knowledge possessed by Toyota.

The reconfiguration of existing process knowledge through new patterns of integration, though complex, may be even more important in creating competitive advantage. In the high-fashion industry, for example, Zara has created quick response capability through innovative approaches to value-chain reconfiguration (discussed in Chap. 1).

3.9 Process Platform

Conceptually, a process platform can be considered the means of operationalizing the core competencies of a company. It defines the processes the company excels in, and has built-in competitive advantage. That is, the company possesses expertise in quickly implementing these processes and can make them function cohesively. To the extent these processes are adequate for exploiting a new business opportunity; the company would perform exceedingly well. However, if that is not the case and the process platform is inadequate, the company would either need to invest heavily to renew the platform or forgo the new opportunity.

Consider a typical university that has built teaching expertise in terms of course design, course delivery, classroom scheduling, faculty scheduling, student enrollment, and student evaluation – all oriented to face-to-face interactions in classrooms. The face-to-face orientation can be considered a platform for university pedagogy. Recently, this platform has come under considerable pressure from the massive open online courses (MOOC) that are very cost effective for both the university and the students (McAuley et al. 2010). Universities are struggling with how to respond to this new challenge: create a separate platform for MOOC, reengineer the current platform to incorporate MOOC synergistically with face-to-face model, and ignore MOOC. While creating a separate MOOC platform would be the simplest solution, it may cannibalize the current face-to-face product.

Creating a single platform to incorporate MOOC, even if it is possible, has the potential to cause disruptions in the pedagogical supply chain of the university for years to come.

A process family, in a process platform, comprises a set of process variants necessary to process all product variants in that family. Component commonality causes the processes and their sequences, related to products, to be similar. Therefore, there exists a common product structure and a common process structure within a product family, and variety is the embodiment of the variants of these common structures.

A typical manufacturing or assembly process, producing a family of product variants, employs a set of machine classes, each class comprising a number of machines, and a number of material handling devices. For example, in moving materials, assemblies, and finished products, the material handling processes variants would be based on conveyor belts, AGVs, robots, and fork lifts.

To select the processes for the process platform, production-knowledge such as rules and constraints act as guidelines. These rules also establish the precedence relations among processes. For each assembly process the sequences of execution and resources requirement for its sub-assemblies are specified. Corresponding to parts at the lowest level of each assembly, the set of machining operations and their execution order are provided.

3.9.1 Process Configuration

The configuration of processes determines how the customization needs of customers are satisfied. Hence, the formal representation of process configuration becomes critical. An important issue in process modeling is the understanding of the characteristics of the processes so as to design or select appropriate modeling tools. The issues associated with process configuration are product variety, process modifications, and constraints.

The number of components designed and produced by a company can be large (in order to satisfy the diverse customer-preferences), resulting in a large number of operations, precedence relations, and needed resources. Thus, the decision on process and product variety should be made simultaneously. Frequent changes in design specifications of products cause corresponding changes in processes. The changes in operations, process sequences, and resources lead to structural changes, and they must be considered explicitly in designing processes.

Next, to ensure compatibility in process configuration several constraints must be satisfied. First, the "inclusion" conditions determine whether specific processes should be included in a configuration. Second, the precedence constraints that determine which processes are completed before the commencement of others must be clearly stipulated. These constraints, therefore, determine the sequence of processing, also known as process-flow. Finally, the execution-rules specify equipment needed. The process sequence together with the assignment of resources to process determine the estimated cycle time for completing the operation.

In the context of a supply chain, process decisions include transportation, contractual relations with suppliers, supplier monitoring, warehousing, distribution, and postponement. Process decisions also include service levels, delivery schedules (e.g., just-in-time), vehicle routing, and crew planning. Product characteristics that may affect these processes are the degree of commonality across components, the way in which the components interact with each other, and the type of the interfaces between the components. For example, supplier relationships, service levels, and delivery frequencies, are all impacted by the number and type of components a product is made of.

3.10 Process Restructuring

A physical process may involve combining two or more tasks, splitting a large process, altering task sequence, relocating tasks to different sites, and maximizing parallel processing. While combining tasks obtains scale economy and better control, splitting a process may enable enhanced agility in the system and increased customization.

For example, a company making a large variety of paints in different colors and textures can split the paint making process into several sub processes (Feitzinger and Lee 1996). It can make a single generic paint and a variety of color pigments, which the retail stores stock. They can then mix the color pigments with the generic paint based on customers' specific requirements. The supply chain must be restructured accordingly to incorporate the split processes. This would involve splitting the paint production process into two: a simple mass volume production of the generic paint, and low volume processes for producing specific color-pigments. The retail stores now assume additional responsibilities: storing the color pigments and the generic paint, investing in a chromatograph and using it to match customer's paint preference with the color pigments, and mixing the pigments with the generic paint. It also alters the customer's processes, as he/she must wait for the completion of the chromatograph analysis and the process of mixing the pigments with the generic paint.

Modularization of a process is a form of process splitting. From a supply chain perspective, the modular approach differs from the traditional approach in two aspects: first, the decisions regarding multiple components can be centralized in a single tier-1 supplier responsible for a segment of the supply chain; second, a portion of the assembly job is moved from the final stage to upstream tiers, reducing the production capacity need at the final stage. In the modular assembly model used by Nissan, suppliers deliver assembled vehicle sections for the final assembly line (Chappell 2001). Volkswagen recently introduced its modular assembly platform hoping to decrease the assembly time by 30% (Bidsar 2012; Feng and Zhang 2013). Boeing's modularization of its 787 aircraft production was discussed earlier.

Compared to the traditional approach, modularization of assembly processes can help integrate sections of the supply chain by allowing a single supplier to manage many suppliers that help build a section of the supply chain. In contrast, in a traditional assembly system many individual component suppliers independently manage their operations. By using modularization, the manufacturer can also outsource an increasing proportion of the assembly job to its suppliers.

In contrast, if agility is not an issue, combining tasks may be beneficial because of the scale economy arising from reduced overhead costs. In scenarios where the processes are complex, combining them may better facilitate coordination. Process combination may also motivate creativity and innovation, as proximity facilitates frequent exchange of ideas among workers.

The underwriting process of Mutual Benefit Life (MBL) involved 40 steps with over 100 people in 12 functional areas and 80 separate jobs (Davenport and Short 1990). To restructure, it merged the data analysis and consolidation tasks under a case manager who coordinated all underwriting tasks centrally, pulling data from all over the company. Other specialists such as lawyers and medical directors in knowledge-intensive fields were also added.

Relocation of the final assembly task closer to a customer may provide fulfillment efficiency. Processes can also be relocated to a low wage region for cost saving. An automatic manufacturing line of a Danish company included four robots and a number of automated manufacturing processes like welding, rolling, turning, grinding and assembly. The manufacturing line was transferred from Denmark to a green field site in Mexico (Madsen et al. 2008). At the new location, a large manufacturing building and a new organization were established to host the line.

Task re-sequencing and parallel processing can help with improving resource utilization and/or reducing the project completion time. Any redesign of a process requires corresponding redesign of the inputs, outputs, controls, and resources. By re-sequencing tasks, Benetton succeeded in customizing its sweaters (discussed earlier).

Exploiting parallelism in business processes can speed up project completion time. This is possible for processes that do not possess sequential dependencies. The considerable time savings accrued by exploiting parallelism may justify additional efforts needed to manage parallelism. However, complex processes may inhibit exploitation of parallelism as it may increase the coordination effort, disproportionately.

Redesign may also involve programmable automation of processes so as to increase flexibility and the scope of outcomes: rapid switch-over between processes, more precise control, reductions in throughput time, and reduction in manpower. Automation is easier for tasks with low complexity. Automation of processes that incorporate semi-structured and unstructured tasks and are performed by high-skill knowledge workers remains challenging. The IT capabilities needed include: storage and retrieval of unstructured and multimedia information, capturing and automating the decision logic, and managing the pool of far-flung and complex data resources.

3.11 Applications of Process and Supply Chain Redesign

The linkages between the manufacturing and product design processes are shown in Table 3.5.

Table 3.5 Product design and manufacturing processes

Product design	Manufacturing
Design-to-market integration	Overlapping engineering processes with design
Integrating suppliers in the design process	Supplier relationship process
Component planning	Manufacturing process and outsourcing
Integrating product launch	Coordinating inventory network
Design compliance	Manufacturing sustainability

With multiple processes, such linkages define the entire supply chain. Therefore, the interaction between processes and the supply chain needs to be understood from two different perspectives: processes altering the supply chain, and the supply chain altering the processes.

3.11.1 Processes Dictating Supply Chain Capability

Companies can increase process reliability and hence improve product quality by redesigning the work method. This provides opportunities for the company to penetrate different market segments, where customers are willing to pay higher prices for the product. However, for this to happen, the company would need to restructure its supply chain to include a fulfillment process for the new market segments. In addition, the segments may require quick deliveries and better after-sales-service.

The choice of technology often determines processes, which can then influence the structure of the supply chain. Consider the technology for making hamburger, which defines the cooking process. McDonald's uses grills while Burger King uses broilers. Typically, a maximum of 12 patties can fit onto a grill. Therefore, McDonald's uses a batch process, cooking 12 patties at a time, and stocking the cooked patties and buns ready for assembly; the buns and patties are mated during the assembly process. Burger King, on the other hand, uses a continuous broiler assembly process – similar to an assembly line in manufacturing – requiring no human intervention. Burger King's "make to order" process requires semi-finished inventory in which mated buns and uncooked patties sit. As the cooked burger and bun pairs exit the broiler one at a time, the hamburger can be customized to customer's preferences. Therefore, Burger King's process is make-to-order and is more customizable, while McDonald's process is make-to-stock and is standardized. It should be noted that BK's mating of the buns and patties before the assembly process creates product variety in the supply chain such as whoppers and burgers as per customer order, which are of different sizes. McDonald's make-to-stock process hinders customization.

3.11.2 Supply Chain Dictating Process Capability

If the processes in a supply chain are not coordinated, they impact multiple companies negatively. It may lead to activities being replicated, inconsistent flows, and erroneous decisions. It may take a great deal of effort to manage the cumbersome interactions between companies. For example, HP's supply chain for computer monitor comprised a contract manufacturer (Solectron), who bought the plastic case for the monitor from an injection molder, shown in Fig. 3.7 (Hammer 2001). The molding-material was purchased from a plastics compounder, who acquired the resin (for compounding) from a resin supplier. The supply chain was linear and hierarchical, with the resin supplier upstream, HP downstream, and the rest in-between.

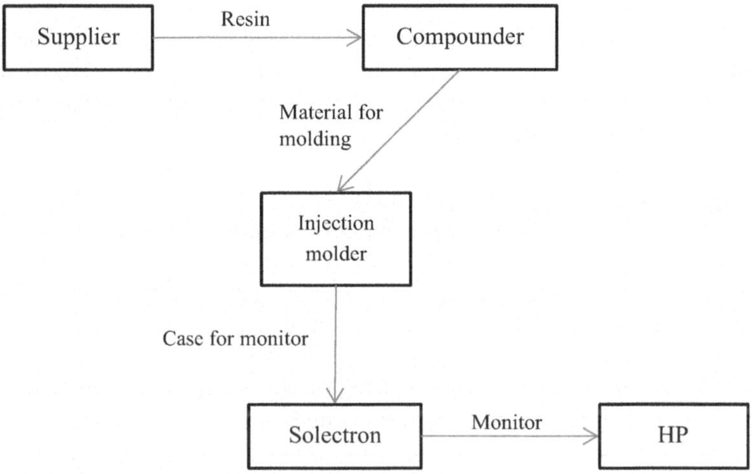

Fig. 3.7 HP's linear supply chain for the monitor

The linear supply chain suffered from a lack of transparency that caused mismatches between production and demand; and led to accumulation of inventory upstream, and loss of customers as well as revenues downstream. The disparity in scale was another problem: HP, Solectron, and the resin supplier are large companies, while most injection molders and compounders are relatively small. Therefore, every HP order for monitor cases was split among many compounders, who bought resin in small volumes and, consequently, at relatively high prices from the resin maker.

HP has transformed the supply chain to a hub-and-spoke architecture to coordinate the supply process by sharing information among the participants, shown in Fig. 3.8.

Fig. 3.8 HP's restructured hub and spoke supply chain

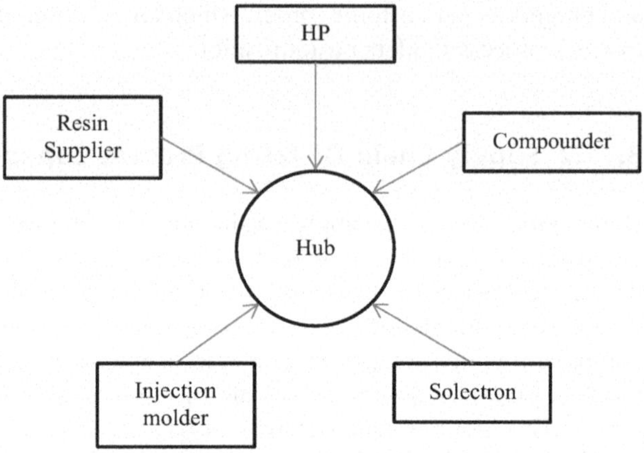

HP posts its demand forecasts and all forecast revisions for its partners to generate their own forecasting, at the hub. The partners post their plans and schedules and use the system to communicate with their own suppliers and customers, exchanging electronic orders, acknowledgments, and invoices. HP owns and manages the processes at the hub: monitoring the performance of suppliers, setting procurement guidelines for quality control, ensuring adherence of schedule deadlines, helping to resolve conflicts between partners, and monitoring the total capacity and demand in the supply chain. As all information must flow through the hub, any change to an order becomes instantly visible, allowing everyone to react quickly. Information visibility permits HP to order all its required resins directly from the resin supplier. HP aggregates orders for resin from all compounders, and obtains a price discount for bulk purchase from the resin supplier. HP arranges for the resins to be shipped directly to the compounders from the resin supplier, using drop shipping. The resin maker gains advantage from this new relationship as well; it benefits from the simplicity and security of dealing with one large customer rather than a host of small ones. Individual partners needed to restructure their processes to be able to access and complete transactions on the hub.

The second example comes from the apparel industry. Several U.S. retail, apparel, and textile companies have linked their business processes to speed up reordering of apparel. When Dillard's (department store) inventory of a particular pants style fell below a specified level, Haggar (apparel manufacturer) was notified electronically (Davenport and Short 1990). If Haggar did not have the cloth to manufacture the pants, Burlington Industries (textile manufacturer) was notified electronically. Thus, process design by these three companies needed to be much broader than simply speeding up the routine purchasing transactions involving invoices or bills of materials. Redesign included processes ranging from demand awareness, to the development of vendor lists, and the delivery and use of the purchased product. Sellers needed to look at all buyer processes in which their products were involved.

Du Pont's uses "effectiveness in use" as the major criterion of customer satisfaction (Davenport and Short 1990). The company is motivated to link its internal processes for creating value in a product, to its customer's processes for using the product, discussed in Chap. 2. Du Pont provides "material safety data sheets", along with the chemicals it sells, to its customers to ensure their safe use.

The fourth example of process redesign comes from Toyota's auto plant in Brazil (Pires et al. 2012). A major Brazilian supplier produces several "lines" such as pipes and hoses for automotive fluids, especially brake fluid, and fuel. The supplier's supply chain includes ten major processes: material purchase, receiving and storage, production planning, cutting the material, flange conformation and inspection, bending, kit assembly, final assembly, packaging, and transportation.

In 2010, Toyota decided to build a supply chain campus at a new location that would include its ten major suppliers. Toyota was motivated by two factors: reducing its logistics cost of doing business with its supplier, and ensuring just-in-sequence deliveries from suppliers (the final assembly schedule is sent to the supplier via EDI, shortly before the actual need for the item).

To meet the needs of the automaker within the campus, the supplier needed to perform a series of changes in its internal supply chain processes. The supplier moved processes (6) to (10) to the campus as they needed to be completed in real time, while processes (1) to (5) remained at their current location. The supplier bore the responsibility for all transportation between its current plant and the new campus. The automaker managed its final assembly line at the campus using a novel system called "jumbiki" which required short replenishment cycles of procurement from its suppliers. The supplier's facility within the campus, in essence, acted as an extension of Toyota's final assembly line. Thus, when the vehicle body was about to leave the paint shop, Toyota notified the vehicle's final configuration, as well its final sequencing on the assembly line, via EDI to the suppliers.

The supplier had at most 36 min to perform processes (6) to (10) and make the assemblies available for Toyota to collect them, through a milk-run within the campus. The parts from Toyota's ten suppliers were picked up in a precise sequence that matched the assembly sequence. The operation of the milk run was placed under the responsibility of a logistics service provider.

References

Balakrishnan, N., & Chakravarty, A. (1996). Managing engineering change: Market opportunities and manufacturing costs. *Production and Operations Management, 5*, 335–356.

Bidsar, K. (2012). Future Volkswagen MQB platform unveiled – All VW cars to be built on this platform. http://www.carblogindia.com/volkswagen-mqb-flexible-platform-revealed/

Carloni, L., Bernardinis, F., Pinello, C., Sangiovanni-Vincentelli, A., & Sgroi, M. (2000). Platform-based design for embedded systems. http://www.cs.columbia.edu/~luca/research/pbdes.pdf

Chakravarty, A. (2010). *Market driven manufacturing*. Boston: Now Publishers.

Chakravarty, A., & Balakrishnan, N. (2001). Achieving product variety through optimal choice of module variations. *IIE Transactions, 33*, 587–598.
Chappell, L. (2001, March 5). Nissan's solution: Modules. *Automotive News*, Issue 5919
Davenport, T., & Short, J. (1990, July 15). The new industrial engineering: Information technology and business process redesign. *Sloan Management Review – Magazine*. http://sloanreview.mit.edu/article/the-new-industrial-engineering-information-technology-and-business-process-redesign/
Diez, J. (2000). Innovative networks in manufacturing: Some empirical evidence from the metropolitan area of Barcelona. *Technovation, 20*, 139–150.
Feitzinger, E., & Lee, H. (1996, January–February). Mass customization at Hewlett Packard: The power of postponement. *Harvard Business Review*, 116–120
Feng, T., & Zhang, F. (2013). *The impact of modular assembly on supply chain efficiency*. Production and Operations Management. doi:10.1111/poms.12182. http://onlinelibrary.wiley.com/doi/10.1111/poms.12182/full
Fine, C., & Whitney, D. (1996). *Is the make-buy decision process a core competence* (Working Paper). Cambridge, MA: MIT Center for Technology, Policy, and Industrial Development.
Fine, C. H., Golany, B., & Naseraldin, H. (2005). Modeling tradeoffs on three dimensional concurrent engineering: A goal programming approach. *Journal of Operations Management, 23*, 389–403.
Fisher, M. (1997). What is the right supply chain for your products. *Harvard Business Review, 75*, 105–116.
Fisher, M., Ramdas, K., & Ulrich, K. (1999). Component sharing in the management of product variety: A study of automotive banking systems. *Management Science, 45*, 297–315.
Grant, R. (1996). Prospering in dynamically-competitive environments: Organizational capability as knowledge integration. *Organization Science, 7*, 375–387.
Gupta, S., & Krishnan, V. (1998). Integrated component and supplier selection for a product family. *Production and Operations Management, 8*(2), 163–182.
Hammer, M. (2001, September). The superefficient company. *Harvard Business Review*, 82–91.
Hoyt, D., Lopez, E., & Lee, H. (2007). *Lucent technologies: Provisioning and postponement, Case: GS-02*. Stanford Graduate School of Business. Stanford University. Palo Alto, California, USA
HP (2002). Inventory optimization for product rollover. HP Invent. ftp://15.217.49.75/pub/services/erp/info/hp_lab_iop.pdf
Krishnan, V., & Ulrich, K. (2001). Product development decisions: A review of the literature. *Management Science, 47*, 1–21.
Lilien, G., Kotler, P., & Moorthy K. (1992). Marketing models, Chapter 2, Prentice Hall: Englewood Cliffs, CA.
MacCormack, A., Newmann, L., & Rosenfield, D. (1994). The new dynamics of global manufacturing site location. *Sloan Management Review, 35*, 69–80.
Madsen, E., Chengand, Y., & Liangsiri, J. (2008). Means for transferring knowledge in the relocation of manufacturing units. http://www2.ifm.eng.cam.ac.uk/cim/symposium2008/08proceedings/10%20-%20Erik%20Skov%20Madsen%20et%20al.pdf
McAuley, A., Stewart, B., Siemens, G., & Cormier, D. (2010). The MOOC model for digital practice. http://www.elearnspace.org/Articles/MOOC_Final.pdf
Muffatto, M., & Roveda, M. (2002). Product architecture and platforms: A conceptual framework. *International Journal of Technology Management, 24*, 1–16. http://www.csun.edu/~rsh12598/415/additional/MSE-415%20Reading3.pdf.
Piezunka, H. (2011). Technological platforms – An assessment of the primary types of technological platforms, their strategic issues and their linkages to organizational theory. *Journal für Betriebswirtschaft, 61*, 179–226.
Pires, S., Guarnieri, E., Neto, M., & daSilva, E. (2012). A supplier's restructuring process to become a "Systemist" within a new automotive industrial condominium. http://www.pom2012.org/fullpapers/add199978_yJtd7bZw0r.doc.pdf
Pisano, G. P. (1994). *ITT automotive: Global manufacturing strategies, Case 9-695-002*. Cambridge, MA: Harvard Business School.

Randall, T., & Ulrich, K. (2001). Product variety supply chain structure, and firm performance: Analysis of the U.S. bicycle industry. *Management Science, 47*, 1588–1604.

Robertson, D., & Ulrich, K. (1998). Planning for product platforms. *Sloan Management Review*, 19–31, Summer

Sanderson, S., & Uzumeri, M. (1995). Managing product families: The case of the Sony Walkman. *Research Policy, 24*, 761–782.

Suh, E. (1990). *The principles of design*. New York: Oxford University Press.

Suh, E., Weck, O., & Chang, D. (2007). Flexible product platforms: Framework and case study. *Research in Engineering Design, 18*, 67–89.

Takeishi, A., & Fujimoto, T. (2001). Modularization in the auto industry: Interlinked multiple hierarchies of product production and supplier systems. *International Journal of Automotive Technology and Management, 1*, 379–396.

Tang, C., & Zimmerman, J. (2009). Managing new product development and supply chain risks: The Boeing 787 case. *Supply Chain Forum: An International Journal, 10*, 74–85.

Ulrich, K. (1995). The role of product architecture in the manufacturing firm. *Research Policy, 24*, 19–440.

Ulrich, K., & Eppinger, S. (1995). *Product design and development*. New York: McGraw-Hill.

Verdouw, C., Beulens, A., Trienekens, J., & Verwaart, T. (2010). Mastering demand and supply uncertainty with combined product and process configuration. *International Journal of Computer Integrated Manufacturing, 23*, 515–528.

Vuuren, V., & Halman, J. (2001). *Platform driven development of product families: Linking theory with practice* (Working Paper 01.06). Eindhoven: Eindhoven University of Technology. http://alexandria.tue.nl/repository/books/551507.pdf.

Williams, C., Allen, J., Rosen, D., & Mistree, F. (2007). Designing platforms for customizable products and processes in markets of non-uniform demand. Georgia Institute of Technology. https://smartech.gatech.edu/bitstream/handle/1853/32890/CBW.CERA07.DRAFT.pdf?sequence=1

Managing Suppliers

4

4.1 Introduction

Supplier management involves multiple partners in the supply chain with evolving expertise. Suppliers are the source of components, raw material, finished products, services and, more importantly, new knowledge creation. They can help companies in improving supply chain performance in cost, quality, revenues, responsiveness, and reliability.

However, not all suppliers are equally capable in all performance dimensions. Individual suppliers must decide what capabilities they would build, given changing customers' needs (discussed in Chap. 2). Examples of such capabilities would be lean-processes, collaboration, procurement, agility, information transparency, responsibilities as tier-1 supplier, and innovation. The buyer, therefore, must carefully build a portfolio of suppliers that maximizes the fit between what it needs and what the suppliers can provide.

It follows that the two factors – supplier's capability and the interface between the supplier and the buyer –determine the supply chain effectiveness. The interface governs the relationships between the supplier and the buyer, and helps operationalize the business models of collaboration. The supplier and buyer must be able to formalize such collaboration by leveraging the individual supplier capabilities. Clearly, the buyer would be effective if s/he can build a dynamic supply network that is responsive to the needs of consumers, including rapid fulfillment of orders. The buyer may also need expertise to manage relationships with offshore suppliers.

The interface and supplier-capability are shown in Fig. 4.1. It should be noted that the interface issues of fulfillment, collaboration, network, and outsourcing are intricately related to the supplier's capabilities of procurement, quality, cost, and tier-1 responsibility.

A.K. Chakravarty, *Supply Chain Transformation*, Springer Texts in Business and Economics, DOI 10.1007/978-3-642-41911-9_4,

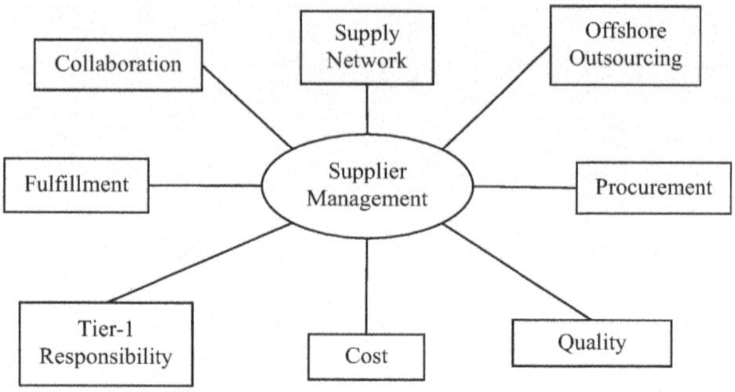

Fig. 4.1 Supplier management

As conceptualized above, supplier management is not just a responsibility of the buyer; it involves multiple partners in the supply chain with evolving expertise. While managing the supply network is more of a buyer's role, quality and procurement are typical supplier roles, and fulfillment may involve third parties. However, these roles continue to evolve.

In the automobile industry, for example, an OEM needs to focus on fewer projects, as the projects are becoming increasingly complex and the OEMcan no longer afford a very broad R&D portfolio (Booz 2013). It must prioritize, which may necessitate dropping of some projects. Some OEMs are now focusing only on new technologies for powertrain such as electric, hybrid, and fuel-cell. Major suppliers are filling the holes left uncovered by the OEMs, making significant investments in key emerging technologies such as batteries, motors, and vehicle positioning & guidance (Booz 2013). New suppliers from other industries such as consumer electronics are emerging in several fields, related to the auto industry.

With the OEMs controlling a narrower portion of R&D, the supply chain is evolving by incorporating more technology partnerships with suppliers; Chrysler purchased diesel engines for its pickup trucks from Cummins (Booz 2013). It follows that suppliers will need to manage complex technology portfolios, operate in uncertain scenarios, develop a sense of consumer needs, and manage risks.

As suppliers take on more responsibility for technology innovation, they will need to become more agile and adept in managing the technology portfolio. This may create opportunities for the entry of more established suppliers from the consumer electronics to create modular systems. In some cases suppliers with high business volumes (because of multiple customers) can advance technology much more effectively than the OEM. The supply chain, as a whole, must determine what kind of partnership would be most effective. In what follows, we describe some of the elements of supplier management (shown in Fig. 4.1).

4.2 Supply Network

A supply network comprises interconnected entities that exchange information, products, and services through procurement and transformation of resources to goods and services (Harland et al. 2001; Choi and Rungtusanatham 2001). The supply network can have multiple tiers and in each tier must focus on compatible objectives to reduce unnecessary activities and duplication of effort. Network design is the process of selecting a structure of the supply chain that maximizes its profitability. It does so by evaluating alternative structures to improve performance at each link in the chain.

The major decisions in a supply network include the design of the network structure, supplier selection, procurement, outsourcing, and logistics, as mentioned earlier. While these decisions are interrelated, there are unique features of each decision. The network design determines which supplier produces what and who can supply whom; other processes such as procurement and logistics create and implement contracts for purchasing and moving goods. There are usually many supplier options for each component. As a component may be procured through multiple channels (e-markets, auctions, catalogs, retail stores etc.), shown in Fig. 1.8 in Chap. 1, the choice of channels would also influence the choice of suppliers. Together, these processes generate a large number of alternatives to choose from. In a market driven environment, with demand and supply uncertainties, the buyer needs to develop the capability to switch from one supply chain alternative to another. If demand is volatile, it would make sense to use a portfolio of suppliers who can individually satisfy different components of demand. If there is an increasing trend in demand, it would make sense to move away from general-purpose suppliers and select focused suppliers so as to exploit the scale economy (Balakrishnan and Chakravarty 2008).

4.2.1 Network Structure

The structure of a supply network can be described in terms of the pattern of connectivity, number of tiers in the network, and the position of the company in the network (upstream/downstream). Burt et al. (2013) discuss how the structure of the network and where a company and its contacts are positioned in the network can influence the company's profitability.

4.2.1.1 Connectivity Pattern

The connectivity pattern determines the paths, connecting any two suppliers, on which the goods, information, and cash may flow in the network. Three different connectivity patterns are usually present in a supply network: dyadic, multiple dyadic and multi-channel (Ring and Van de Ven 1992). Dyadic network (one-to-one) refers to the interaction between exactly two companies; the path connects a supplier to another supplier or to a customer. Multiple-dyadic network refers to the interaction of one company with several other companies (one-to-many or many-to-one). This can take the form of N suppliers and one service provider, or one supplier and N retailers. An example is the relationship between an airline and several independent travel agents, or between a car manufacturer (such as GM) and its multiple dealers. A multi-channel network denotes relationships in which several companies interact with several other companies (many-to-many). Possible interactions include M suppliers linked to N retailers. These connectivity patterns, in a 2-tier network, are shown in Fig. 4.2.

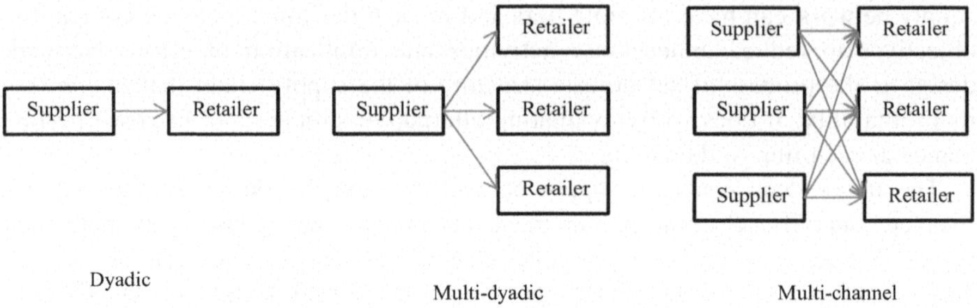

Fig. 4.2 Connectivity patterns

4.2.1.2 The Tiers

Two three-tier networks are shown in Fig. 4.3. In the simplest form of a multi-tier network, a single company is connected to one other company at the next downstream tier in a chain, and it is known as the multi-tier dyadic network or a tree-structure.

Fig. 4.3 Multi-tier structure

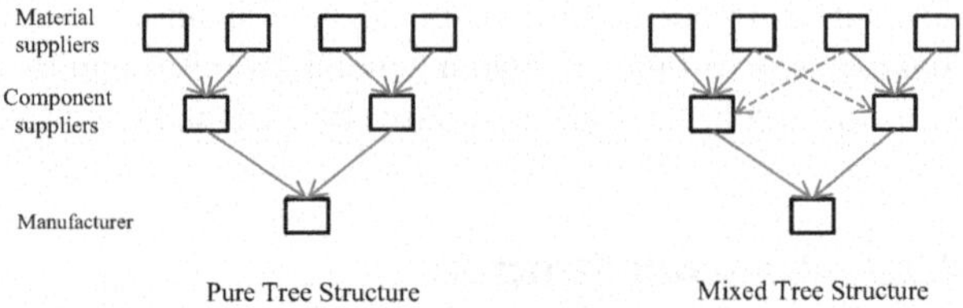

Fig. 4.4 Cyclic network

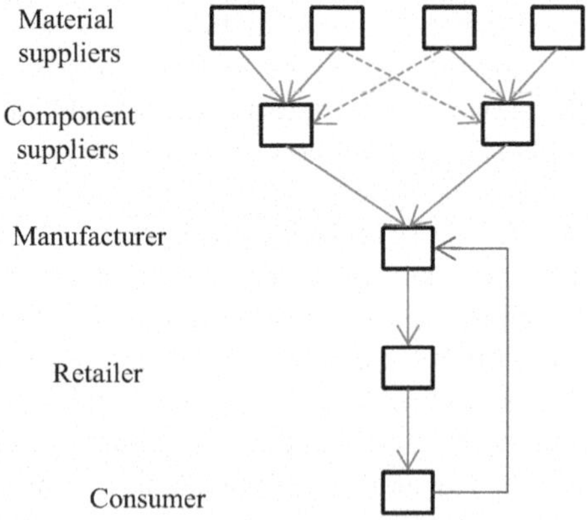

In the pure tree-structure (Fig. 4.3), one upstream supplier *feeds* exactly one company at the next downstream tier. In the mixed tree structure, relationships can exist between one upstream company and multiple downstream companies – the case of common suppliers. In this pattern there is both a vertical and a horizontal dimension, as one upstream supplier can feed outputs to two or more horizontal downstream companies (Fig. 4.3). The horizontal dimension depicts the number of suppliers at each stage. The vertical dimension depicts the number of tiers extending from the initial raw material located furthest upstream to the manufacturer, or to the consumer. As shown in Fig. 4.3, the chain can extend from the initial raw material supplier to the end consumer.

Note that there are no cycles present in the structures shown in Fig. 4.3. In contrast, supply networks that include return of goods from customers or recycling of used products contain cycles, as shown in Fig. 4.4.

The level of integration between the upstream suppliers and downstream buyers depends on their perceived importance in the network.

4.2.1.3 Upstream/Downstream Position of the Company in the Network

The third factor in the supply network configuration is the position of a firm in the network. Each supplier has a specific position in the network that can range from the initial source of raw materials to the consumer. A company's position in the supply network can affect its experiences and consequently its interactions with others in the network. For example, the demand volatility in a network is at its highest at the raw materials tier and the least at the retailer tier (bullwhip effect). Amplification of demand volatility can be reduced through information transparency all the way from the upstream supplier to the downstream customer. As we shall see in Chap. 6, upstream suppliers can be shielded from demand volatility by decoupling him from the downstream companies.

4.2.2 Coordination Scheme

In a supply network, a critical issue is how to coordinate the decisions made by multiple companies. The coordination scheme plays an important role in integrating the network, and it requires sharing of information across boundaries. Two of the key determinants of coordination structure are the decision rights and information architecture.

4.2.2.1 Decision Rights

Decision rights determine who makes what kind of decision. The two extremes cases are centralization and decentralization. In a centralized structure all decisions are made at a single point, while in a decentralized structure the decision making authority is dispersed throughout the network. Each supplier in the decentralized structure can independently make its own decision. As the supply network evolves and becomes complex, centralized decision-making can become arduous because a single decision maker cannot effectively handle all the nuances of local decision-making activities. That notwithstanding, decentralized decisions can result in sub-optimization because of the presence of double-marginalization (Sanders 2008).

4.2.2.2 Information Architecture

The second determinant of coordination structure is the type of information that is available for making decisions. Decisions can be based on local information, global information or a hybrid of the two. A supply network with centralized authority is predicated on the availability of global information for decision making, while a decentralized authority relies on local information. A decentralized decision structure would be more appropriate when there are characteristics that are unique to a particular location or the company, such as local nuances that must be incorporated in the decision. It will also be appropriate when the local setting is prone to rapid changes. In contrast, centralized coordination would be desirable for reasons such as protection of intellectual property, protection against disruptions, use of standard products and procedures, and economic efficiency.

4.2.3 Network Taxonomy

Cravens et al. (1996) identify four types of supply networks based on two dimensions: market volatility and the type of relationships between suppliers. The relationship may span from being collaborative to being transactional in nature as shown in Figure 4.5.

Fig. 4.5 Network Taxonomy

	Market Volatility	
	Low (Functional products)	**High** (Innovative products)
Transactional	Value-added Network	Hollow Network
Collaboration	Virtual Network	Responsive Network

The value added network is usually associated with markets where customer preferences are stable. Companies that complement each other in adding value to products or services, work as a part of a network to create significant customer value. The stable market allows companies to invest in long term innovations and creative product designs, as they can be confident that relationships with suppliers and intermediaries would not come under stress. While retaining the leadership roles for innovations, companies may outsource production and distribution to global suppliers and service providers on a transactional basis.

A virtual network is formed for a specific project, and replaced by another network for a new project. In the garment industry, a project typically implies the production and delivery of a fashion line of clothing for the following year. The required processes may need an assortment of competencies. Therefore, the focal company lines up suppliers, who collaborate to deliver the project. When a new design of clothing (a new project) must be produced, it may require formation of a significantly different network. As production must be committed to almost a year in advance, the market must remain relatively stable for the year. In addition, in a project mode, all suppliers need to be on the same page. Hence they must be aware of their individual roles in the context of the total project. Such a network relies on incremental changes in the collaborative system.

The Hong Kong based Li & Fung has perfected an effective way of working with suppliers who possess capabilities that overlap. It enables the company to switch effortlessly between different garment-production projects. Li & Fung have analyzed supplier relationships and outsourcing in a network context.

The hollow network becomes relevant in a highly volatile market, where the networked companies possess limited capabilities. They use other suppliers to perform functions in response to individual transactions. The network is similar to a virtual network, except that it is difficult to develop collaboration among suppliers. This is common in the software industry where technical specialists are contracted to perform structured tasks for limited durations. Cravens et al. (1996) discuss the case of The Registry Inc. It maintains a database of 50,000 technical specialists who can be used on a contract basis.

The core companies in a responsive network maintain a long term relationship with other networked members. Each member company is adaptable; they can switch products and processes quickly in view of the market volatility. The network coordinator is typically an entrepreneur, who manages a team that identifies customer needs, designs products, and secures the sources of supply. Because of the market volatility, product development cycles need to be very short.

The characteristics of the four networks are summarized in Table 4.1.

Table 4.1 Network characteristics

Characteristics	Responsive network	Hollow network	Virtual network	Value-added network
Relationship	Long term	Short term	Short term	Long term
Network coordination	Collaborative with flexibility	Transactional	Collaborative	Transactional
Market structure	Diverse customers	Segmented customers	Segmented and dynamic	Stable preferences
Core competency of the focal company	Market knowledge and process design	Marketing	Product innovations and production	Product design, marketing, and production

Source: Cravens et al. (1996)

4.2.4 Network Design

Supply network design or redesign can be driven by many factors that include: emergence of new competitors, introduction of new products, entering new geographical markets, incorporating new technologies, mergers and acquisitions, cost escalation, customer service expectations, and new business models such as disintermediation. Such design decisions are driven by the need to realign the network with the changing supply and market conditions or to satisfy new strategic objectives of the buyer.

If a company grows through a merger or acquisition, network re-design addresses rationalization of the supply base, elimination of redundancies, and use of revised flow paths. Introduction of new products may require incorporation of new suppliers with different competencies without disrupting existing supply lines. It may also be possible to develop some of the current suppliers to possess new capabilities. Entering new geographical markets would require new logistics service providers and cultivation of relationships with new suppliers with unfamiliar business-norms. Breakthrough technologies such as the internet typically require reengineering of the entire supply/demand network. New business models such as Dell's "direct to customers" (DTC) required suppliers to assume new responsibilities, and customers to be more knowledgeable. For example, Boeing's modularized supply network for the Dreamliner empowers the tier-1 suppliers. Finally, to satisfy customers' fulfillment preferences, the supply network may need to incorporate drop shipping. Companies may also want to design a "greenfield" network, if the existing network becomes unwieldy due to frequent unplanned changes.

The supply network may use mathematical and simulation techniques to support optimal decisions. Typical decisions include: the optimal number of suppliers to use; whether to source globally and, if so, where; and how to manage various risks.

4.2.4.1 Design Alternatives

In general, a supply network can be defined as comprising $n(i = 1, n)$ value-adding tiers with $m_i(j = 1, m_i)$ options to choose from, at tier i, so that the number of possible design alternatives for the supply network would be $\prod_{i=1,n} m_i$, some of which may not be feasible. Assume that $K(k \in K)$ is the set of feasible design alternatives (a path in the network). It is clear that some of the m_i options available at tier i would not be used in the supply network set K, if it is not cost effective to do so. We denote option j at tier i by the pair ij, and the set of chosen options in tier i as $S_i(j \in S_i)$ and $S = \sum_{i=1,n} S_i$. Clearly, $S \subseteq T$, where T is the set of all ij pairs. The simultaneous choice of products and the supporting supplier base, using optimization and simulation techniques, is discussed in Balakrishnan and Chakravarty (2008).

4.3 The Supply Base

In a typical supply network, many decisions are made that are outside the managerial purview of specific buyers. For example, supplier X may decide to use a different raw material supplier or it may adopt a new manufacturing process; neither of these changes may be known to the company that buys from X. At the same time, there are other suppliers that are closely monitored by the buyer, and this group of suppliers is known as the supply base of the buyer, as shown in Fig. 4.6. These suppliers are interrelated through the flows of materials, services, information, and cash.

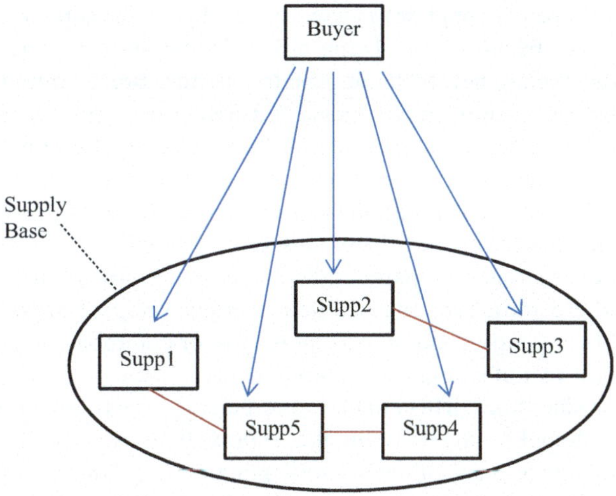

Fig. 4.6 The supply base (Source: Choi and Krause 2006)

The performance metrics of such a supply base includes transaction cost, supply risk, supplier responsiveness, and supplier innovation (Choi and Krause 2006). These performance measures are largely determined by the architectural characteristics of the supply base and include, supplier diversity, number of suppliers in the base, and the linkages between suppliers, as shown in Fig. 4.7.

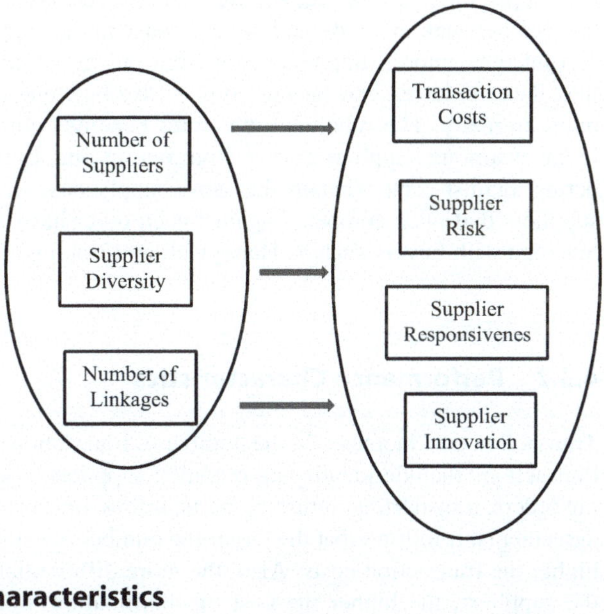

Fig. 4.7 Supply base performance

4.3.1 Supply Base Characteristics

The size of the supply base has a major impact on its performance. For example, a reduction in the number of suppliers in the supply base causes the purchase volume per supplier to increase, benefitting the suppliers due to scale economy. The buyer benefits from being able to focus on longer term strategic issues such as developing new technologies; freeing themselves from protracted negotiations with multiple suppliers for price and delivery. Large OEMs such as General Motors (GM) and General Electric (GE) maintain small supply bases, and that enables them to leverage their respective supply bases very effectively (Krause 1997). On the flip side, with fewer suppliers in the supply base the buyer's operations can become entwined with the operations of the suppliers, increasing suppliers' power and buyer's risk. To maintain leverage, many buyers invest in supplier development. These initiatives may include reengineering of suppliers' processes to be more buyer-centric, and increasing the use of common components to cut cost. Automobile manufacturers use common parts such as fasteners across several different product lines. This enables specific suppliers to obtain large volume orders.

Supplier diversity in the supply base (Fig. 4.7) stems from differences in organizational cultures, work practices, technical capabilities, and geographical distances between suppliers. Supplier homogeneity in terms of cultural and geographical proximity can help in coordinating the flows in the network. A good example is the Japanese keiretsu that functions like an extended family in terms of shared norms, communication styles, and business practices. Diversity, on the other hand, acts in the opposite direction. Diversity may be caused by cross-border barriers such as languages, humanitarian values, and business ethics; informal transaction procedures; and differentials in technology expertise. While diversity increases the effort needed in coordination, it often leads to increased innovation in the supply base. Therefore, through supplier development, maintaining the right balance between diversity and homogeneity in the supply base becomes critical.

Linkages among suppliers result from business transactions. As described earlier, these linkages can be one-to-one (dyadic), one-to-many, many-to-one, and many-to-many. The relational dynamics becomes much more complex when the linkages among suppliers evolve from one-to-one, to two or more suppliers competing against each other in the same supply base. In the aeronautical industry, suppliers that once competed against each other have formed coalitions for better leverage with buyers such as Honeywell and Boeing (Choi and Wu 2009).

4.3.2 Performance Characteristics

Transaction cost increases in the number of transactions in the supply base. Typical transactions include identifying qualified suppliers, contracting, monitoring, placing orders, transporting, returning parts, follow-up, correcting orders, and enforcing agreements. It follows that the larger the number of suppliers in the supply base, the higher the transaction costs. Also, the more differentiated and yet inter-dependent the suppliers, the higher the cost of coordination. For instance, if the operating procedures of some suppliers at the supply base are different from others, the transaction costs incurred by the buyer will be high. Moreover, if the suppliers interact closely, the suppliers' opportunistic behaviors could hinder optimization of buyer's decisions.

The buyer's risk increases if one or more suppliers acquire the technological knowledge of the buyer, as those suppliers now becomes potential competitors. Thomson Consumer Electronics, a supplier of JVC, eventually moved into the JVC's market space as a competitor after acquiring the technologies from JVC (Bleeke and Ernst 1995). In addition, if the technologies that suppliers bring to the supply base are critical for the buyer, the suppliers may threaten to hold it back; it becomes a structural risk for the buyer.

As mentioned earlier, with single sourcing, the buyer can be exposed to a high risk because of over dependence on that supplier; the buyer becomes vulnerable to negative events at the supplier's plant. The buyer is further exposed to the risk of not having access to new and varied technologies that other suppliers may possess. On the other hand, a very complex supply base with a large number of suppliers may introduce significant delivery risk.

Thus both the differentiation and relationship dimensions can increase the difficulty in coordinating with the supply base, increasing the probability of undesirable events occurring. Indeed, close relationships among suppliers in the supply base may lead to a higher supply risk.

In addition, information sharing among suppliers may increase supply risk. When suppliers become too close the buyer loses control over the flow of information between the suppliers that it could otherwise have leveraged for its own benefit, and the cooperative relationship may lead to collusive activities (Choi and Hong 2002). Therefore, information exchanges among suppliers in a supply base increases the potential for collusion among them.

Supplier responsiveness can increase in the size the supply base because a larger supply base increases the competitive pressure on suppliers, causing them to be more responsive. At the other extreme, a close relationship between the buyer and its suppliers also motivates supplier responsiveness (Liker and Choi 2004). Therefore, optimal supplier responsiveness can be achieved with a supply base that is neither too large nor too small. It follows that by working with a limited number of preferred suppliers the buyer can be more effective in communicating its needs and can better induce the suppliers to be responsive.

The linkages in a supply base help in absorbing, sifting, and classifying new technology developments. Therefore, low levels of linkages would be expected to lead to less innovative improvements in suppliers' operations.

In summary, a supply base with a small number of suppliers decreases buyer's transaction cost. It also makes it harder for the suppliers to form coalitions for contractual decisions, and decreases the occurrence of opportunistic behavior by suppliers. Thus, it helps increase buyer-seller collaboration. With a small supply base, the probability of a supplier being chosen a preferred supplier increases. However, to remain a preferred supplier, suppliers must improve performance such as responsiveness. A small supply base increases the frequency of repeated transactions, reducing the supplier's risk. However, it may increase the supply risk due to over dependence on a small group of suppliers who may exploit the situation.

It may also prevent the use of more cutting edge technologies from other suppliers. Differentiation between the suppliers in the supply base is reduced, buyer can easily substitute one supplier with another, increasing the supplier's risk; conversely, increased supplier differentiation increases the buyer's risk.

4.4 Supplier Development

Increasingly, companies are leveraging the specialized competencies of their suppliers to generate customer value. Buyers benefit from suppliers' expertise through joint development of products and processes, and through sharing resources with key suppliers. The frequent exchange of information on product design, production, and logistics enables the buyer to run its production processes without interruptions (Wagner 2010).

However, not all suppliers are sufficiently prepared or capable of supporting the buyers in accomplishing their objectives. As a consequence, buyers often focus on developing those suppliers that are either too critical for the buyer or have sufficient preparation for further development.

4.4.1 Supplier Management Options

The buyer has three possible recourses in working with a deficient supplier: replace the supplier, bring the needed product in house (insource), and assist the deficient supplier in improving its operations. Switching suppliers involves searching for alternative sources of supply and picking the one that has a better fit. This option, however, might not be justifiable if the switching costs of developing relationship with a new supplier are too high. For the second option of bringing the needed product in house, the company may have to acquire the supplier or build in-house capability to manufacture the product. Vertical integration may require substantial investments by the buyer, and it may become a distraction from the company's core competencies.

The third option of assisting a deficient supplier implies investment in upgrading the supplier's capabilities (Wagner 2007). Strategically, supplier development implies enhancing the long-term capability of the supplier – upgrading the technical, quality, delivery, and cost capabilities – and to foster a culture of continuous improvement. The development processes may range from simply informing the supplier about the deficiencies in its product and delivery performance, to on-site consultation and support at the supplier's premises.

John Deere relies on a strategic approach for upgrading the just-in-time capability of its suppliers. Working with John Deere, suppliers can achieve significant cycle time reductions (Golden 1999). In a similar vein, a large number of suppliers of Toyota receive assistance from Toyota's consulting teams in building up lean manufacturing capabilities. These organizational capabilities benefit Toyota and its suppliers in the long run (Dyer and Hatch 2006).

There are two approaches for implementing supplier development projects: direct and indirect. In the direct approach the buyer commits relationship-specific resources to a supplier. In the indirect approach the byer offers incentives and communicates with the suppliers how they might improve.

4.4.2 Direct Development

The buyer dedicates capital resources such as financing for machines, facilities, and tools to a specific supplier. In addition, direct development may include activities that transfer knowledge to the supplier's organization. Some examples would be education and training programs, on-site consultation, temporary personnel transfer, and inviting supplier's personnel for on-the-job learning. The buyer often invests in suppliers' processes, such as product-customization, to tailor them for seamless integration with the buyer's processes. Clearly, the buyer will make such investments only if they add value or reduce costs above what could have been achieved through supplier switching or vertical integration.

In contrast to other options, development through knowledge sharing involves processes that evolve continuously in time. Knowledge transfer from the buyer to the supplier is a common method of sharing knowledge. The means of knowledge transfer include transmission of codified knowledge (electronic or otherwise), and transfer of tacit knowledge through co-location of employees. Buyers transferring explicit knowledge must have well-structured documentation for processes such as quality control, work organization, scheduling, and process-plans. Toyota is a good example. The transferred knowledge can help suppliers in developing, producing, and distributing products more efficiently. It is harder to transfer tacit knowledge, as it is embedded in and carried through multiple entities including organizational culture and identity, policies, routines, and story-telling. Inter-organizational project teams can help in transferring tacit knowledge, even if partially.

4.4.3 Indirect Development

The indirect development approaches are less invasive and include assessing suppliers, communicating recommendations, setting performance goals, and instilling competition through goal attainment. The buyer may use the communication process and/or market forces to achieve performance improvements.

Indirect supplier development does not require the buyer to become actively involved in the supplier's processes, and it relies more on the buyer's coercive and non-coercive powers than on collaborative joint actions. The supplier is motivated by its desire not to lose the buyer's business, especially if the buyer possesses significant market power. It responds by improving quality, cost, and fulfillment through a better management of time, labor, inspection, and sourcing. Wagner (2010) outlines six initiatives for indirect development of suppliers, as shown in Table 4.2.

Table 4.2 Indirect development of suppliers

Initiative	Description
Information exchange	The buyer supplies product or process related information without specifying any actions on the part of suppliers
Recommendation	The buyer suggests a specific supplier action for improving performance
Request	The buyer asks the supplier to act without mentioning any subsequent positive or negative sanctions
Promise	The buyer offers a specified reward to the supplier if the supplier complies with the buyer's stated desires
Threat	The buyer informs the supplier that failure to comply will result in negative sanctions
Contract	The buyer requires supplier's compliance using a formal agreement

4.4.4 Development Strategy

Clearly, there is a need of structuring a strategy for applying the indirect and direct development initiatives. One possible strategy would be to build a portfolio of applications, and apply only a subset of the initiatives to a supplier. As shown in Table 4.3, while supplier 1 is subject to the threat of negative sanctions and coerced into compliance, supplier 6 is offered training and on the job learning. Thus, only a subset of development initiatives is applied to a supplier.

Table 4.3 Supplier development portfolio

	Suppliers					
Development initiatives	1	2	3	4	5	6
Information exchange				X		
Recommend			X			
Request						
Promise					X	
Threat	X					
Contract	X					
On-site consultation		X				
Training						X
Personnel transfer				X		
On-the-job learning						X
Invest in machines and tools			X			

To construct such a portfolio, the buyer would need to determine the fit between the development initiatives and the suppliers' current capabilities. The buyer would also need to assess the strategic value of individual suppliers to the buyer.

While the buyers might apply indirect or direct supplier development to different suppliers in the above supplier portfolio, the indirect supplier development initiatives usually serve as the key enablers of direct supplier development (Krause et al. 2000). Therefore, the buyer should first attempt to apply the indirect development options, before engaging in direct supplier development activities. This would imply that the indirect and direct supplier development activities should be combined in a certain sequence. The buyer may utilize indirect strategies to *influence* suppliers to improve and, while committing resources to the direct development of deficient suppliers (Wagner 2006).

Many companies often adopt the above strategy of employing indirect supplier development to precede direct supplier development activities (Wagner 2006). This would be the case for suppliers 3, 4, and 6 in Table 4.3. It would make sense for the buyer to first implement indirect supplier development activities such as information exchange and recommendation for improvement, before starting knowledge transfer and investment in machines and tools for the supplier. Such a sequence is consistent with the fact that information acquisition initially occurs indirectly from documents and reports and, later, directly through observation of supplier's processes. The indirect supplier development activities by the buyer enable the supplier to focus on capabilities that help achieve improvement targets. For example, if the buyer communicates that the supplier's delivery performance is unsatisfactory the supplier can work on improving its logistics capabilities.

This implies that the indirect supplier development activities must be transparent, clear, unambiguous, and possess measurable goals. In contrast, as Wagner (2006) points out, the goals of direct development such as the degree of process improvement at the supplier are hard to quantify. Therefore, simultaneous engagement in direct and indirect supplier development may cause distraction to the supplier from a blurring of goals; it may dampen the supplier's intensity, direction, and persistence towards improvement goals.

4.4.5 Integration of Suppliers

Supplier integration is achieved through the blending of supplier activities with the buyer activities, using both *buyer and supplier resources* (Wagner and Johnson 2004). Such integration during the product development and manufacturing phases are of considerable interest, as the development process is iterative in nature; products need to be innovative while market-driven and manufacturing processes need to be consistent with the products and cost-effective. For effective integration, the buyer should involve the suppliers early in activities such as product-conceptualization, design, engineering, production, and sourcing.

With early integration, suppliers become stakeholders in the development process, and assume greater design and manufacturing responsibility. It also helps in building trust between the parties and enhancing inter-organizational learning (Primo and Amundson 2002). For the Boeing Dreamliner project, the tier-1 suppliers were integrated starting from the design phase. Tier-1 suppliers assumed responsibility for specific segments of the aircraft in terms of R&D, design, financing, and manufacturing (discussed in Chap. 3). Through such integration companies can gain new competencies, conserve resources, and move into new markets.

Buyers can set priorities and allocate resources to supplier based on their roles in terms of risk and the degree of responsibility assumed (Wynstra and Pierick 2000). Because of the many inherent risks the parties must agree in advance on how to minimize risk with respect to technology, completion time, and commercialization.

Integrating and overlapping tasks across company boundaries increases interdependence and adds complexity to the management processes. Buyers must coordinate suppliers' efforts closely. As discussed earlier, training programs, co-location of suppliers, presence of on-site supplier representatives, supplier participation in buyer project teams, shared physical assets, and reliable ways of sharing R&D information with suppliers, all help in integration. Note that supplier integration in the development phase is different from integration in the manufacturing phase. The former emphasizes comprehensive project-oriented activities, whereas the latter involves transaction-oriented improvements in the supply network.

4.4.6 Supplier Consortium

The supply network can become more resilient to disruptions, if there are multiple suppliers per component. To exploit this resiliency, the network design should generate the optimal set of business-activity options, for each supplier scenario (choice of suppliers). If one or more suppliers are disabled, the contingency plan would trigger the next best supplier scenario and the corresponding option set (Balakrishnan and Chakravarty 2008). However, switching from one supplier scenario to another would require that new relationships are developed. It would be prudent to develop such relationships, before contingencies arise. For example, each supplier scenario may be conceived of as a supplier consortium. One way of managing such a consortium would be to develop a lead supplier who would be responsible for coordinating other suppliers in the consortium (discussed in Chap. 2). It would also make sense to alter product design to a modular form, with one module assigned at a time to a consortium. The lead supplier can now be made responsible for acquiring components from other suppliers in the consortium and assembling them into modules. The success of the consortium, to a large degree, would depend on the ability of the lead supplier to monitor the production capacity and process capability of consortium members. Clearly, information transparency within the consortium coupled with appropriate incentives would be critical.

4.4.7 Monitoring Suppliers

Establishing a continuing bilateral relationship requires careful scrutiny of other parties in a relationship. Therefore, supplier-monitoring is an important step in ensuring supplier's ability to perform as expected. It may range from financial checks to examining a supplier's manufacturing operations, production capacity, personnel, and technological capabilities. As financial data may be confidential, suppliers may be asked to submit financial statements, balance sheets, and cash flow, to a third-party (such as an accounting firm) with a confidentiality agreement. The accounting firm can provide the buyer with a scorecard for each supplier, based on financial ratios. To ensure supplier participation, award of new business contracts can be made contingent on their submitting required data. Such information helps the buyer identify suppliers that might be in trouble or headed for trouble.

Another useful strategy is to study suppliers' performance metrics based on data on delivery, quality, and agility. With this data it would be easy to create monthly supplier scorecards to assess supplier's future delivery and quality trends.

4.5 Supplier Selection

In choosing suppliers to be included in the supply base, the buyer needs to be concerned about not only the transaction costs but also the long term strategic fit. Typically the buyer identifies, evaluates, and contracts with suppliers. Unreliable suppliers, in terms of quality and fulfillment, increase the buyers cost. In many cases, because of the suppliers' capacity constraints, the buyer may need to allocate its purchase order to more than one supplier.

In consumer products, many product safety issues have been traced back to suppliers failing to meet the buyer's requirements, resulting in dangerous lead paint in toys, unsafe car tires, and pet food containing poisonous chemicals (Spencer and Casey 2007). Production delays due to parts shortages and recalls of faulty products produced by deficient suppliers cost millions of dollars to buyers through recalls, warranty cost, cost of inventory adjustments, and damage to reputation.

There are many business related factors that differentiate suppliers. It includes new technology and new management initiatives that can help reduce a supplier's cost (Beil 2009). Location endows structural cost advantage to the supplier; it can gain competitive advantage from low labor costs, availability of raw materials, and favorable import/export regulations. Other dimensions of competitive advantage are reduced supply disruption risks and enhanced supplier diversity. Therefore, the buyer must ensure that the supply base comprises the right number of viable suppliers for each product or service procured.

The evaluation factors used most frequently are price, quality, and delivery. Other possible factors would be production capacity and flexibility, technical capabilities and support, information and communication systems, financial status, innovation and R&D, corporate social responsibility (CSR), safety awareness, political stability, cultural congruence with the buyer, and terrorism risk. Other frequently used factors include organization structure, training and development policies, performance history, reputation, amount of past business, warranties and claim policies, procedural compliance, and return policies.

The buyer must combine the relevant factors in a meaningful way to come up with a rank-ordering of the suppliers. Very often a linear rank ordering may not be adequate, in which case the buyer must construct multidimensional assessment procedures. For example, if the buyer wishes to evaluate suppliers on the dimensions of price and lead time, the buyer must construct a tradeoff between these two dimensions to determine whether it prefers, say, a supplier with a high price and short lead time to a supplier with a low price and long lead time. The challenge of supplier evaluation lies in constructing this tradeoff in a way that accurately reflects the buyer's preferences.

4.5.1 Supplier Evaluation Models

A variety of tradeoff methodologies have been proposed but the overall approach is usually one of three types. The first approach is to build a simple linear tradeoff (such as the weighted average) between the dimensions. However, the simplicity of this approach has a downside, as it may be difficult to find the weights that truly reflect buyer's preferences. To overcome this problem more intricate mathematical models have been suggested. The list includes data envelopment analysis (DEA), clustering techniques, analytical hierarchical process (AHP), mathematical programming, fuzzy sets, and artificial intelligence.

The AHP divides a complex decision problem into a hierarchical system of decision elements. A pairwise comparison matrix of these elements is constructed, and then the normalized principal eigenvector is calculated for the priority vector, which provides a measure of the relative importance (weight) of each element. Consider the example in Nydick and Hill (1992), where four suppliers are evaluated based on the factors of quality, price, service, and delivery.

First, the buyer carries out a pairwise comparison of the four factors as shown in Table 4.4, which shows that the buyer considers supplier quality twice as important as price, four times as important as service, and three times as important as the delivery performance. Price is three times as important as both the service and delivery performances, and service is two times as important as the delivery performance. The buyer then determines the adjusted weights by dividing each column in Table 4.4 by the column total. The adjusted weights of the factors are shown in Fig. 4.8.

Next, the buyer repeats the above procedure to determine the pairwise rankings and the adjusted weights of the suppliers (last row in each table), for each of the four performance factors. These are shown in Tables 4.5, 4.6, 4.7, and 4.8.

Denoting w_i as the weight of factor i, and s_{ji} as the weight of supplier j with respect to factor i, we can express the factor-moderated weight of suppliers as,

$$W_j = \sum_i w_i s_{ji}, \quad \text{for all } j$$

$$\text{Thus, Supplier Weight} = \sum_{factors} \{(\text{factor weight})(\text{supplier weight for the factor})\}$$

Table 4.4 Pairwise weights of the four performance factors

	Quality	Price	Service	Delivery
Quality	1	2	4	3
Price	0.5	1	3	3
Service	0.25	0.33	1	2
Delivery	0.33	0.33	0.5	1
Column total	2.08	3.66	8.5	9

Fig. 4.8 Adjusted weights of the four performance factors

	Quality	Price	Service	Delivery	Row Average
Quality	0.48	0.54	0.47	0.33	0.46
Price	0.24	0.27	0.18	0.33	0.30
Service	0.12	0.09	0.12	0.22	0.14
Delivery	0.16	0.09	0.06	0.11	0.10
				Total	1.00

Table 4.5 Supplier weights with respect to quality

	S_1	S_2	S_3	S_4
S_1	1	5	6	0.33
S_2	0.2	1	2	0.17
S_3	0.17	0.5	1	0.13
S_4	3.0	6.0	8.0	1
Adjusted weights	0.3	0.09	0.05	0.56

Table 4.6 Supplier weights with respect to price

	S_1	S_2	S_3	S_4
S_1	1	0.33	5	8
S_2	3	1	7	9
S_3	0.2	0.14	1	2
S_4	0.13	0.11	0.5	1
Adjusted weights	0.3	0.57	0.08	0.05

Table 4.7 Supplier weights with respect to service

	S_1	S_2	S_3	S_4
S_1	1	5	4	8
S_2	0.2	1	0.5	4
S_3	0.25	2	1	5
S_4	0.13	0.25	0.2	1
Adjusted weights	0.6	0.14	0.21	0.05

Table 4.8 Supplier weights with respect to delivery

	S_1	S_2	S_3	S_4
S_1	1	3	0.2	1
S_2	0.33	1	0.13	0.33
S_3	5	8	1	5
S_4	1	3	0.2	1
Adjusted weights	0.15	0.06	0.64	0.15

Table 4.9 Supplier weights, adjusted for factor weights

	Quality effect	Price effect	Service effect	Delivery effect	Supplier weight W_j
S_1	(0.46)(0.3)	(0.30)(0.3)	(0.14)(0.6)	(0.10)(0.15)	0.33
S_2	(0.46)(0.09)	(0.30)(0.57)	(0.14)(0.14)	(0.10)(0.06)	0.24
S_3	(0.46)(0.05)	(0.30)(0.08)	(0.140)(0.21)	(0.10)(0.64)	0.14
S_4	(0.46)(0.56)	(0.30)(0.05)	(0.14)(0.05)	(0.10)(0.15)	0.29

In Table 4.9, the factor-moderated supplier weights W_j are shown. Clearly, the supplier 1 is most valuable to the buyer, and supplier 3 is the least valuable.

The DEA approach is built around the concept of the efficiencies of decisions that are evaluated on benefit (output) and cost (input) criteria (De Boer et al. 2001). The efficiency of a supplier is defined as the ratio of the weighted sum of the performance factors pertinent to the supplier to the weighted sum of the costs of using the supplier (see Chap. 10 for additional details of DEA formulation). Each supplier is assigned the most favorable set of weights so as to maximize the supplier's efficiency rating without making its own or any other supplier's rating exceed one. The optimal set of weights determines the relative efficiency of each supplier. The fact that the DEA approach ignores the interrelationships between performance attribute is its inherent weakness. In the real world, decision makers must rank the criteria in order of importance based on their expertise or experience, even though they cannot tell the exact weights.

Cluster analysis is a statistical technique that uses an algorithm to group a number of items (De Boer et al. 2001). The items are described by a set of numerical attributes. The clustering objectives are to minimize the differences between items within a cluster and maximizing the differences between clusters. It follows that cluster analysis can be applied easily to a group of suppliers that are described by performance scores.

4.5.2 Integration of Supplier Selection with Order Allocation

The supplier ranking is not sufficient for allocating purchase quantities, as it does not consider other purchasing related issues such as supplier capacity and the fixed cost of using a supplier. In addition, while the supplier's unit price may be included in the ranking in relative terms, it does not allow for the buyer's purchase budget constraint. Mathematical programming formulation is one of the ways of combining supplier-ranking with order allocation (Tektas and Aytekin 2011).

Consider a buyer with a purchase budget B who needs to allocate a purchase quantity of R units of a product to n suppliers. Assume that using the AHP ranking procedure, discussed earlier, the weight of the supplier i ($i \in I$) is determined to be W_i. This supplier charges a unit price p_i and has a production capacity u_i ($\sum_{i \in I} u_i \geq R$). Assuming Q_i to be the amount bought from supplier i, the value of purchase from this supplier can be expressed as $W_i Q_i$. Define $y_i = 1$ if supplier i is used, 0 otherwise.

The buyer's optimization problem can be expressed as,

Maximize $\pi = \sum_{i \in I} W_i Q_i$

Subject to

$Q_i \leq u_i y_i,\ i \in I \rightarrow$ implies that the Q_i must not exceed the supplier's capacity, which will exceed zero only if the supplier is used ($y_i = 1$)

$\sum_{i \in I} Q_i = R \rightarrow$ Requires the total quantity bought to be equal to the needed amount

$\sum_i p_i Q_i \leq B \rightarrow$ Ensures that the total payments does not exceed the purchase budget

$\sum_i y_i \leq m \rightarrow$ Implies that number of suppliers used cannot exceed the number available

This is an optimization problem with some binary variables, which is a special case of integer program. However, the special structure of the problem lends itself to an easy solution procedure. Essentially, the procedure will be a "cherry picking" algorithm, where suppliers are chosen after they are ranked in the descending order of W_i/p_i, with $n(i)$ being the rank of supplier i, $n(i) < n(j)$ if $W_i/p_i > W_j/p_j$. Suppliers are chosen one at a time starting from $n(i) = 1$, and allocated a quantity $Q_i = u_i$. As m is the rank of the last supplier chosen, we would have $Q_j = u_j$ for $1 \leq n(j) < r$, and $Q_r \leq u_r$; $\sum_{n(j)=1}^{n(j)=r} Q_j = R$.

Consider the example shown in Table 4.10, with $R = 50$ and $B = 1000$.

The rank ordered suppliers with allocated purchase quantity shown as in Table 4.11.

The fixed cost of using a supplier can be modeled by setting a minimum order quantity q_i (q_i increasing with the fixed cost), and incorporating additional constraints $Q_i \geq q_i y_i,\ i \in I$

4.5.2.1 Cost Minimization

Chakravarty (1979) suggests a cost minimization approach, where cost includes the cost of purchase $p_i Q_i$ as well as additional inventory carrying cost to cover supplier

Table 4.10 Supplier data

Supplier i	W_i	p_i	u_i	W_i/p_i
1	.3	10	50	.03
2	.5	5	10	.1
3	.8	20	30	.04
4	.4	12	25	.033
5	.2	8	15	.025
6	.3	15	45	.02

Table 4.11 Supplier ranking and order allocation

Rank	Supplier i	W_i	p_i	W_i/p_i	u_i	Q_i	Cumulative R
1	2	.5	5	.1	10	10	10
2	3	.8	20	.04	30	30	40
3	4	.4	12	.033	25	10	50
4	1	.3	10	.03	50	0	
5	5	8	15	.025	15	0	
6	6	.3	15	.02	45	0	

deficiencies. The author considers three types of deficiencies: short supplies, delayed supplies, and supplies with defective items. Assuming $F(\tau)$ and $G(\xi)$ to be the distributions of delivery time τ and quantity delivered ξ, the shortage relative to the ordered quantity Q and delivery time T are established as in Chakravarty (1979). In addition, there would be shortages due to defectives in the delivered consignments. Assuming the buyer holds adequate inventories to cover all expected shortages, the order quantities can be determined to minimize the total cost.

4.6 Procurement

The dynamic supply environment, the increasing level of competition in the worldwide market, the uncertainty of global business environment, and the corresponding changes in firm's purchasing function indicate that procurement must play a significant role in a company's strategic decision-making process. In a strategic sense, buyers need to use a differentiated procurement approach for products and suppliers based on their characteristics. This is the so called portfolio approach, first proposed by Kraljic (1983).

4.6.1 Procurement Portfolio

The core idea is to minimize supply risk and make the most of the buying power, where appropriate (Gelderman and VanWeele 2005). It classifies products in two dimensions: profit impact and supply risk with two measures (high and low) for each dimension, as shown in Fig. 4.9.

Figure 4.9 shows classification into four categories: bottleneck, non-critical, leveraged and strategic items. Each of the four categories requires a distinctive

Fig. 4.9 Kraljic's portfolio model

	Low Supply Risk	High Supply Risk
High Profit	Leveraged	Strategic
Low Profit	Noncritical	Bottleneck

approach. Non-critical items (low profit, low risk) are usually the standard products that are amenable to efficient processing and bulk purchase. Solutions such as e-procurement can be used to reduce transaction costs. Leveraged items (high profit, low risk) are the "cash cows" where the buyer possesses significant purchasing power; enabling it to use RFQ for tendering, and product substitution as needed. Bottleneck items (low profit, high risk) are the ones that the company must buy for reasons other than profitability. The high risk is mediated through insurance, and low profits suggest supplier monitoring and holding safety stocks. In some cases, a search for alternative suppliers or products would be advisable. Strategic items (high profit, high risk) need to be managed through collaboration with the supplier. By plotting the buying strengths against the strengths of the supply market, Kraljic identifies three procurement strategies: exploit (buyer power), diversify (supplier power), and balance (no dominant party).

4.6.2 Restructuring for Market Volatility

Consider restructuring purchasing strategies with a third dimension: market volatility (Chakravarty 2010), in the context of Kraljic (1983). While for a low market volatility scenario Kraljic's framework is appropriate, it would need modification for high market volatility. For each of the four scenarios shown in Fig. 4.9, we discuss the implications of high market volatility as in Fig. 4.10.

Fig. 4.10 Purchasing strategy

	Low supply uncertainty	High supply uncertainty
High purchase cost	High unit cost Stable supply	High unit cost Unstable supply
Low purchase cost	Low unit cost Stable supply	Low unit cost Unstable supply

Scenario A: High Unit Cost and Stable Supply

The buyer would have several options to counter market volatility: purchase-postponement, buy-back contract, and use hedging options such the "option-to-purchase". The buyer gains by postponing purchase of the items to a time when demand becomes known. This would require the supplier to hold inventory that would, otherwise, have been owned by the buyer and held at buyer's warehouses. However, suppliers would require incentives, such as a price premium, to carry such inventory.

Purchase-postponement is comparable to supplier owning the inventory in the buyer's warehouse until needed by the buyer, as is the case in specific applications of vendor-managed-inventory. However, to implement purchase postponement the supplier must modify its logistics system to be able to transport large quantities of the item at short notice, controlled by the buyer. With a stable supply, the supplier's market power will be limited, and he would be receptive to changes in the procurement system.

To protect against the downside market risk, the buyer may implement a buy-back contract to share the risk of unused items with the supplier. This would permit the buyer to return unsold or unused goods to the supplier at a unit price negotiated by the two parties.

To protect against the upside demand risk, the buyer may buy an option to purchase the needed quantity of the item at an agreed upon price, which is exercised by the buyer at some future date when demand is high. The option price is determined by the stock market based on the extent of market volatility.

One way of implementing one or more of the above three options would be for the buyer to include the specifics of the options in the RFQ used in a tender and have the suppliers bid. That is, the supplier bids not only on the specifications of the product, but also on the level of difficulty in implementing the stipulated option(s).

Scenario B: High Unit Cost and Unstable Supply

Both unstable supply and volatile market "combine" to increase the uncertainty of being able to satisfy the demand. Therefore, if the buyer chooses to hold safety stock, its size needs to be sufficiently high. However, because of the high unit cost, too high a safety stock will be uneconomical.

The second option would be to buy insurance against uncertainty. However, such insurance would be complicated as it would need to indemnify against two sources of uncertainty – supply and demand – with a possible correlation between them.

The third option would be for the buyer to diversify by earmarking multiple suppliers for the same item, and purchasing online from electronic marketplaces. One way of managing demand volatility is to place a small order and then follow it up with additional orders if demand picks up. The availability of multiple suppliers greatly simplifies placement of follow-up orders. Multiple suppliers also help protect against supply uncertainty.

Vendor managed inventory, where the supplier is responsible for replenishment of inventory with an agreed upon service level, would be the fourth option for the buyer. Because of the unstable supply, suppliers' price premium for VMI would be high. The buyer has protection against demand volatility, as it only pays for the items used, not the items inventoried.

In addition, the buyer might pursue spot market purchases more aggressively.

Scenario C: Low Unit Cost and Stable Supply

With a volatile market, it would be optimal to order in small lots, more frequently. With a stable supply, it would be possible to implement a JIT replenishment policy that incorporates small orders. Another possible approach would be to commit to a total dollar value of purchase for a planning horizon, with actual orders placed more frequently. Order details such as product type, order quantity, and delivery time are then communicated when the demand becomes known. Although the supplier faces uncertainties in order size, he safeguards a business volume for the duration of the contract.

Scenario D: Low unit cost and unstable supply

The buyer may pursue JIT purchasing. With unstable supply, it would require close relationships with suppliers that enable information transparency between the parties. The supplier, for example, could be involved in the buyer's product design and market research. The supplier will have advance information not only on the type of components the buyer is likely to order but also on possible order-sizes. The buyer has visibility of the supplier's inventory and his component production plans, and it helps him decide which orders, and how much of it, to direct to particular suppliers.

4.6.3 The Buying Process

Procurement spans several activities of the buyer and the supplier, and therefore it is of crucial strategic importance. It includes the supplier's processes such as production and fulfillment, in addition to the buyer's activities such as product-design, development, and operations. To ensure that procurement policies are effective, the underlying process-architecture must be well integrated.

The procurement process begins with the buyer identifying potential suppliers. The buyer then formally request that the suppliers provide information about their goods and/or services, which typically takes one of three forms: request for information (RFI), request for proposal (RFP), and request for quotation (RFQ). RFI seeks to obtain market intelligence on available products and services that satisfy the buyer's needs; the suppliers get the opportunity to differentiate themselves from the competitors. The RFI does not commit the buyer to award a contract.

For complex products, requiring suppliers' inputs, the buyer issues an RFP, formally stating the work and performance requirements such as product configuration and/or delivery schedule. Suppliers respond to the RFP with details on how they would satisfy the requirements and the price they would be willing to accept. In response, the buyer negotiates final terms with suppliers. Thus, the process is generally iterative.

The RFQ is used in a tendering process where the buyer is able to state the exact specifications of standard goods or services needed. The suppliers submit closed bids on the tender and the buyer selects the best bid based on some objective criteria. The information received through RFP, RFQ, and the supplier-bids are translated into formal contracts.

4.6.4 Contracts

A contract is a formalization of the terms of agreement between two parties. It usually stipulates how the partner is expected to perform. In some contracts, it also specifies corrective actions if the partner's performance is below expectation. Contracts that are based on measurable values such as quantity, time, and price, are easy to implement. Contracts based on partner-specific attributes such as cost and effort, suffer from information asymmetry and are hard to implement.

Supply chain contracts can be of three types: cost-plus, fixed price, and incentive-contract (Bolton and Dewatripoint 2004). Many procurement contracts are in fact a combination of the three, specifying incentives on some aspects and fixed-price on others. The contracting environment determines how contracts should be combined. While a contract is a convenient way of moderating partner behavior, it is of little value without a mechanism for monitoring conformance.

In a cost-plus contract the buyer agrees to pay a certain percentage above the supplier's actual unit cost. This insulates the supplier against all cost escalations, and provides little motivation for him to improve productivity. In projects such as R&D a cap on unit price is applied, as the buyer may be unable to estimate how much work is needed. This contract has no built-in incentive for the supplier to cut cost.

In a fixed-price contract, price is determined through a bidding process (reverse auction). The supplier bids a fixed price but must satisfy the buyer's stipulated specifications in the RFQ. As the supplier bears all the risk of cost escalation, s/he is motivated to control cost. However, the fixed contract-price provides no incentive for the supplier to attempt cost reduction. Such contracts can be effective if markets do not suffer from unexpected shocks.

Between the two extremes of cost plus and fixed price, lies the incentive contract that specifies a mechanism for sharing gains/losses from cost reduction or overruns. While the fee payable is related to the supplier's performance, the supplier does not bear the entire risk of fluctuations in costs. For such a contract the unit price p paid to the supplier can be expressed as $p = c_1 + \alpha(c_2 - c_1) + m = \alpha c_2 + (1 - \alpha)c_1 + m$; c_2 and c_1 are the actual and targeted costs respectively, m is the supplier's margin, and α is the cost sharing parameter. While c_1 is known, c_2 is uncertain at best. The supplier will participate only if $p \geq c_2$, implying $m \geq (1 - \alpha)(c_2 - c_1)$. However, as c_2 is not known *ex ante*, the value of m would be set through negotiation between the two parties. The buyer, on the other hand, would expect $p \leq c_1$, implying $m \leq \alpha(c_1 - c_2)$. It follows that the buyer's constraint cannot be satisfied, if $c_2 > c_1$.

In a volatile market, contracts must reflect the uncertainties in the marketplace. This can be included in a contract in multiple ways. For example, the buyer can incorporate a delivery window instead of a fixed delivery time; permit a small variation in the quantity delivered; permit the wholesale price to vary within some bounds; and buy insurance or financial hedges to insulate against market volatility. Examples of creative risk sharing contracts would be, capacity reservation, build-operate-transfer, forming supplier consortium, and forecast revision.

Contracts can be sole-award or multiple-award types. Sole award contracting may be favorable if the scope of work is best accomplished by a single supplier. For example, the contract may require significant capital investments on the part of the supplier and/or buyer. Sole award contracts would also be preferred if dealing with multiple suppliers is very costly. This would be the case for items with intellectual property value that require close monitoring. Suppliers of sensitive back-office operations also need close monitoring to ensure client security. Multiple-award contracting can be useful if the buyer wishes to diversify its supply sources to mitigate disruption risks, or if the suppliers have insufficient capacity. The buyer may also be motivated to prevent formation of supplier monopolies. The buyer may favor incumbent suppliers to foster trust and loyalty or, for example, to avoid the administrative costs of training a new supplier on invoicing and payment procedures.

4.6.5 E-procurement

E-procurement is the automation of an organization's procurement processes using web-based applications. It enables widely dispersed buyers and suppliers to interact, and execute purchase transactions. Each step in the procurement process is captured electronically, and all transaction data is routed automatically, reducing the time and cost of procurement. Properly deployed, e-procurement can deliver tremendous value to enterprises in different ways. A procurement platform is shown in Fig. 4.11.

Fig. 4.11 E-procurement exchange

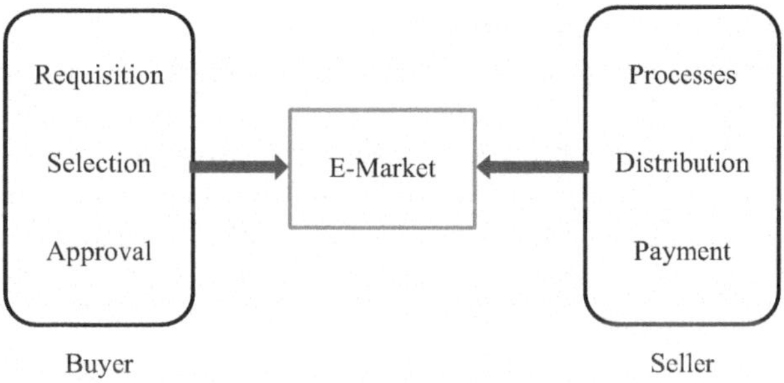

The most common e-procurement model involves an electronic exchange linking buyers and sellers using a web-based procurement application. The online exchange manages transactions, facilitates communication, aggregates and maintains catalog content, and provides the general infrastructure for the exchange. Through this network, buyers may compare products from multiple suppliers in a single electronic catalog or check an item's price and availability in real time before creating a purchase order. The electronic purchase-order is sent to the supplier through the exchange, which also enables suppliers to track the real time status of the purchase order, after an order ships.

Creation of invoice and authorization of payment are also handled by the exchange. The buyer can make a purchase in different ways: through the catalog based on fixed prices, through an established contract on specific products, and through an RFQ-based bidding and negotiation process for one or more product items. Typically, the buyer browses an aggregated catalog on the procurement exchange, adds items to his/her list for purchasing, and places an order specifying the shipping preferences. The exchange captures the order, splits it into multiple supplier-specific orders, and routes them to the relevant suppliers. The components of e-procurement include: catalog management, procurement processes, user management, billing management, price establishment, and data transmission.

4.6.5.1 Catalog Management

The electronic catalog that contains detailed information on products or services available is central to any e-procurement system. Data is aggregated from multiple suppliers in the centralized catalog, using import and other aggregation tools. Buyers can retrieve the catalog content in a user friendly format using tools such as browsing, search, sort, and alert.

Three types of catalogs can be conceptualized that address buyer needs. Product catalogs contain data on tangible items such as furniture, equipment, and medical supplies. Service catalogs offer professional service such as maintenance services, temporary personnel services. Commodity-specific catalogs offer specific product families or groups such as chemicals, paper, or other raw materials, and are useful to manufacturers and assemblers.

4.6.5.2 Procurement Processes

Procurement processes establish the workflow in completing a purchase from the start to finish. They include requisition, order management, real-time tracking, order fulfillment, receiving shipment, invoicing, payment processing, returns management, and generating reports. Note that an effective e-procurement platform must support both the buyer's and supplier's business processes. It should also offer functionality that can easily be customized and configured to meet specific e-procurement requirements.

4.6.5.3 User Management

This component serves as the foundation for managing the complex buyer–supplier relationships that occurs within the marketplace. It defines the user profile: authorization to use the e-procurement system, enrollment procedure, access to the trading community, catalog filters, and workflow. It allows stipulation of unique pricing and contractual relationships between a buyer and a supplier. It creates a supplier profile to identify sellers, and creates shipping and other parameters for suppliers. It establishes authorization levels to address questions such as who may access catalogs, who may create or edit requisitions, and who has access to purchase orders and the shipping/billing information.

4.6.5.4 Billing Management

E-procurement revenues are generally based on transaction fee. A billing management system calculates the usage charges, generates documents, and distributes statements and invoices to the buyers and sellers. Suppliers use the billing system to calculate ordering charges or to distribute operating costs for specific orders. These functions interface with the back office invoicing systems to automatically generate bills.

4.6.5.5 Data Transmission

The key issues in transmitting data over the internet include messaging agents and data security. Data and messaging tools enable the exchange of transactional data between buyers and suppliers. Transactions are sent as "messages" through the internet and then integrated into a supplier or buyer's back-office system, enabling financial postings that coincide with the receipt, payment, and invoicing processes. Security is an important aspect of any internet transaction. Protecting the buyer and supplier confidential financial information is critical to ensuring confidence in any e-procurement system. In certain transactions it also becomes necessary to protect the supplier's intellectual property information.

4.6.5.6 Advantage of E-procurement

Buyers can lower the cost of goods through comparative shopping and aggregated purchases. In addition, bundled purchases through group auctions and promotions result in substantial savings. To enable bundling, buyers must be able to access detailed information on product descriptions, technical specifications, support services, and order status.

Through enhanced matching of inventory with business volumes and stocking levels, both the buyers and suppliers can lower their respective stocking levels. It enables buyers in developing improved capabilities for searching products and supplier-capabilities. Using search engines and online catalogs, it enables products to be organized and cross referenced in a variety of ways to facilitate rapid product identification.

Supplier sites can be individually configured and profiled to generate buyer-specific views of product needs and interests (avoiding product clutter). Better communications and support services (e-mail confirmations, order status, data mining, product tracking) can provide more favorable pricing, offerings and related activities. Online purchasing sources are available 24/7.

E-procurement systems tend to drive down the prices of goods and services in multiple ways. First, they foster purchasing from preferred vendors, who in turn offer volume discounts and contract pricing. A typical e-procurement system can boost preferred vendor transactions by 25% (eJiva 2001). Second, as volume shifts to preferred vendors low-volume vendors are eliminated, causing consolidation that enable preferred vendors to increase consumer discounts. Third, because of the dynamic nature of on-line pricing, vendors can offer limited-time spot discounts on excess inventory. Even a 1-% reduction in purchasing expenses can yield significant savings for large organizations.

Table 4.12 E-procurement savings

Process step	Manual or EDI	E-procurement
Product selection	20	3
Availability/price check	10	1
Requisition creation	11	2
Requisition approval	21	3
PO generation and approval	14	0
Send PO to vendor	18	0
Status check	11	1
Receive shipment	12	2
Match invoice receipt	8	5
Process exceptions	8	3
Payment approval	4	3
Payment generation	8	5
Total minutes	145	28
Cost/cycle time	$75	$15.50

Source: NAPM

Table 4.12 shows the approximate time (minutes) it takes to complete each step in a single corporate purchasing cycle, comparing manual or electronic data interchange (EDI) purchases to e-procurement (eJiva 2001). It demonstrates the dramatic, nearly fivefold reduction in time and nearly 80% reduction in cost directly attributable to e-procurement.

4.6.6 Sustainable Procurement

Sustainable procurement includes the economic, environmental, and corporate social responsibility (CSR) dimensions. CSR aspects of sourcing often deals with assuring ethically sound relationships, establishing codes of conduct, and human rights. Environmental aspects facilitate recycling, reuse, reduction of toxicity, and reduction of resource use. In many cases, as in reduced resource use, sustainable practices can, in fact, decrease the cost in the supply chain. These issues are discussed at length in Chap. 9.

4.7 Restructuring Outsourcing

Outsourcing is the process of leveraging a supplier's expertise and capacity in a way that benefit both parties. It establishes and manages a contractual relationship with an independent supplier. In an outsourcing arrangement, companies rely on other independent companies to run some of their business functions. Outsourcing is a key part of many companies' supply management strategy. The strategy has proven to be effective but brings with it significant risks that must be recognized and managed. If not managed well, the company may negatively affect its operations and those of its customers.

4.7.1 The Pros and Cons of Outsourcing

Outsourcing can create competitive advantage for a company in multiple ways. First, the company can take advantage of the partner's scale economy, as the outsourcing partner can cut her unit cost by aggregating orders from all customers into a large volume production. The partner generates scale economy by specializing in a limited set of tasks and sharpening her expertise through innovation and learning. Second, the company gains agility, as its investment burden is reduced, having outsourced many tasks. Third, it can use the outsourcing partner to quickly fulfill customer orders through drop-shipping. For example, Dell gets UPS to pick up monitors and CPU from Dell's contract manufacturer, match them according to customer orders, and deliver them. The company may also arrange for the outsourcing partner to set up its final assembly plant close to the company's customers, thereby leveraging the partner in implementing its postponement project (discussed further in Chap. 6). Cisco and FedEx have relied on a similar arrangement.

In any outsourcing practice there may be loss of knowledge, as the principal is not involved in performing the outsourced tasks. This causes a degradation of critical capabilities, which the company may find very hard to rebuild. A related issue is the over dependence on the supplier causing a lock-in; it may become hard to switch out once lock-in happens, leading to economic inefficiency. Second, the project related intellectual property (IP) information shared with a supplier may find their way to competitors, leading to a competitive disadvantage. Third, as the company may not be fully able to monitor the performance of outsourced tasks, it may lose control of how the tasks are completed. Fourth, there may be coordination and expediting problems if the terms of agreement between the partners are ambiguous. Fifth, poorly designed contractual agreement can lead to unnecessary litigation. Sixth, outsourcing, especially to offshore locations, can make the supply network vulnerable to disruptions due to distances involved and a reduced transparency between the partners (information asymmetry).

4.7.2 Scope and Importance

The strategic value of outsourcing can be understood in terms of its scope and importance, as shown in Fig. 4.12. Scope is defined in terms of the nature of relationship between the two parties: transactional, shared responsibility, and supplier leadership. Importance of outsourced tasks in terms of the company's operations and competitive advantage can be categorized as low, medium, and high.

Scope	Importance	Low (Non core)	Medium	High (Core)
	Supplier or Buyer Leadership	**Contracts** Electrical system of a car		**Alliance** Nokia and Microsoft devloping new mobile products
	Shared responsibility		**Partners** Joint manufacturing	
	Transactional	Outsource specific tasks (payroll processing)		

Fig. 4.12 Outsourcing strategy

A car manufacturer may consider the engine of the car to be of high importance, brake system of medium importance, and electrical systems of low importance. The underlying issues in this context are knowledge and production capacity. The company may possess both design-knowledge and production capacity for engines, it may possess design-knowledge but no capacity for the transmission system, and possess neither design-knowledge nor capacity for electronics.

In outsourcing a complex task, such as design of a new product, either the buyer or the supplier assumes a leadership role. However, both parties must be jointly responsible for designing, prototyping, testing, and market-positioning (1st row in Fig. 4.12). It may also require the supplier to coordinate the procurement of components and material from other suppliers. This type of outsourcing relationship will have to be based on a great deal of trust between the parties. Nokia and Microsoft, for example, have entered into a broad global strategic alliance where they plan to combine assets and develop innovative mobile products on an unprecedented scale (Microsoft 2011). By integrating their complementary strengths and expertise, these potential competitors would ensure their mutual survival in the new global mobile ecosystem and marketplace.

If the product is not very important to the OEM such as electrical system in a car, it will be outsourced. Toyota outsources both design and manufacturing of electronics on a contract basis (1st row, 1st column, Fig. 4.12).

If the tasks are somewhat complex, information exchange would need to be a two-way process (2nd row in Fig. 4.12). This would call for shared responsibility for task transference, monitoring, expediting, and fulfillment. Renault-Nissan joint venture is a strategic partnership between automobile manufacturers Renault (based in Paris) and Nissan (based in Yokohama), which together sell more than one in ten cars worldwide (Wikipedia 2013). The companies have nearly 350,000 employees and control seven major brands. In a similar vein, YouTube and NBA have joined forces to develop a special NBA Channel on the video platform, which is essentially an individualized micro site providing primarily NBA material and offering user recordings of NBA games (Jessop 2012).

If the tasks are very specific and standardized, the instructions for suppliers can be easily codified and transmitted. For example, highly structured work such as a payroll processing system that requires standard technical skills would be an ideal candidate for arms-length transactional outsourcing. Such tasks are of relatively low importance to the company in creating a competitive advantage. A large number of tasks in a typical company are of this type and hence the number of suppliers managed through transactional relationship is quite high.

4.7.3 Offshore Outsourcing

The worldwide spending for offshore outsourcing has increased significantly since 2000. Using advanced technology, a global service or manufacturing network can operate as a single virtual unit, irrespective of where the services are created and used. By leveraging operations in other countries, such networks create competitive economies of scale for offshore facilities, while satisfying customer orders in the client's home countries. However, it may expose the client company to the risks of intellectual property protection, uneven quality, and political uncertainty. Therefore, the company must decide whether to contract with global suppliers and, if so, to what extent.

The major dimensions in such decisions are ownership (insource or outsource) and location (onshore or offshore) of servicing facilities. Chakravarty (2010) discusses taxonomy of offshoring, denoting a domestic facility as onshore and a foreign facility as offshore, as shown in Fig. 4.13.

	Onshore	Offshore
In-source	**Domestic in-sourcing** Full control Complex products Customization High cost	**Offshore in-sourcing** Cost advantage Knowledge vacuum Quality and productivity risk Reduced operations control
Outsource	**Domestic outsourcing** Low cost Low operations risk Simple products	**Offshore outsourcing** Low investment risk Very low procurement costs Quality risk

Fig. 4.13 Offshore outsourcing choices

The multiple tradeoffs involving factors such as cost, quick response, and risk in moving from onshore to offshore operations are shown in Fig. 4.13. It suggests many choices for a company in terms of various combinations of onshore and offshore facilities. An example would be EDS's portfolio of onshore and offshore facilities – based on uncertainty, cost, and language requirements (Kolakota and Robinson 2004). Such a portfolio can be interpreted, in the context of Fig. 4.13, as the choice of cells where the company likes to position itself.

The offshore facilities can be of different types, based on ownership and operations. A *dedicated* offshore facility is owned and managed by the supplier and provides products or services geared to the onshore buyer. The buyer seeks to leverage the expertise of an established offshore supplier in building and managing an offshore facility. The onshore buyer minimizes her risk by cutting investment in onshore facilities, in an uncertain environment.

A *captive* offshore facility is owned by the onshore buyer. An owner-operated facility becomes desirable when offshore operations are very critical in terms of quality and fulfillment, or when the proprietary information must be safeguarded. However, to operate such a facility, the onshore buyer would have to train and manage employees, comply with the laws and regulations of foreign governments, and bear the risk of a high start-up cost.

4.7.4 Build, Operate, Transfer

To minimize her exposure to offshore operations risk, the onshore manufacturer may opt to acquire the facility only if the facility is sufficiently productive. This is known as the build-operate-transfer (BOT) model (Chakravarty 2013). The offshore facility is transformed from a dedicated center to a captive center. A company that stands to benefit from a captive center but lacks business experience in a specific region of the world would find BOT appealing. An example is J.D. Edwards (Kalakota and Robinson 2004) that reduced the cost of its offshore operation by leveraging the infrastructure of Covansys, a technology services company in India. JDE (the principal) signed a 6-year contract with Covansys with the option of buying the facility at the end of the contractual term, or walking away. A related issue, therefore, is whether the onshore buyer should make the acquisition of facility contingent on the state of its productivity. In addition, the buyer may consider providing incentives in order to motivate the offshore supplier to enhance its productivity.

Chakravarty (2013) establishes that the BOT possesses a special structure that ensures that the buyer's decision to purchase the facility (yes/no) is not impacted by demand uncertainty, although the uncertainty is central to the determination of the size, productivity, and the price of the facility. While decisions are sensitive to the respective reservation values, the two parties react differently – principal buys a smaller facility if supplier desires a higher profit, and the supplier cuts price if principal wishes to reduce her cost. The author also establishes that the principal and supplier can both be better off if the principal is more experienced in offshore operations.

4.7.5 Capacity Strategy

Both demand and supply uncertainties play vital roles in determining the capacity of the onshore buyer and her offshore supplier. The onshore buyer would be interested in acquiring a portfolio of onshore and offshore facilities. For example, the buyer may maintain an onshore presence for rapidly responding to demand-changes, while exploiting the offshore cost advantage, through a BOT arrangement.

4.7.5.1 Griffin Manufacturing

The Griffin Manufacturing (Griffin 2013), an apparel supplier in US supplying to a design house in US, was concerned of losing manufacturing to low cost producers in Honduras (Stratton and Warburton 2003). That not withstanding, the facts working in Griffin's favor were the extended logistics line to Honduras that necessitated 8-month advance orders from the US customer, and dependence on long-term forecasts. In addition, the Honduran supplier lacked the capability of responding rapidly to changes in real time demand. During the selling season, the customer needed very short delivery windows. Therefore, orders that deviated from forecasts resulted in losses from markdowns and/or resulted in lost sales.

Griffin positioned itself in the manufacturing niche with its rapid response capability. The customer saw the benefit in splitting her orders into two: an *early* order of "basic" products for the forecast-amount with the Honduran suppliers at a low price, and a second order *after* the start of the selling session with Griffin to make up the variations from forecast at a premium price. This arrangement worked well for all the parties: the design house, the Honduran supplier, and Griffin.

4.7.5.2 Restructuring for Market Volatility

As shown in Fig. 4.14, in a market driven environment, the source of volatility may reside in a domestic or a foreign market.

Fig. 4.14 Onshore/Offshore capacity portfolio

	Domestic market response	**Foreign market response**
Onshore facility	Reactive capacity (Griffin)	Generic core (in-house)
Offshore facility	Speculative capacity (Honduras)	Customized add-ons (outsource?)

For a volatile domestic market, the supply chain response should be as in the case of Griffin Manufacturing discussed earlier. However, if the source of market volatility happens to be in an important foreign market, an appropriate supply chain strategy would be to redesign the product in a modular fashion with a large core and add-on peripherals for customization (discussed in Chap. 3). If quick-response is the market driver, the generic core can be produced in an onshore facility (in advance of the demand-knowledge) and shipped to the offshore facility where peripherals can be added quickly, based on actual demand. Lucent Technologies adopted this strategy to win a large Saudi order, which had a 3-week delivery lead time from the time the contract was awarded.

In a similar vein, HP used a postponement strategy to add peripherals to its core printer at a distribution center in Europe; the objective was to customize the printer to the different needs of European customers (language, power supply etc.).

4.8 Supplier Relationship Management

Buyer–supplier relationships are a function of the extent to which the two parties are able to obtain a share of the surplus. Both parties seek to maximize their share of surplus, whether in the long or short term. However, the relationship may be skewed by the power relationship between the two parties, which impacts how the surplus is shared (Cox et al. 2003). Power can be defined as the ability of A to make B act in a manner that B would not otherwise have done. Power comes from dependency, and possession of information.

Buyers and suppliers, while interested in obtaining the largest possible share of surplus, may collaborate to increase the surplus created by their interaction – either through cost reduction or utility enhancement. These interactions can be based in four dimensions: information exchange, operational alignment, co-operative norms, and relationship-specific adaptations (Cox et al. 2003).

Information exchange includes the transfer of technical information, cost information, demand information, and proprietary information. Operational linkages include procedures that facilitate the flow of goods, services, payments, and information. Examples include use of ERP, scheduling, and tracking. Co-operative norms are the agreed standards of conduct such as corporate social responsibility, and business sustainability. Relationship-specific investments refer to the non-transferable investments that are often made in business relationships. These could be modified technology platforms for information exchange, training for managing the interface, and relocation of facilities for proximity.

Hartman et al. (2001) explain how the major characteristics of a relationship can be defined in terms of trust, commitment, involvement, and satisfaction. Trust is the basis of any relationship and can be describes as the degree of confidence in how the partner would behave. While trust is not a major factor in transactional relationships, it becomes critical in a strategic relationship, such as developing a new product with a partner in an iterative fashion, where formal contract documents do not cover all eventualities. Commitment defines attitude towards a long term relationship. Toyota's business practice is based on making long term commitments to its suppliers. This commitment stipulates enduring business relationship, while improving performance through learning continuously. Involvement can be defined as getting to know each other's business practices, and it provides opportunities to learn from each other. Therefore, involvement can be considered a mechanism for knowledge transfer in an ongoing basis. Last but not the least is satisfaction with partner's performance. It is a proxy for the effort the partner makes relative to what is expected. The challenge would be how to integrate satisfaction expectation with commitment because expressing a lack of satisfaction should not be construed as a lack of commitment.

As partners may possess varying degrees of sophistication and as relationship with them may not be equally valuable to the company, the optimal strategy would be to think in terms of a portfolio of relationships with partners. For example, to a manufacturer, collaboration with suppliers and retailers would be of more value than that with trading companies (Clark and McKenney 1994). Such relationships can be conceptualized as exploratory, occasional, significant, and intensive. The explorative format is a pre-relationship exchange of ideas. An occasional relationship signifies a low frequency of interaction that may be adequate in scenarios where the partners are only interested in access and information visibility. A relationship is called significant when it requires interfacing of demand with the supply chain processes. Finally an intensive relationship indicates joint decisions in a project management environment. This enables partners in coordinating their activities and updating real time resource decisions.

The company would like to decide in which relationships to invest so as to build the portfolio of relationships. The two major determinants in such decisions are the value and cost of relationship. The factors that impact value creation and costs in any relationship are the frequency of interactions with a partner, revenue and cost per interaction, and the cost of setting up a relationship. If a partner is technologically unsophisticated but can create significant values in the supply chain, the manufacturer may wish to subsidize the partner for it to attain technology savvy. The value (and cost) of relationship would obviously depend on the frequency of interactions between the parties – higher the frequency, higher the value and the cost.

4.8.1 Implementing Collaborative Relationship

To implement collaboration in a relationship, partners must agree on the purpose of collaboration and means of achieving it. Collaboration may necessitate certain changes in how the companies operate. It should be tailored to specific business processes such as, order placement, order fulfillment, logistics, and payment. Cisco Systems coordinated supply and demand across its supply chain with an intelligent-software for early detection of conflicts and their resolutions (Lee and Whang 2004).

One of the success stories of collaboration is the sharing of forecasts. Collaborative Planning Forecasting and Replenishment (CPFR) is based on the idea of setting up rules for resolving conflicts in forecasts (Aviv 2001), and strengthening partner's operations. It has a built in metrics for performance evaluation, accountability, and corrective actions (VICS 2001). Another technology called workflow (Basu and Kumar 2002) can be used for crafting collaboration. It can track supplier's production related events to keep the customer informed. If the customer is concerned of supplier's ability to deliver an order on time, he/she may run the electronic workflow that links the ordering event to the order fulfillment processes of the supplier. Simultaneously, the workflow updates all documents to ensure that the customer has the most recent information at all times. Collaboration must allow alignment of incentives (of partners) with the overall objectives. This requires an equitable sharing of both gains and risks among the members.

4.8.2 Supplier Coalition

The motivation for suppliers to form coalitions in supply chains can be strategic or operational. IBM teamed up with Siebel to provide hosted CRM software service that customers can deploy at the click of a button (Chakravarty and Zhang 2007). Boulton (2003) describes this collaboration arrangement as "the latest fruit of a partnership in which a leading applications provider and leading infrastructure provider teamed up to reach more customers, as neither makes what the other specializes in".

In general, if a customer order contains a set of different products, the supplier must be able to produce all elements of the package, or else form a coalition with suppliers (Oshkai and Wu 2010). For example, consider General Motors' car seats. The seat itself may contain many different parts, which could be produced by different suppliers and later assembled either by the one of the suppliers or by a separate assembler.

The larger the package, the easier it becomes to involve multiple suppliers into production of parts for the package with subsequent assembly. Therefore, by forming a coalition, suppliers hope to increase their bargaining power and market share, and as a result, increase their profits. Coordinated supply is a major incentive for buyers to encourage formation of supplier coalitions.

Operational goals of forming coalition include cost savings through shared resources and economies of scale, risk pooling, and improved capacity utilization. Granot and Sosic (2005) study the benefits of coalitions and explore conditions under which the formation of alliances is optimal. They also provide insights on the stability of these coalitions. However, the dynamics of such coalitions are not very well understood (Cattani et al. 2004) in terms of their stability.

References

Aviv, Y. (2001). The effect of collaborative forecasting on supply chain performance. *Management Science, 47*(10), 1326–1343.

Balakrishnan, N., & Chakravarty, A. (2008). Product design with multiple suppliers for component variants. *International Journal of Production Economics, 112*, 723–741.

Basu, A., & Kumar, A. (2002). Research commentary: Workflow management issues in e-business. *Information Systems Research, 13*(1), 1–14.

Beil, D. (2009). Supplier selection. University of Michigan, Stephen M. Ross School of Business. http://www-personal.umich.edu/~dbeil/Supplier_Selection_Beil-EORMS.pdf

Bleeke, J., & Ernst, D. (1995). Is your strategic alliance really a sale. *Harvard Business Review, 73*, 97–105.

Bolton, P., & Dewatripoint, M. (2004). *Contract theory*. Cambridge, MA: The MIT Press.

Booz (2013). 2013 Automotive industry perspective. Booz and Company. http://www.booz.com/global/home/what-we-think/industry-perspectives/display/2013-automotive-industry-perspective?pg=all

Boulton, C. (2003). Siebel, IBM take CRM on demand. http://www.internetnews.com/ent-news/article.php/3086451

Burt, R., Kilduff, M., & Tasselli, S. (2013). Social network analysis: Foundations and frontiers on advantage. *Annual Review of Psychology, 64*, 527–547.

Cattani, K., Gilland, W., & Swaminathan, J. (2004). Coordinating traditional and internet supply chains. In D. Simchi-Levi, D. Wu, & M. Shen (Eds.), *Supply chain analysis in the Ebusiness era*. Boston: Kluwer. http://public.kenan-flagler.unc.edu/faculty/swaminaj/research/paper/CGS.pdf

Chakravarty, A. (1979). Supplier performance and order allocation. *Journal of Operational Research Society, 30*, 131–139.

Chakravarty, A. (2010). *Market driven supply chains*. Boston: NOW Publishers.

Chakravarty, A. (2013). *Productivity driven offshore outsourcing*. Boston: Northeastern University.

Chakravarty, A., & Zhang, J. (2007). Collaboration in contingent capacities with information asymmetry. *Naval Research Logistics, 54*, 421–432.

Choi, T., & Hong, Y. (2002). Unveiling the structure of supply networks: Case studies in Honda Acura, and DaimlerChrysler. *Journal of Operations Management, 20*, 469–493.

Choi, T., & Krause, D. (2006). The supply base and its complexity: Implications for transaction costs, risks, responsiveness, and innovation. *Journal of Operations Management, 24*, 637–652.

Choi, T., & Wu, Z. (2009). Triads in supply networks: Theorizing buyer–supplier-supplier relationships. *Journal of Supply Chain Management, 45*, 8–25.

Choi, T., Dooley, K., & Rungtusanatham, M. (2001). Supply networks and complex adaptive systems: Control versus emergence. *Journal of Operations Management, 19*, 351–366.

Clark, T., & McKenney, J. (1994). *Campbell Soup Company: A leader in continuous replenishment innovations*. HBS Case 9-195-124, Harvard University.

Cox, A., Lonsdale, C., & Watson, G. (2003). The role of incentives in buyer–supplier relationships: Industrial cases from a UK study. http://www.impgroup.org/uploads/papers/4320.pdf

Cravens, D., Piercy, N., & Shipp, S. (1996). New organizational forms for competing in highly dynamic environments: The network paradigm. *British Journal of Management, 7*, 203–218.

De Boer, L., Labro, E., & Morlacchi, P. (2001). A review of methods supporting supplier selection. *European Journal of Purchasing & Supply Management, 7*, 75–89.

Dyer, J., & Hatch, N. (2006). Relation-specific capabilities and barriers to knowledge transfers: Creating advantage through network relationships. *Strategic Management Journal, 27*, 701–719.

eJiva (2001). eJiva creates eProcurement development practice using commerce one and PeopleSoft Solutions. http://www.dynamicimages.net/special/ejiva/news/releases/pr_012501.htm

Gelderman, C., & VanWeele A. (2005). Purchasing portfolio models: A critique and update. *Journal of Supply Chain Management, 41*, 19–28

Golden, P. (1999). Deere on the run: Quick response manufacturing drives supplier development at John Deere. *IIE Solutions, 31*, 24–31.

Granot, D., & Sošić, G. (2005). Formation of alliances in internet-based supply exchanges. *Management Science, 51*, 92–105.

Griffin (2013). Griffin manufacturing. http://www.griffinmanufacturing.com/

Harland, C., Lamming, R., Zheng, J., & Johnsen, T. (2001). A taxonomy of supply networks. *The Journal of Supply Chain Management, 37*, 21–27.

Hartman, E., Ritter, T., & Gemuenden, H. (2001). *Determining the purchase situation: Cornerstone of supplier relationship management*. 17th Annual IMP Conference, Norwegian School of Management BI, Oslo.

Jessop, A. (2012). NBA and YouTube partner to present NBA D-League games live on YouTube. http://www.forbes.com/sites/aliciajessop/2012/11/20/nba-and-youtube-partner-to-present-nba-d-league-games-live-on-youtube/

Kalakota, R., & Robinson, M. (2004). *Offshore outsourcing: Will your job disappear in 2004*. Inform IT, February, http://www.informit.com/articles/article.aspx?p=169548

Kraljic, P. (1983). Purchasing must become supply management. *Harvard Business Review, 61*, 109–117.

Krause, D. (1997). Supplier development: Current practices and outcomes. *International Journal of Purchasing and Materials Management, 33*, 12–19.

Krause, D., Scannell, T., & Calantone, R. (2000). A structural analysis of the effectiveness of buying firms' strategies to improve supplier performance. *Decision Science, 31*, 33–55.

Lee, H., & Whang, S. (2004). E-Business and Supply Chain Integration, In the practice of supply chain management: Where theory and application converge. *International Series in Operations Research & Management Science, 62*, 123–138

Liker, J., & Choi, T. (2004). Building deep supplier relationships. *Harvard Business Review, 82*, 104–113.

Microsoft (2011). Nokia and Microsoft announce plans for a broad strategic partnership to build a new global mobile ecosystem. http://www.microsoft.com/en-us/news/press/2011/feb11/02-11partnership.aspx

Nydick, R., & Hill, R. (1992). Using the analytic hierarchy process to structure the supplier selection procedure. *International Journal of Purchasing and Materials Management, 28*, 31–36.

Oshkai, S., & Wu, D. (2010). Supplier coalition formation in buyer-oriented markets. Lehigh University. http://www.lehigh.edu/~sdw1/Oshkai1.pdf

Primo, M., & Amundson, S. (2002). An exploratory study of the effects of supplier relationships on new product development outcomes. *Journal of Operations Management, 20*, 33–52.

Ring, P., & Van de Ven, A. (1992). Structuring cooperative relationships between organizations. *Strategic Management Journal, 13*(7), 483–498.

Sanders, N. (2008). Surplus effects of vertical integration with and without double marginalization – examples. Graduate Department of Economics, UC Davis. http://njsanders.people.wm.edu/121B/Double_Marginalization_Handout.pdf

Spencer, J., & Casey, N. (2007). Toy recall shows challenge China poses to partners. *Wall Street Journal*, August 3, http://online.wsj.com/news/articles/SB118607762324386327

Stratton, R., & Warburton, R. (2003). The strategic integration of agile and lean supply. *International Journal of Production Economics, 85*, 183–198.

Tektas, A., & Aytekin, A. (2011). Supplier selection in the international environment: A comparative case of a Turkish and an Australian company. *IBIMA Business Review*, 2011, 1–14, http://www.ibimapublishing.com/journals/IBIMABR/2011/598845/598845.pdf

VICS (2001). CPFR: Collaboration planning, forecasting, and replenishment. www.cpfr.org

Wagner, S. (2006). Supplier development practices: An exploratory study. *European Journal of Marketing, 40*, 554–571.

Wagner, S. (2007). Supplier switching decisions. *European Journal of Operational Research, 183*, 700–717.

Wagner, S. (2010). Indirect and direct supplier development: Performance implications of individual and combined effects. *IEEE Transactions on Engineering Management, 57*, 536–546.

Wagner, S., & Johnson, J. (2004). Configuring and managing strategic supplier portfolios. *Industrial Marketing Management, 33*, 717–730.

Wikipedia (2013). Renault–Nissan Alliance. http://en.wikipedia.org/wiki/Renault-Nissan_Alliance

Wynstra, F., & Pierick, E. (2000). Managing supplier involvement into new product development: A portfolio approach. *European Journal of Purchasing and Supply Management, 6*, 49–57.

Logistics and Fulfillment Services 5

5.1 Introduction

Logistics is responsible for moving goods and services such as materials, components, finished products, people, information, documents, messages, consumables, and money. An oil company uses logistics to deliver gasoline; a television company uses logistics to transmit programs to customers; and a hotel uses logistics to provide food, bed, and entertainment to its guests. Thus, logistics is a necessity, irrespective of the industry.

In the context of a supply chain, logistics enables the movement of raw material to production, and products from production to consumers through wholesalers and retailers. It is also responsible for moving waste to disposal and the reverse flow for recycling/remanufacturing. The activities include storage, inventory control, packaging, transportation, and freight forwarding services. Logistics services may be owned by the user such as a manufacturer or they can be outsourced to 3rd party service providers (3PL). Physical assets include warehouses, transportation fleet, and the electronic system for tracking the flow of goods. The functions are shown in Fig. 5.1.

The warehousing function must determine the number of distribution centers (DC) to own and their location. The optimal location of the DCs would clearly be a function of the locations of customers, sources of material, and the locations of manufacturing plants. To be responsive to customers, the DCs must store a mix of products that would vary by location, and therefore the product-mix must be chosen carefully. Material handling equipment enable efficient stacking and retrieval of items in a warehouse. They include trucks, forklifts, side loaders, platform trucks, conveyor systems, automatic guided vehicles (AGV), and robotic delivery systems.

The packaging function is driven by multiple objectives such as protection of the product, visual effect, environmental sustainability, and minimization of weight and volume. Packaging should be considered as an integrated approach, where packaging and logistics complement and adapt to each other.

A.K. Chakravarty, *Supply Chain Transformation*, Springer Texts in Business and Economics, DOI 10.1007/978-3-642-41911-9_5,

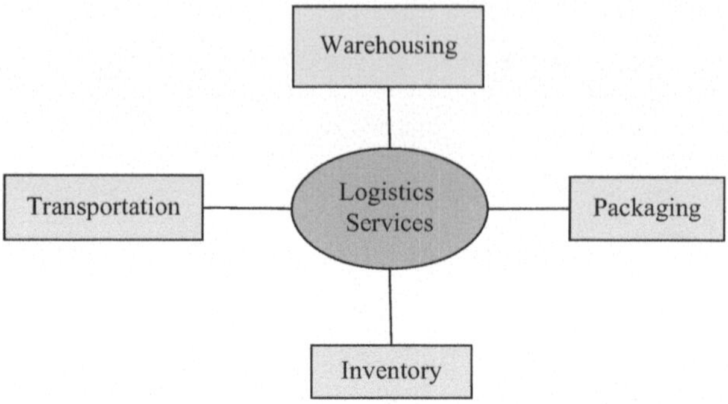

Fig. 5.1 Logistics services

Managing inventory in the warehouses is a critical function, as the investment in inventoried items does not earn any return. In addition, the inventory items must be accounted for with visual inspection or using advanced technology such as RFID. To match customer's changing needs, inventory items transported and stocked in pallet loads may need to be re-palletized. Special storage must be provided for hazardous material, and perishable products.

Of the four logistics functions in Fig. 5.1, transportation has the highest percentage of investment. It entails four distinct modes: road, rail, ocean, and air (Brogan et al. 2013). Almost 90% of the global movement of goods use ocean shipping. The primary domestic transportation modes are rail and road. In general, raw material, mineral, and crude oil are transported in railway wagons, whereas, finished products are trucked using the roadways. The major issues in road transportation are route planning, and scheduling of the moving stock.

The traditional logistics functions, described above, are evolving in view of the emerging economic scenarios, summarized next (described in depth in later chapters). The logistics needs for humanitarian relief is driven by saving lives, while the participating service providers and retailers may be driven by profit. Warehouses for relief effort may carry large inventory, a significant part of which may go waste if disasters do not occur. The last mile delivery in emerging economies works in a reactive mode, and is mostly informal and unreliable. Logistics for sustainable operations needs to emphasize reverse logistics and reduction of green house gas emission. Logistics for the service sector such as healthcare is undergoing major changes, as healthcare is mutating in terms of extended span of network, segmented patient markets, remote connectivity, decentralized centers of excellence, patient-pulled services, and the payment and ownership structures. Supply chain security is posing major challenges for logistics operations in terms of protection of assets, and rerouting to avoid broken links in the supply chain.

The strategy for logistics services, whether owned or outsourced, needs to include the above concerns associated with economic and business turbulences. It could take various forms such as increased transshipment between distribution centers; rapidly switching the mode of transportation and rerouting; using air-drops for humanitarian relief; linking the informal and formal components of logistics; and enabling the technology for telemedicine, wearable devices, and deployment of appropriate treatments.

5.2 Key Drivers

The major drivers of a logistics system include products, customer preferences, technology innovation, globalization, sustainability, infrastructure, and cost, as shown in Fig. 5.2. A logistics service, whether internal or outsourced, can carve out a competitive niche in terms of how it responds to these key drivers. It can try to responsive to all the drivers or it can focus on a subset of them.

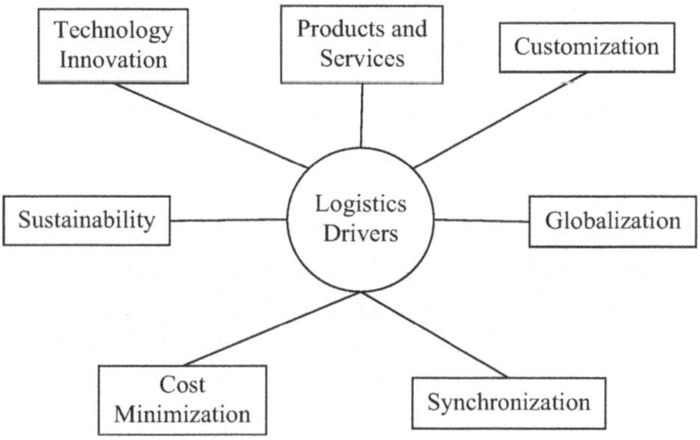

Fig. 5.2 Logistics drivers

5.2.1 Product Portfolio

Products impact logistics in terms of transportation, storage, ordering, and packaging. They can be classified as commodity or customized, durable or perishable, and hazardous or safe. Individual logistics companies may decide to focus on certain types of products such as the long-haul commodity. It also implies that companies must carefully examine the synergy between its product portfolio and its logistics operations. It may need to redesign some of its products and/or modify the logistics functions appropriately.

Commodities include grains, metals, iron ore, and crude oil. As these products are transported in bulk with long delivery times, the best fitting transportation mode would be ships, barges, and freight trains. Non commodity goods such as cars, computers and television sets come with a large variety and must be delivered in small shipments, as specified by the dealers and retailers. If delivery flexibility is not an issue they can be transported in scheduled freight trains; otherwise, they would be transported in long-haul trucks. For unique or delicate products the transportation mode and carrier should be chosen carefully – door- to-door trucking with special handling for the valuable objects. It can be expensive.

Hazardous materials, especially toxic materials, must be routed away from populated areas to reduce risk. They must be routed to minimize accident (derailment) probability, likelihood of toxic emission, and population exposure. Rerouting may increase the delivery time and cost. Stringent protections measures must be taken in packaging and storage of hazardous material, increasing cost. As explained in Chap. 1, Dow Chemical had to redesign rail cars to avoid disasters during transportation of chlorine gas. The cars were made thicker so that they would not be punctured upon impact, if derailed.

Transportation of perishable products such as food and blood require containers equipped with atmosphere-control technologies that can extend the shelf life of perishable products and thus expand the types of perishables that can be shipped in refrigerators without spoilage. Another technology is radio frequency identification (RFID) with wireless sensors that detect the condition of products (such as food) and transmit the information to the RFID, and the RFID communicates with the computer system. It ensures the products' quality during transportation from farms to consumers. Once products with low quality or in a deteriorating condition are detected, appropriate actions such as removing the bad products can be taken immediately in the distribution centers.

In transporting products with moving parts, such as a printer, adequate care must be taken to secure the product. In contrast, the core printer with no moving parts, to which location-specific customization can be added, can be transported in bulk like a commodity. But the add-on parts must be transported separately to the sales locations. Hewlett Packard reports that the total cost with core printer plus add-ons actually decreased in their case, while customization increased.

If a logistics company chooses to focus on (say) iron ore, it would have little use for trucks, vans, and aircraft. It would choose a mass transportation mode such as an ocean liner or freight train. In contrast, a company specializing in computers may design logistics to support merge-in-transit, and drop shipping (discussed later).

Two major trends in supply chain – declining product life cycles and decreasing time-to-market cycles – are increasing the flow of products through the logistics network. Thus the probability, of not being able to satisfy customers, has risen in many industries. However, the increase in flow is not being felt equally by all companies. Individual companies, some significantly more than others, would thus need to invest more in logistics and create innovative expertise.

5.2.2 Customization

Companies can create competitive advantage by tailoring products and services to customers' specific needs. However, customization preferences and service expectations can make filling of orders more complex. Not only must the order have the right quantity of items, they may need to be configured rapidly.

Speed of delivery is critical for most customers, unless it is a commodity product. Reduced lead times enable faster response to variations in demand as well as the short life cycles of products, and increase the reliability of delivery. There are several ways the delivery can be speeded up such as using a faster mode of transportation, and deferring product customization to distribution centers. Some customers may want the deliveries to be very explicit such as splitting the total order and delivering specific quantities by specific dates and at specific locations. This would require rescheduling of freight movement to enable nominated-day delivery and timed-delivery (OECD 2002). Delivery distance can become a critical factor when product-freshness is the issue. For example, fresh flowers must be delivered within a few hours to a customer's location.

Order Tracking enables a customer trace his/her orders. It is a common expectation in the logistics of parcel services (UPS and FedEx, for example). The supplier and customer can obtain real-time information on the position of shipments in the network that enable fast and dedicated interventions in case of any problem.

The *return* option protects the customer as it enables him/her to return defective or unsatisfactory products. It entails a reverse logistics service provider transporting the product back to the supplier. Many e-business companies, realizing that online shopping cannot provide the same tactile feeling as in a physical store, have developed fast and effective processes for product-returns with pre-paid labels.

Customers are sensitive to shipping cost in varying degrees. As mentioned earlier, shipping costs can be high for special products such as hazardous material, products that are fragile such as glass, and fine art such as paintings, carvings, and antiques. Some customers may require special packaging for items such as gifts, bulky products such as the living room furniture, and special machine tools.

While using a low-cost transportation mode can reduce shipping cost, the associated longer delivery time may increase the probability of damage from toxicity, fragility, and pilferage, in addition to causing delays. Insurance is an option, but that increases cost.

Customers may insist on consignment delivery as in VMI, where the supplier owns the product at a customer's facility until it is sold. This is a daunting logistics issue for the supplier, as it must monitor the inventory level at customer's premises. On the positive side, the supplier can better estimate when deliveries need to be shipped out.

To overcome the mismatch between centralized production and product customization, companies are postponing product customization to the DCs at destinations. Thus the number of inventory items is minimized until the point of customization, minimizing transportation cost, inventory risk and costs, and lead-times.

5.2.3 Technology and Automation

Technology creates value in logistics through automatic identification of products and equipment, communication with service providers and customers, automation of materials handling, and ordering and tracking of products. It enables firms to monitor inventories, improve utilization of transportation and warehouse assets, and eliminate duplication in performing logistics activities.

Communication technology is important to logistics because it makes the right information available at the right time and at the right place. It requires a messaging architecture that exchanges business data while customizing information flows and formats. Real-time communication allows dynamic routing and scheduling when the vehicles are on the road. The technology includes global positioning system (GPS), smart phones, internet, electronic data interchange (EDI), point of sale (POS), optical scanning, and the traditional telephone. Automatic identification technologies can be used for tracking rail cars, and cartons moving on a conveyor belt, among others. It includes technologies such as bar coding, RFID, data-communication, and voice recognition to enter data into computer without human intervention.

Material handling technology enables automated loading, unloading, storing of goods in warehouses, using technologies such as programmable conveyor belts, automated guided vehicles (AGV), robots, and automated storage and retrieval system (ASRS).

5.2.4 Globalization

In recent years the flow of goods and components between countries has risen significantly for reasons that include increased outsourcing, and increased global trade. More and more, companies source or manufacture components at many locations around the world, assemble them at several other locations, and warehouse them in numerous distribution centers. The requisite logistics network has become complex for many reasons: a large number of sources and destinations arising from geographic fragmentation, extended travel distances, barriers to flow between countries (tariffs, import restrictions, customs formalities, fluctuating exchange rates), and country-specific variations in logistics performance.

Geographical fragmentation causes problems in the coordination of shipments. The travel distances are not only long, the available transportation modes are either too slow (ocean liners) or too expensive (airplanes). While the liberalization of trade barriers has given impetus to global trade, unnecessary formalities in customs clearance can delay shipments. The country specific variations in logistics performance add to the uncertainty. As Hausman et al. (2005) point out, to export a full container load (FCL) of cotton apparel it may take 93 days in Kazakhstan, 67 days in Mali, and only 6 days in Sweden. While the transaction cost (including inland transportation) for the same shipment in Sweden is $500, it can exceed $3,000 in Namibia.

While 90% of the world's goods are moved by ships, the widespread infrastructural deficiencies cause inefficiencies in linking the shipments to overland and air routes. The lack of real-time communication with banks, transport providers (including ports) and cargo owners is crucial. The bill of lading must be processed by banks, and ports facilitate customs clearance. In addition to the internet, private communication channels such as VPN, are being used increasingly to connect the logistics community of ship operators, port operators, the customs authority, suppliers and customers, trading agencies, land-carriers, and logistics companies.

Major logistics clusters have emerged around the world. Leading clusters include Singapore, Rotterdam, Panama, Dubai, and Los Angeles. Because of the economy of scale, it is possible to develop and maintain specialized knowledge bases at the clusters. Clusters also enable the use of large ships for consolidation of shipments.

5.2.5 Sustainable Practices

The drive for sustainability has direct consequences for the logistics industry, as logistics and transportation services are responsible for a significant percentage of all global greenhouse gas emissions. Actions that the logistics industry can take include: minimizing energy use by improving the operations procedures of trucks, ships, and aircrafts; switching to a fleet that use less carbon-intensive fuels and renewable fuels; and partnering with manufacturers of trucks, vessels, and aircraft to improve vehicle design and energy consumption.

Some of the actions that companies might take include: use of sensors that control the speed, direction, braking, and the performance of the engine and drive train; optimal routing incorporating traffic congestion and number of stops; minimizing less-than-truckload and small freight transportation; using renewable fuels in vehicles (propane, compressed natural gas, liquefied natural gas, electric engines, and liquid petroleum gas). They should switch to locomotives that run on hydrogen-fuel cell whose only effluent is water vapor. Ships can implement the use of waste heat recovery system to convert heat into electrical energy (Hansen and Cooper 2012).

Logistics companies can work with truck manufacturers to design lighter, more aerodynamic vehicles that run on alternative fuels. Companies can partner with ship manufacturers to redesign ships to incorporate a two-engine/two-propeller system, which is more fuel efficient than the one-engine/propeller system. In addition, cargo ships should be designed to operate at slower speeds to reduce energy consumption, and increase total cargo loading capabilities to increase energy efficiency (Hansen and Cooper 2012).

5.2.6 Cost Minimization

Logistics requires heavy investment in fixed assets in transportation fleet, distribution centers, and warehouses. Materials handling equipment, AGVs, and storage racks can also add considerably to the total sum of fixed assets. In addition, there are many logistics related operating costs: crude oil prices, prices of key commodities, interest rates that impact inventory costs, shipping charges, and currency valuations.

The pressure to reduce cost has implications for logistics functions. Firms need to focus on warehouse locations, make critical decisions about inventory, and work with a reliable third-party logistics provider. In recent years, trucking costs have been falling and reliability has been improving. Firms have responded by using faster modes of transportation and reducing other components of logistics costs such as using fewer warehouses and holding smaller inventories. As ICF (2002) points out, Ford Motors managed to decrease the total logistics cost by using faster and reliable means of transportation.

Selective use of international transportation modes is now common in the personal computer industry, in which items with low added values are normally carried by sea in order to reduce transportation costs (scale economy), whereas key parts with higher added values are selected according to demand shifts and are transported by air.

5.2.7 Synchronization

Logistics synchronization helps resolve conflicts between partners, so each can perform specific tasks and assume specific responsibility. It eliminates duplication and collusion that negatively impact system performance. The road-map for strategic initiatives in the logistics network helps determine the roles of partners. Synchronization creates value in acquisition, consumption, and disposition of products and services. It enables a variety of products reaching the marketplace as and when the customer needs. Inventory management, facility management, and transportation help create rapid response to customer requirements, lowered inventory costs, improved product availability, minimum obsolescence and minimize delays that disrupt chain performance.

To be competitive, the logistics industry would need to be more global, more segmented around customers, products, and geography. The companies would need to create better visibility in the logistics network, and be adept at integrating with partners. They would need to excel in moving and storing goods as well as in executing business processes.

5.2.8 Capabilities

It follows that the company needs to build a set of logistics capabilities to respond to the external and internal drivers, discussed above. As shown in Fig. 5.3, the four first tier capabilities are warehousing, transportation, packaging, and managing customers. The cluster of capabilities, labeled "managing customers" refers to external drivers, whereas the other three clusters labeled "warehousing", "transportation", and "packaging", are related to the internal drivers.

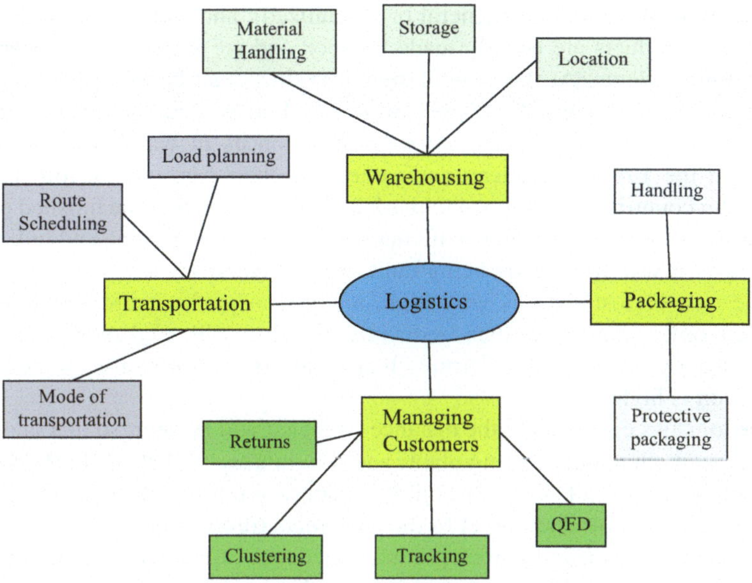

Fig. 5.3 Logistics capabilities

5.3 Warehousing

Traditional warehouses were conceptualized as a facility for storing goods. Today, warehouses are used to carry out a multitude of tasks, and may be customized for manufacturers, wholesalers, and exporters/importers. They can also be customized by products such as commodity, customer-returns, and high value products; and customized for functions such as storage, cross docking, distribution center, and value-adding customization.

Warehouses are designed to have loading docks for trucks, and vast spaces fitted with cranes for moving goods. Goods are stored in standard racks or on pallets. Warehousing has several functions: moving goods in and out of the facility; storing them on appropriate racks; regrouping and picking products for each customer-order; packaging; and shipping the orders to customers. Information flow serves to manage the warehousing activity. In an automated warehouse, products are moved on programmable conveyor belts, automated guided vehicles (AGV), and on automated storage and retrieval systems (ASRS).

5.3.1 Warehouse Location

Warehouse location is determined by many factors that include: the strategic priority for competition (made to stock or assemble to order); transportation cost and time; type of warehouse (general or specialized); and warehouse productivity. Inexpensive products are usually made-to stock and are warehoused closer to the manufacturing plants. At the other extreme products can be assembled based on specific orders; HP ships the "core" printer to Europe and then customizes the printer based on the European customers preferences in warehouses located in Europe. In the 1980s, American companies had their cars manufactured in low wage Asian countries and shipped to markets in the US, so that the finished product warehouses were in the US. Now with markets opening up in China and India, more finished goods warehouses are being built in those countries.

If the cost of lost sales is very high, a firm can justify more warehouses. If customers order small quantities frequently, the firm will need more warehouses located closer to the markets. A firm will also have more warehouses if the level of competition is high.

If the logistics cost is high, the distances traveled will be short so that one would expect to see both manufacturing plants and warehouses built closer to the markets. The other option would be to use contract manufacturer at customer locations and have them ship the product direct to the customers (drop shipping).

A warehouse for made-to-stock products needs the capability of handling a large number of finished products of different types. For such a warehouse efficiency in storing and retrieving the items would be imperative. An automated warehouse with RFID tags on pallets or containers would be a good fit in terms of creating value for the customer.

Some of the important factors affecting the size (sq ft) and capacity of a warehouse are: customer service levels, size of markets served, number of products, bulkiness of individual products, material handling systems, and the pattern of demand. Materials handling decisions include space-occupied and efficiency, economy of scale, aisle requirements, and the types of storage racks and shelves used.

5.3.2 Storage

A storage system in a warehouse holds products until they are shipped to customers. Products such as TVs, medicine, iron ore, and grain come in different forms and sizes. While most storage systems are stationary, some are moveable such as an intercity long-haul truck or an ocean liner. A storage system has three main functions: placing products into racks or bins; holding them in a stationary position; and removing them for shipping (order picking). The performance of a storage system depends on several factors: storage density, ease of access, pick and place procedure, and information technology. Other factors such as safety, structural soundness, energy efficiency, and environmental impact, become important from a long term perspective.

There are many designs of racks used for storage. The most common are flow rack system, single deep rack, double deep rack, and push back rack. Some other storage systems are carton flow rack, cantilever rack and drive-in rack. *Flow racks* use rails on an incline that permit loads to move by gravity from the loading bay to the racks. It reduces the extent of handling needed, minimizes the cost of equipment such as fork-lifts, and saves time and labor. The *Single-deep Rack* that places all loads on a single aisle is the most commonly used in industry today. Loads do not need to be stackable, and may be of varying heights and widths. *Double Deep Rack* is simply a single-deep rack that is two load positions deep. The advantage is that it requires fewer aisles, resulting in a more efficient use of floor space, though the access to the rear load could pose a problem. The *Push-back Rack* provides a rail-guided carrier for each pallet load. Loads are placed on carts on inclined rails, which are then pushed back into the rack with subsequent loads. Each storage lane operates independently.

Dedicated storage policies require that a particular storage location be reserved for units of a single product during the entire planning horizon. In such a situation, the storage assignment problem becomes a simple case of assigning products to slots according to an appropriate criterion such as the CPO index, which is the ratio of the volume of a product to its demand rate.

Shared storage policies permit different products to be stored in the same location at different times. Products can be assigned storage space randomly or based on product class. In a random storage assignment every product has equal chance of being assigned to any of the open locations. In the class-based storage, storage racks are partitioned into a few clusters, and the class of items having the highest turnover rate is paired with the cluster of rack locations with the shortest travel time. It can be shown that the shared storage policy provides flexibility in the use of space and reduces the effective storage area needed. Both reduce the average travel time.

5.3.3 Materials Handling

The Materials Handling Industry of America defines materials handling management as "The movement, storage, control and protection of material, goods, and products throughout the process of manufacturing, distribution, consumption and disposal. The focus is on the methods, mechanical equipment, and related controls used to achieve these functions".

Material handling (MH) employs several types of equipment, depending on factors such as customer demand (volume uncertainty), product type (regular, hazardous, perishable), type of warehouse (commodity, finished goods, value-adding), speed of handling, and the order-picking procedure. It includes handling of unit loads (collection of materials, stored and moved as a single unit) that require standard handling equipment, and bulky loads at shipyards and airports. Examples of unit load are pallets, trucks, and shipping containers. A typical material handling operation involves unloading the inbound material/product, moving it to an assigned location in warehouses, picking it per customer order, conveying it to a packing station, and loading it on trucks for shipping.

Material handling has implications for the efficiency and responsiveness of order-picking and determines the order completion cycles. A continues flow of material clearly minimizes idle time and increases the efficiency of material handling. Cost of material flow can obviously be minimized by incorporating the pull of gravity as much as possible. While the use of standard equipment can enhance efficiency, special equipment may be needed for handling product variety.

Material handling equipment can be classified as conveyors, cranes and hoists, trucks, containers, automated storage and retrieval system (AS/RS), robots, and automated guided vehicles (AGV). Conveyors facilitate continuous movement on fixed paths along which the conveyor belt or rollers is laid out. They can move a large amount of load rapidly, but the hardware cost can be high. Cranes are overhead devices that facilitate vertical movement. Robots are programmable with multiple degrees of freedom in movement. AGVs are small, independently powered vehicles that are guided by cables that are buried in the floor or they use an optical guidance system (Sople 2007). They can be used as mobile workstations to replace the more traditional conveyor system.

5.3.4 Cross Docking

Cross Docking is a process where pallet lots of products arriving at a distribution centre are broken up by individual products and sorted into palletized store orders or consolidated to make full truck loads. Next, the reassembled pallets are loaded onto trucks for delivery to stores. The products never go into storage because they are placed in shipping trucks, soon after they are unloaded from arriving trucks and consolidated. The benefits from cross docking include rapid fulfillment of customer orders, reduced inventory, increased flow of goods, and reduced consolidation effort by suppliers.

Wal-Mart- The world's largest retailer, delivers about 85% of its merchandise through a cross-docking system. Home Depot serves more than 100 stores in the Northeast US through a cross dock (Bartholdi and Gue 2004). Each store may place orders with many vendors who send truckloads of products to Home Depot's cross dock. The products are consolidated into store-specific truck loads bound for individual stores. If the number of vendors exceeds the number of stores supplied, the number of receiving doors (bays) at the cross dock would exceed the number of shipping doors. Transportation cost is reduced because of the FTL shipments.

Toyota uses cross docking to directly feed components to its assembly lines. It places JIT orders with its suppliers, who deliver the components to a cross dock managed by a third party logistics service provider. The 3PL sorts the components and configures them as pallet loads, as per demand of specific assembly lines; and delivers them just in time.

Many retail chains negotiate with their vendors to have the vendors prepare the product for cross-docking before shipping it to the retailer's distribution center. This is known as pre allocation or pre distribution. The supplier would be expected to label products by the destination store for automatic sorting upon arrival at the retailer's DC. As an example, consider a supplier producing three products – A, B, and C – supplying five stores of a retailer. The supplier would stack five units of

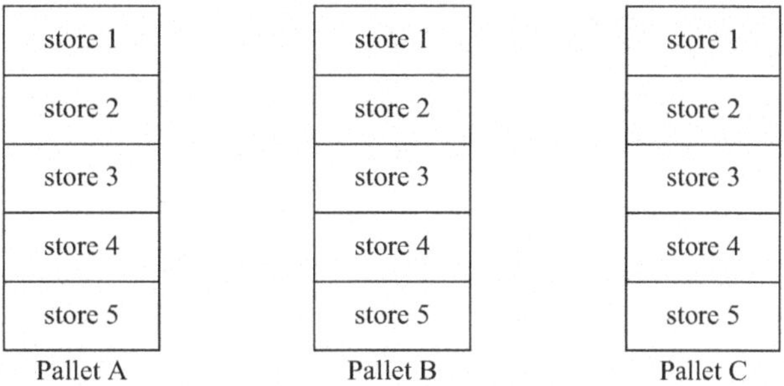

Fig. 5.4 Pallets by the supplier

product A on the first pallet, five units of B on second pallet, and five units of C on the third pallet; and label the products on each pallet by the store number, as shown in Fig. 5.4.

At the buyer's cross dock, the units of the products will be sorted on five store-specific pallets as shown in Fig. 5.5. The retailer then loads the store-specific pallets on trucks bound for the specific stores. Alternatively, the supplier may be required to sort and palletize cartons for each of the retailer's stores, as in Fig. 5.5, for quick and easy cross-docking. It follows that the supplier would need additional effort, manpower, technology, and expertise, to sort the products. In addition, the supplier would need more detailed demand information – which stores need how many units of each product.

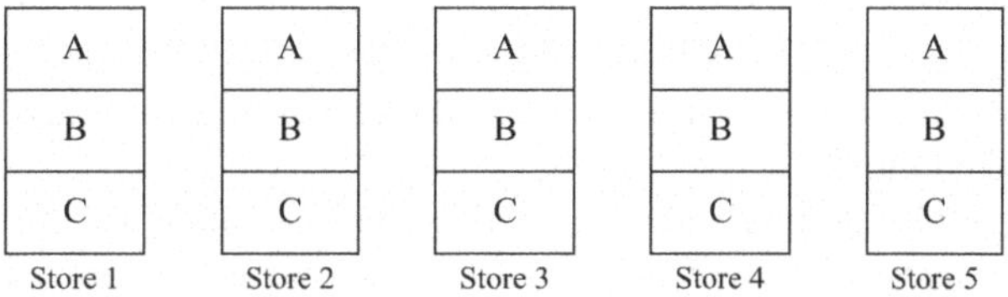

Fig. 5.5 Sorted pallets after cross docking

If the number of stores involved is large, the supplier would need a different level of effort in terms of software and automation. Note that in this case no sorting is needed at the cross dock, and the retailer can load the pallets directly on the store-bound trucks. A cross docking consolidation advantage would accrue by being able to consolidate LTL from several suppliers to full truck loads for shipping.

A third option would be to have the retailer perform all sorting and consolidation tasks. This would reduce supplier's investment in sorting and labeling and, hopefully, its wholesale price. The incoming pallets from the supplier would look as shown in Fig. 5.6.

Now a specific pallet may contain several untagged loads. The retailer's cross dock will be responsible for all sorting and consolidation tasks. This form of cross docking essentially postpones the sorting and consolidation tasks and thus it benefits from the latest demand information from the stores.

For cross docking to operate seamlessly the system must possess certain capabilities (Ertek 2012). First, information flow between the suppliers and the

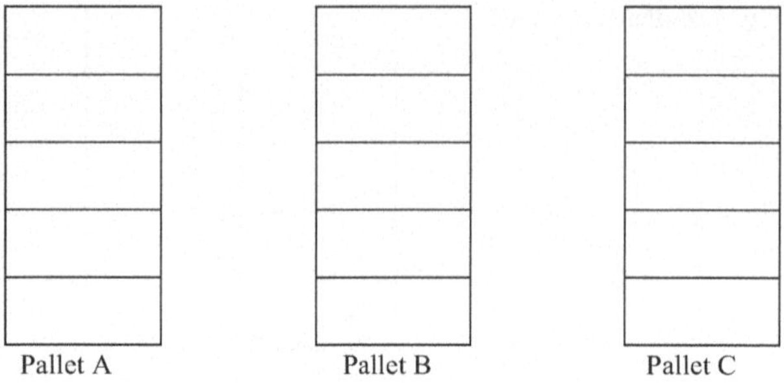

Fig. 5.6 Pallets by the supplier

cross docking station must be smooth, and must be monitored continuously by all parties concerned. This entails a considerable investment in information technology. The Wal-Mart stores, for example, are connected by a satellite system that transmits the point-of-sale (POS) data from the stores directly to Wal-Mart's suppliers (Stalk et al. 1992). Second, the absence of inventories, as in a JIT system, makes it critical to have a perfect coordination in material flows. Third, many interrelated decisions by the suppliers and retailers must be made under resource and time constraints.

Note that the suppliers deliver to the cross dock in a JIT mode. Therefore, in relation to a traditional warehouse operated by a retailer, cross docking would increase the delivery lead time to the stores by the amount of time it takes the suppliers to deliver to the cross dock. As a consequence, individual stores may have to hold additional inventory to cover for the lead time. Therefore, for the total system inventory to decrease, the additional lead time inventory at stores must not exceed the inventory that would accrue at a centralized warehouse without cross docking.

The delivery lead time to stores can increase further if shipping trucks are unavailable when delivery trucks arrive at the receiving doors of the cross dock. To minimize this possibility, the supplier, retailer, and the trucking company must coordinate their activities. For example, the retailer may need to consider an incentive scheme for the trucking company to have a sufficient number of trucks waiting at the shipping doors.

5.4 Freight Transportation

Transportation is the crucial element in moving freight; it links shippers to carriers, and customers. Transportation accounts for approximately one-third of the logistics costs and therefore it has a significant impact on the overall logistics performance. It is estimated that overall freight demand in the US will double, from 15 billion tons today to 30 billion tons by 2050. Freight carried by trucks and railroads will increase by 41% and 38%, respectively, from today's quantities. The number of trucks on the road compared to today will also double. By 2020, the U.S. trucking industry will move three billion more tons of freight than it hauls today. To meet this demand, the industry will put another 1.8 million trucks on the road (AASHTO 2010). The industry would need to restructure itself to be ready for the new scenario.

5.4.1 Modes of Transportation

Shippers rely on carriers such as vans, trucks, railcars, ocean shipping lines, and airplanes. While railcars hauling freight in mass quantities are less expensive per unit weight or volume, they are not open to flexible scheduling, as are the trucks. Trucks can deliver customized loads right at the premises of the customer and they can be rerouted easily. Vans are ideal for short haul (last-mile delivery) because of low operations cost. Ships, used for global transportation, possess massive freight capacity but the lead times can be long. Shipments are assigned to standard containers, and the containers are treated as unit-load. They are ideal for transporting commodities, and products that are not needed immediately (low-value products with steady demand). Airplanes are ideal for swift delivery but at high cost, whether domestic or international.

Unlike railcars, ships, and airplanes, truck schedules are negotiated between the shipper and the carrier. While this provides additional flexibility in the use of trucks, it can lead to unproductive fuel consumption if trucks must return empty to the base. Vans are used for delivering smaller packages to multiple customers. In this case, the shipper is interested in optimizing a complete tour – sequencing deliveries so as to minimize the total distance traveled in visiting all customers.

As ships ply only between seaports and airplanes fly only between airports, use of ships or airplanes imply the need of multimodal transportation; shipments must be picked up from the shipper's premises and delivered to the seaport/airport and, at the other end, they must be picked up from the seaport/airport and delivered to the customer. As an example, consider loaded containers that leave a shipper's facility by truck either directly to port or to a rail yard, from where a train delivers them to port. Next, a ship moves the containers from this port to the destination port. From there they are delivered to the customer by a single or a combination of "land" transportation means: truck, rail, coastal or river navigation. Several intermodal terminals are part of this chain: the initial and final sea-port container terminals where containers are transferred between the ocean navigation and land transportation modes, as well as in-land terminals (rail yards, river ports, etc.) providing transfer facilities between the land modes.

The mode of transportation impacts inventory cost in different ways. First, a slower mode creates higher in-transit inventory, increasing the safety stock needed to protect against the uncertainties in supply and demand. It also increases the order lead time, so additional safety stock may be needed to protect against uncertainty in demand during the lead time. Second, larger shipment sizes may create cycle inventory, which happens when the batch shipment size exceeds the current demand. Thus smaller shipments using faster modes can reduce three types of inventories, but the associated transportation costs would increase.

5.4.2 Hub and Spoke Architecture

A hub-and-spoke system is characterized by an organizational structure in which a central distribution center (hub) is connected by logistic channels (spokes) to multiple depots. A depot covers a region with specific collection and delivery points for goods, hard and soft. A soft good can be software, or orders. In some systems, links between depots permit transshipment (Grahovac and Chakravarty 2001).

Transportation in a hub and spoke system works as follows: At the first stage, suppliers' products are picked up and brought to an appropriate depot. The depots execute their pick up plans. At the second stage, shipments coming from these depots are consolidated at the hub. At the third stage goods are transported from the hub to receiving depots. At the fourth stage, receiving depots distribute the products to customers. The hub and spoke serves an important function with LTL (less than truck load) transportation.

Consider the hub and spoke system in Fig. 5.7. Assume that depots A and B are LTL sending depots, and depot C is a receiving depot, all located in different cities. Both depots A and B can send their shipments to C directly (dotted lines) or through the hub (solid lines). The advantage of using a hub in this case is that the LTL from A and B can be consolidated and a full truck load (TL) can be sent to C, and the total distance traveled from source to destination via the hub will be lower than that for direct shipment between depots. However, additional handling in bulk-breaking and sorting will be necessary at the cross dock. Thus there will be a tradeoff between the cost of additional handling and the savings from distance traveled and capacity utilization of transportation vehicles.

Fig. 5.7 Hub and spoke

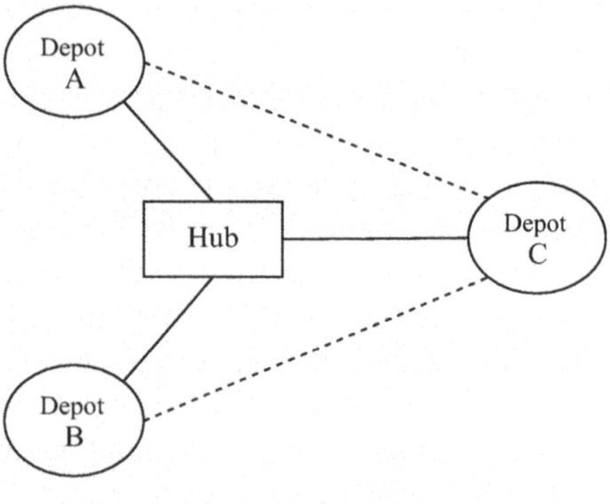

Setting up and operating a hub-and-spoke network involves many decisions, both at the strategic and operational levels. The company must first decide how many hubs, depots, and vehicles it would need in the long run. It needs to specify which depot should serve which retailers (customers) or suppliers. At the operational level the decisions would be allocation of storage and transportation resources to handle current demands, routing of trucks and other vehicles for pickup and delivery at the depots, rapid routing to manage exceptions, and performance improvement.

Each depot can be linked to a single hub, called single-allocation, or it can be linked to more than one hub, called multiple-allocation. Both situations occur in practice. As seen in Fig. 5.7, LTL trucking networks have each depot assigned to a single break bulk terminal (hub) for load consolidation. Passenger airline networks, on the other hand, have flights scheduled from many non-hub cities to a few hubs. Also note that capacity limitations at hubs may force multiple-allocation, as seen in many facility location problems.

In the service sector, the hub and spoke model is used in the context of multi location operations (Parakala and Udhas 2012). In this architecture, a central consolidator called the hub interacts with major customers, while seamless links with delivery centers are leveraged to provide services that are distributed across multiple locations. In the service sector, the depots are also known as spokes. The hub assumes responsibility for quality, risk and performance management, training & development, manpower management, and regulatory compliance. Typical responsibilities of a delivery center include meeting a hub's performance expectations, project management, supporting the hub on legal & tax issues, and operating as an extension of the hub.

Benefits of hub and spoke architecture include, reduced costs enabled by the scale economy in processing at the hub; minimizing duplication of activities among spokes; improved service quality through specialization of activities at the hub; enhanced knowledge sharing, with the hub acting as a knowledge aggregator and distributor; faster innovation cycles; shortened time to market by using standardized products and operations (developed at the hub) at all spokes; and enhanced flexibility enabled by the operations-platform that can facilitate new market entry rapidly (Wyman 2009).

Telecommunication systems use a hub and spoke network to provide for the movement of electronic data (Campbell and O'Kelly 2012). The links are either wired (cables) or wireless. Transmission cost does not increase much with distance traveled. Facilities such as switches, routers, and concentrators are located at the hub to enable communication among a set of nodes, analogous to depots. In relation to a transportation network, the operations, costs, service measures, and constraints are often quite different in a telecom hub network because of the differing natures of objects – freight or passengers versus electronic signals in "packets".

5.4.3 Fleet Management

An important task in freight distribution design is deciding the size of the transport capacity, whether owned or outsourced. Typically, large freight carriers manage multiple fleets comprising diverse resources. For example, truckload trucking firms also manage fleets of tractors and trailers to serve point-to-point full truckload (FTL) demand. Railroads serve point-to-point full carload traffic, but face additional complexity as railcars may be detached from one locomotive and attached to another at a sorting station, before reaching its destination.

The cost of owning a fleet includes both the fixed and variable costs for both vehicles and personnel. Fixed costs are incurred in the acquisition and depreciation of vehicles, purchase of equipment such as GPS and routing software, and taxes and insurance. Fixed costs are also incurred for the salaries of drivers and dispatchers, and equipment for drivers such as mobile phones. Variable costs relate to fuel and spare parts such as tires, vehicle damages, and overtime payments to drivers. Fleet management may become complex for several reasons: vast expanse of the network of destinations covered; demand may not be known with certainty before dispatching the shipment; and the uncertainty in time needed to make full truck load at a hub.

An extended network with multiple hubs and intermodal transfer locations increases the complexity in coordinating the transportation assets (trucks, railcars, containers etc). If the assets are not available as scheduled at specific locations, they must be procured, incurring additional cost. In addition, goods being carried might be labeled erroneously for wrong destination. Recent advances in technology, such as global positioning systems (GPS) and radio frequency identification (RFID) enable real-time visibility of freight operations and a better coordination of freight movement.

Customer demand (pick-up, delivery, split delivery) is often the major source of uncertainty. Customers may book transportation weeks in advance, but the information about customer locations, demand quantities, and demand timing may not be known with certainty in advance of execution. Decisions must be made with the best information available, while accounting for the uncertainty of the parameters to minimize the expected risk. A related issue is what prices should be charged for different lanes (origin–destination pairs) at different time periods (Topaloglu and Powell 2007).

Efficient fleet management must build in real-time updating of decisions, as the information that is used to make decisions may change unexpectedly. For example, a customer may cancel his/her order, delay it, or ask it to be delivered at a different location, while the shipment is on its way. In such cases the trucks would need to be rerouted to the closest hub or to a different destination. In local pickup or deliveries, certain pick-ups cannot be executed, if one or more of preceding drop-offs could not be completed.

5.4.4 Load Planning

A typical less-than-truck-load (LTL) shipment occupies less than 10% of the truck's capacity. Hence consolidation of LTL loads at hubs would be a more economical way of freight transportation, as explained earlier. However, the savings are partially offset by other costs such as the handling cost for consolidation, and increase in the total time and distance in reaching destinations because of hub to hub routing (Erara et al. 2012).

Today, customers often require delivery times not to exceed a few days, and these tighter service needs must be enforced when planning shipment paths. Shorter delivery times reduce opportunities for consolidation. As a result, carriers need methods for designing load plans that capture short delivery time constraints.

Consider a shipper's point to point LTL shipments from Los Angeles, as shown in Fig. 5.8.

Fig. 5.8 LTL Shipments

	Lincoln	Chicago	Pittsburgh
Los Angeles	15,000 lbs 1500 miles		
Los Angeles		35,000 lbs 2000 miles	
Los Angeles			20,000 lbs 2400 miles

Assuming the shipping rate for transporting 1,000 lbs over a distance of 1 mile to be \$1.20, the total cost to the shipper will be (22,500 + 70,000 + 48,000) (1.20) = \$168,600.

The shipper can consolidate the shipments on a 5-axle truck with a capacity of 80,000 lbs (40 tons) and dispatch it to Pittsburgh with stops at Lincoln and Chicago. As shown in Fig. 5.9, the truck will transport 70,000 lbs from Los Angeles to Lincoln, 55,000 lbs from Lincoln to Chicago, and 20,000 lbs from Chicago to Pittsburgh.

Fig. 5.9 Consolidated Shipment

	Lincoln	Chicago	Pittsburgh
Los Angeles	70,000 lbs 1500 miles		
Lincoln		55,000 lbs 500 miles	
Chicago			20,000 lbs 400 miles

We assume that the consolidated shipment is charged \$1.00 for 1,000 lbs over 1 mile, with a minimum of \$10,000 between any two cities.

The total cost of shipment will now be computed as (105,000 + 27,500) (1.00) + max (8,000, 10,000) = \$142,500. Thus the consolidated shipment with the altered routing will generate \$26,100 in savings.

The routing architecture for transportation over long distances usually comprises a backbone routing (line-haul), and delivery routing (last-mile delivery). The backbone routing is used for hub-to-hub transportation, and it may use multiple modes of transportation such as land, rail, ocean, and air. With multiple hubs (for multi modal transportation), the time of arrivals at a hub and transfer of load to the next mode must be controlled tightly. Transportation modes have their own time windows that must be adhered to.

Delivery routing is also known as the last-mile delivery. Typically vans are used to pick up packages from the arrival hub, and distributed to customers in the vicinity. The objective is to cover the largest number of customers, traveling a shortest distance from the hub and return to hub.

5.4.5 Dynamic Routing

Unforeseen events may occur during the execution of planned routing. These include unexpected traffic congestion, vehicle breakdown, blocked roads, and new customer orders. These events may cause disruptions in the scheduled routes and services of customers, and must be addressed appropriately. Dynamic routing facilitates incorporation of add and drop stops as needed, adjustments to traffic and weather conditions, and minimizing operating costs.

Dynamic routing can be implemented in multiple ways: local updates, re-optimization, and advanced approaches. Local updating is usually limited to a simple insertion of new requests. Re-optimization involves solving the entire routing problem that may change the sequence of pick-ups and deliveries to customers.

Advanced approaches include waiting strategy, diversion, and anticipation of future requests (Couriel 2011). Waiting strategies permit vehicles to wait at specific points when they finish service or for a limited time during the servicing of a customer. Diversion permits altering the destination of a moving vehicle to service a newly arrived request. Anticipation of future requests take into account demand patterns by using historical (statistical) information. For example, it may permit keeping the vehicle in an area that has a high probability of arrival of new requests.

5.4.5.1 Tour Creation

Clarke and Wright (1964) designed a heuristic, known as the savings heuristics, for dynamically creating tours that often lead to shortest total distance traveled. It begins by assuming that one vehicle travels from the depot directly to a node and returns to the depot (shown as double headed arrows). As an example (adapted from Couriel (2011)), consider the initial tours (with solid lines), shown in the Fig. 5.10. The dashed lines show paths that may be used but are not in use currently.

Savings is a measure of how much the trip length or cost can be reduced by "linking" a pair of nodes and creating the tour which can then be assigned to a single vehicle. In general, with current deliveries from node *i* to *j* and node *i* to *k*, the saving from linking nodes *j* and *k* can be expressed as $S_{jk} = c_{ij} + c_{ik} - c_{jk}$, where c_{ij} is the cost of one-way transportation from *i* to *j*.

In Fig. 5.11, by linking nodes 2 and 3, a tour can be created for a single vehicle to cover nodes 1, 2, 3, 1 in that sequence. By linking nodes 2 and 3, we add 7 miles (the

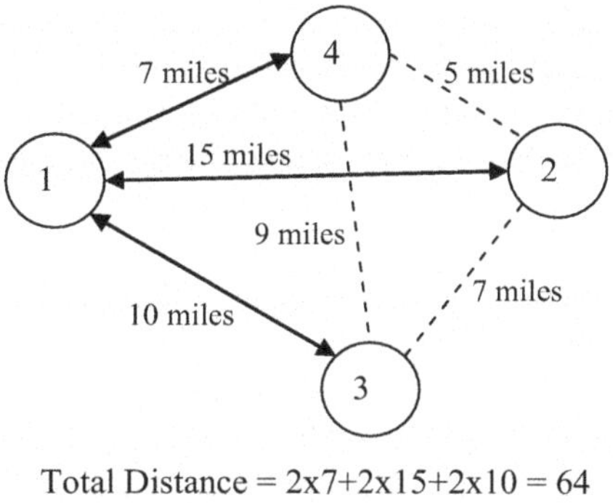

Fig. 5.10 Initial tours

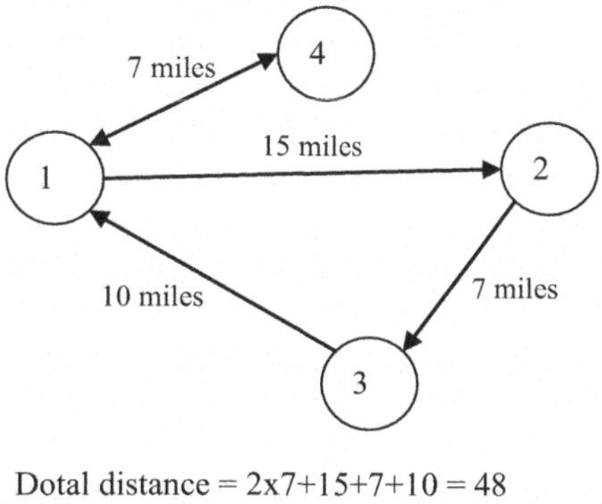

Fig. 5.11 Revised tour

Table 5.1 Savings in tour construction

Nodes (i, j)	Savings	S_{ij}	Ranking
(2,3)	15 + 10−7 = 18 miles	$S_{23} = 18$	1
(2,4)	7+ 15−5 = 17 miles	$S_{24} = 17$	2
(3,4)	7 + 10−9 = 8 miles	$S_{34} = 8$	3

distance from node 2 to node 3), but we save 15 miles for the trip from node 2 to node 1 and 10 miles from the trip from 3 to 1. The saving equals 15 + 10−7 = 18 miles, as shown in Table 5.1. The savings obtained by linking nodes 2 and 4, and nodes 3 and 4, are also shown in Table 5.1.

The tour created by linking nodes 2 and 3 is shown in Fig. 5.11.

Total savings = 64−48 = 18 miles.

5.5 Packaging

Packaging has a significant impact on the efficiency and effectiveness of logistics. As the product moves through the packaging value chain, the costs for handling, warehousing, transportation, and damages are modified. For example, the packaging processes may alter the transportation plan in terms of the choice of freight rates, transportation mode selection, transit time, and service level (Reddy and Sterling 2009). Other implications would be utilization of storage racks in the warehouse, floor space utilization, and handling and distribution costs. Packaging cost is determined by the raw material and its conversion to an appropriate form, cost of printing labels, and the supplier's payment terms. The packaging raw material includes paper, fiberboard, plastic, glass, steel, aluminum, wood, jute, cork, and ceramics.

Packaging satisfies four primary objectives: protection, product-promotion, identification, and handling. Appropriate packaging can prevent damage, spoilage, contamination, tampering, and theft (Lockamy 1995; Saghir 2004). Spoilage can occur from moisture, gases, light, heat, cold, and contamination. Packaging may be used to promote the product with descriptions of the list of ingredients in the product, product features and benefits, and the promotional messages and branding. It can provide product identification information such as product preparation and usage, nutritional data, safety warnings, and opening/closing instructions.

Handling ensures efficient logistics through load unitization. Unitization (unit loads) minimizes unloading time, facilitates material handling, rapid positioning of inventory, and reduces in-transit damage. In essence, it modularizes the content being shipped. There are two forms of unitization: rigid containers, and flexible containers. Rigid containers are formed by rigid enclosure of master cartons (secondary packaging), often using standard-size containers for the shipping mode being used. Flexible containers usually stack master cartons on pallets using block, brick, row or pinwheel shapes. Flexible containers increase damage potential.

Increased packaging protection decreases damage and theft in transit, but increases packaging weight and transportation costs. It decreases insurance cost, increases product availability, and increases product value and carrying costs (Lambert et al. 1998). Enhanced packaging information decreases delays in transportation of shipments and decreases tracking of lost shipments; decreases order filling time and labor cost in warehousing; and minimizes the need for other forms of communication such as telephone calls to track down lost shipments. Packaging unitization increases cube utilization (stacking) in warehouse, but also decreases facility utilization by increasing the size of the product dimensions. Increased standardization decreases handling costs, decreases waiting time for the loading and unloading of vehicles; increases modal choices for shipper and decreases need for specialized transport equipment; and decreases material handling equipment costs in warehousing.

Generally, packaging is classified into the following types: primary, secondary, and tertiary (Saghir 2004). Primary packaging, also known as "sales" packaging, forms a sales unit for the user or final consumer, for example, a box of soap powder. It aims at making the product attractive and readily available to the customer, while ensuring adequate protection for its contents. It enables easy stacking and fit with the storage shelf space. It is also designed for ease of identification (at POS), and handling (opening and resealing). The secondary packaging comprises several primary packages, for example, a cardboard outer containing a number of boxes of soap powder. It consists of outer wrapping that assist in displaying, storing, shipping and protecting products. Tertiary packaging, also known as transport packaging, is used to group secondary packaging together to aid handling and transportation and prevent damage to the products. The most common type of tertiary packaging is the corrugated brown carton. For rigid and stable products a simple shrink-wrap may suffice, while less stable products may need "transport box" for stacking. A good example would be the pallet and shrink wrap used to transport a number of cardboard outers, containing boxes of soap powder.

5.6 Fulfillment

The ability to fill orders reliably and efficiently has never been so important, as suppliers must compete not only on price and quality but also on quick delivery. They must deliver the right merchandise on time and at the right place, while keeping costs low. In addition, as customers can switch suppliers using electronic exchanges, suppliers need to be vigilant in capturing and tracking orders. Increasingly, order-configuration must be customized for specific customers.

5.6.1 Fulfillment Processes

Fulfillment logistics is responsible for three key tasks: order management, product customization at the warehouse, and distribution (Ricker and Kalakota 1999). Restructuring of the fulfillment process is discussed in Chap. 6.

5.6.1.1 Order Management

Order management coordinates all processes starting from receiving of orders from customers and ending at committing to deliver. It involves the coordination of sales, credit checking, accounts receivable, and relationships with suppliers (Davenport 1993). For order commitment it must ensure that sufficient inventory is on hand, and that the fulfillment deadlines can be met. It must incorporate relevant constraints such as truck capacity, use of alternate transportation modes, and availability of resources such as loading docks. Specific processes involved can be categorized as order capture, validation, decomposition, and distribution.

Order capture processes enter customer's request for products. It does so using a standard interface platform. *Order validation* creates the library of business and technical rules that can be used to determine if an order is valid and should be accepted for further processing. For example, business rules define possible actions on customer's credit limit and/or the contractual terms. Technical rules define availability of services needed to complete the order in the customer's premises. *Order decomposition* provides a mechanism to generate multiple sub-orders based on the product definition (product decomposition) or based on the delivery location (work order decomposition). For example, a customer may order a broadband service as well as TV service, triggering the order management processes in the broadband and TV service platforms, as well as the billing platform. Thus, decomposition creates sub-orders that can be represented as a tree-diagram. Note that each of the service platforms needs only a part of the product information, while the billing platform requires all billable services. *Order orchestration* coordinates order management by resolving conflicts (such as inter dependencies) among decomposed orders. The product data is transferred to the customer's financial account in the billing system. Orchestrating the fulfillment across multiple platforms requires transaction management capabilities to be implemented in the order orchestration module. The order decomposition and the order orchestration modules lead to a dynamic fulfillment plan.

The order management process can be automated, and made transparent. The components of order management process facilitate better communication based on the electronic product catalog. As discussed, order management enables the customer to purchase products from multiple service platforms.

5.6.1.2 Customization Centers

Increasingly, product customization functions are being performed at distribution centers. The typical activities include light assembly, kitting, merging, consolidation, packaging, and labeling. The finishing tasks must be scheduled for specific customer orders, necessitating a real-time resource allocation at the distribution center. As the capacities of finishing operations are pre determined, it may imply delays in delivering certain orders. The supplier may also use 3rd parties. For example, Dell assembles its computers to the configuration specified by the customer and then readies them for shipment.

Different strategies for maintaining or outsourcing these processes range from fully outsourced supply and fulfillment to complete in-house control. For example, Calyx and Corolla (calyxandcorolla.com) outsources order fulfillment to its partner suppliers while carefully managing the supplier relationships and continuously developing new suppliers. eToys started by outsourcing order fulfillment to Fingerhut but has since built its own warehouses to gain control of these processes. Under an agreement reached in 2000 by Toys "R" Us and Amazon (now defunct), Toys "R" Us managed the supply processes while Amazon managed order fulfillment (Pyke et al. 2001).

5.6.1.3 Distribution

The distribution process includes packing, document preparation, freight forwarding, transportation, and inventory management, both for forward and reverse logistics. It aims at providing users easy access to data on shipping, tracking, and deliveries. Reverse logistics tracks and delivers products returned by customers, damaged products, and used products for remanufacturing.

5.6.2 E-Fulfillment

The business models of e-fulfillment are shaped by two factors: customer preferences (home delivery, store pick up), and supply sources (warehouse, stores, 3rd party, consolidators). As shown in Fig. 5.12, home delivery can be fulfilled from all the four sources of supply, whereas, store pick up would rule out the warehouse and the 3rd party service provider. The supplier structures the fulfillment system by creating a new logistics infrastructure, or by adapting the existing processes.

Fig. 5.12 E-Fulfillment Models

		Customer	
		Home Delivery	Pick up at the Store
Supplier	Warehouse	X	
	Store	X	X
	Use 3rd Party	X	
	Consolidate from Multiple Sources	X	X

The trend for order fulfillment among most online e-grocers is to establish large, automated distribution centers for home delivery in each major market they serve, while some brick-and-mortar chains have employed only in-store order fulfillment.

5.6.2.1 Direct Home Delivery

Direct home delivery can be completed by the supplier, a 3rd party service provider, or a consolidator (Lang 2010). Before initiating the fulfillment, products have to be picked, packed, and labeled, which can be done either in a central warehouse, a distribution center, or in a store, or in some combination thereof.

Direct Delivery from Central Warehouse

Picking, preparation and packing of ordered items can be done from a central warehouse, where the products are in inventory (Fig. 5.13). This central warehouse can be built specifically to serve exclusively the online channel. The existing warehouse for the store operations can be used, in which case the processes in the warehouse must be reengineered.

Fresh Direct uses a 300,000 square feet central warehouse which hosts 12 different temperature zones to keep produce fresh and avoid contamination.

Webvan (now defunct) used a hub-and-spoke architecture distribution centers to deliver its products to consumers (Hays et al. 2004). In the DC, customer orders were grouped and sorted by destination, loaded onto trucks and shipped to the final

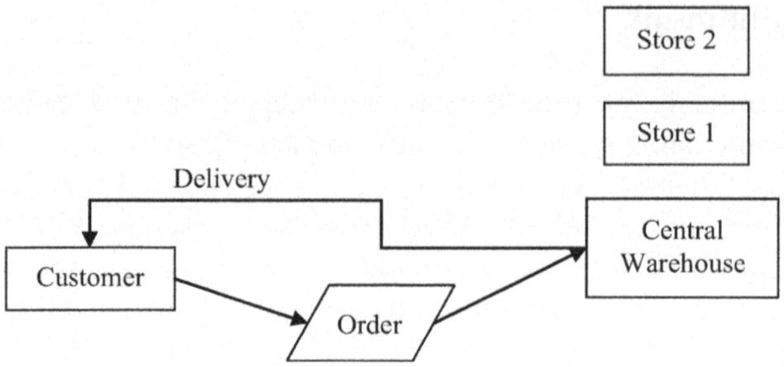

Fig. 5.13 Direct delivery from central warehouse

delivery locations. Cross-docking was used before vans delivered the orders to customers.

Direct Delivery from a Store

Picking and packing of orders can be done inside a physical store out of the shelves. In this case, operators pick the products into a specific order preparation cart from the shelves and then pack and send the completed order to the customer (Fig. 5.14).

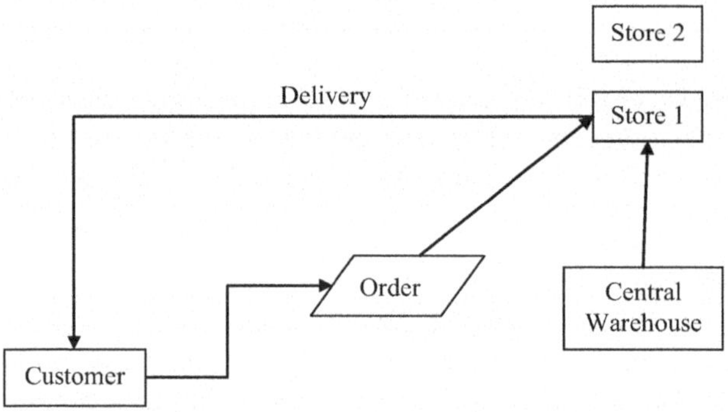

Fig. 5.14 Direct delivery from a store

Instead of building highly automated new warehouses, Tesco's fulfillment and delivery model exploits its network of stores. Dedicated workers pick the items on customers' shopping lists from the shelves of the nearest supermarket, and have the orders delivered as scheduled. When an order is received at the Tesco Direct website, it is routed to the nearest physical outlet. For assembling an order, employees use special carts mounted with screen guides and software instructing them where to pick the items in the list (Hays et al. 2004).

Direct Delivery After Consolidation

If the products ordered are not available in one facility, they must be picked from different locations. They are then consolidated in one site, where they are packed

(Fig. 5.15) and shipped out. Consider a customer ordering a book and a toy. Assume that the book is available in the store, but the toy is in the warehouse. The two items must be picked separately and consolidated at one location (store or the warehouse). They are packed in a single box and shipped out.

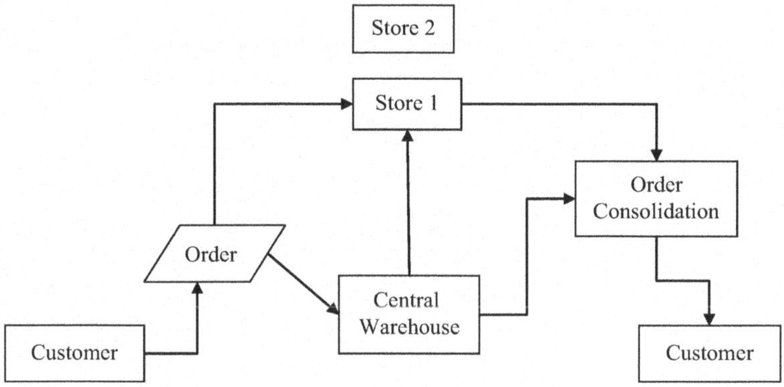

Fig. 5.15 Direct delivery after order consolidation

Use 3rd Party Fulfillment Centers

With increased demand uncertainty, companies are increasingly using 3rd party fulfillment centers, which can be thought of as virtual warehouses (not owned by the company). There are multiple ways a 3rd party service can be utilized, as shown in Fig. 5.16. For example, they can be logistics companies (FedEx and UPS), who pick up the products from different suppliers, provide finishing operations (packing, labeling etc.), and ship them to customers. They could also be assemblers who procure components from suppliers, assemble the products, and ship them. Third party fulfillment facilitates inventory reduction at the company's warehouses, but it may also lead to ceding direct control of fulfillment operations.

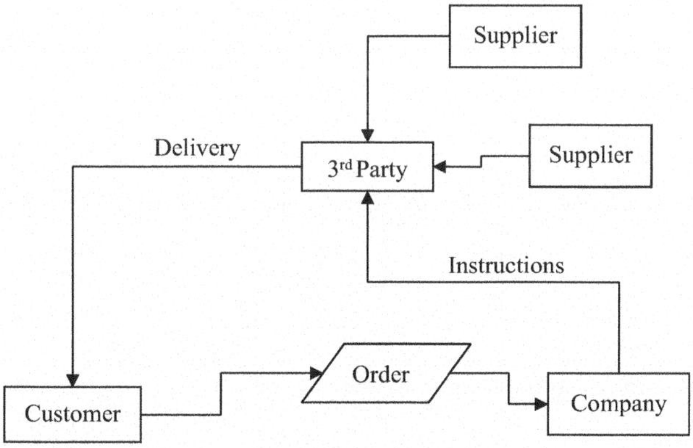

Fig. 5.16 Fulfillment through a 3rd party

5.6.2.2 Customer Pick-up at Store

Customers ordering online may choose to pick up the products in one of the retailer's stores. In such a case, products must be picked, packed, and kept ready for pick-up in the store.

Direct In-Store Order Preparation

If all the products in a customer's order are available in the store chosen by the customer for pick-up, the picking, preparation and packing of the order can be done immediately in the store (Fig. 5.17).

Fig. 5.17 Direct in-store order-preparation

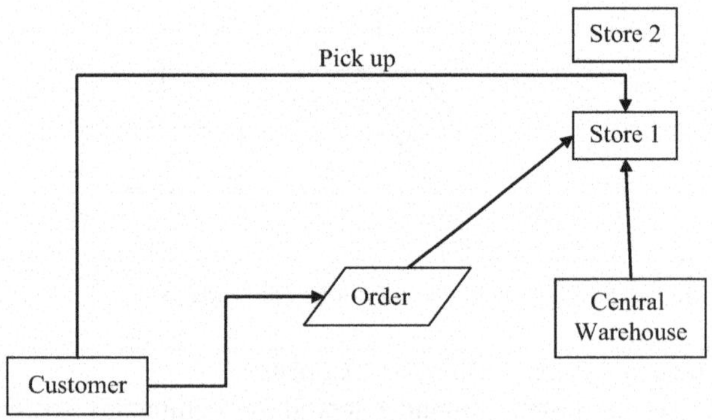

In-Store Order Consolidation from Multiple Sources

If some of the products of the customer's order are not available in the store chosen by the customer, the unavailable items must be ordered from different locations to complete the order. The preparation and packing of the order can be done only after consolidation of all items in the store (Fig. 5.18).

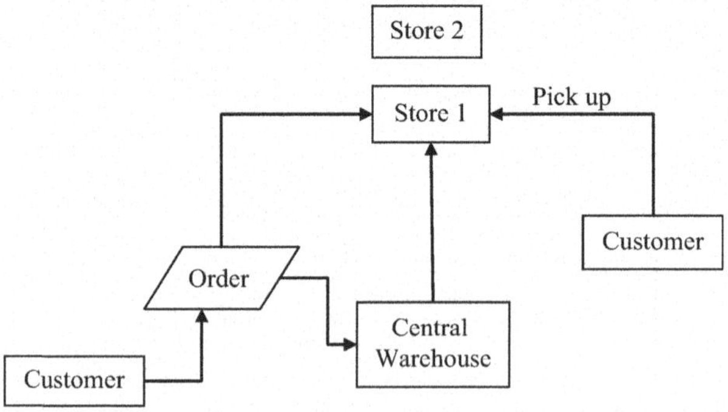

Fig. 5.18 In-store order consolidation from multiple sources

5.6.3 Fulfillment of Bundled Orders

A customer's order with multiple items is known as a bundled order. The supplier might not be able to ship the whole order from one location or at the same time. The supplier will need to split the order into multiple shipments that pick the items from multiple warehouses and/or at different times. For example, a bundled order of a laptop and a TV may have to be split, if both items are not available at a single warehouse. If the choice of warehouses for order picking is made as the orders arrive, the choice will be myopic and hence suboptimal as it does not account for subsequent customer orders and inventory replenishments. In a typical scenario with a mix of bundled and single orders, the strategy of holding-back picking for a bundled order, until the entire order can be satisfied from a single warehouse, may help reduce shipping costs.

As Xu et al. (2009) point out, there is always some time delay (as long as several days) between order placement and its shipment. In some cases, customers may choose delayed shipping for a discount. This time delay can be used to make better choices of warehouses for order picking. Xu et al. (2009) have constructed an effective heuristic for assigning customer orders to warehouses that minimizes the total number of shipments.

5.7 Drop Shipping

Drop-shipping permits an entity to satisfy customer orders without it holding any inventory. Cisco Systems and Dell obtain customer orders, get contract manufacturers to assemble the products and ship them directly to their (Cisco's and Dell's) customers. Suppliers and wholesalers, such as Ingram Entertainment, provide drop shipping services by shipping products directly to customers at the retailer's request. Thus, the intermediary acquires customers and accepts orders, while the wholesaler owns and holds inventory and fulfills orders. Intermediaries benefit by not having to invest in fulfillment and not holding inventory, and the suppliers benefit from higher wholesale prices. If the supplier serves multiple retailers, the whole supply chain may benefit from risk pooling. In addition, information transparency (in e-fulfillment) between the wholesaler and the intermediary permits direct placement of orders in real-time.

Because of the above tradeoffs, the retailer (intermediary) must decide the optimal mix of fulfillment: by itself, and though drop shipping. This is somewhat similar to the optimal mix of make to stock and assemble to order. Netessine and Rudi (2006) provide a newsvendor model of the optimal mix of fulfillment. The authors cite the example of BlueLight.com that uses the mix strategy for CDs. It stocks the top 40 CDs with high profit margins and fulfills orders in its stores, and utilizes drop shipping for the remaining CD titles. Some retailers provide the option of having products shipped to the customer, if they are not available on the shelves in its physical store (Netessine and Rudi 2006).

5.7.1 Merge In-Transit

In-transit merge combines pieces of the order coming from different locations so that the customer gets a single delivery. For example, when a customer orders a PC from Dell along with a Sony monitor, FedEx picks up the PC at the Dell factory, the monitor at the Sony factory and merges the two together as per specific customer orders, before delivering to the customer.

Transportation costs will usually be lower than drop shipping because any order with products from three manufacturers would require only a single delivery to the customer compared to three, otherwise (Chopra 2003). Performing the in-transit merge requires additional capability of doing so. Customer's transaction cost will be lower because a single delivery is received. The required information infrastructure would be complex, as operations at the retailer, manufacturers, and the carrier must be coordinated.

5.8 Logistics Service Providers (Outsourcing)

A logistics service provider (LSP) is an external service provider who manages, controls, and delivers logistics activities on behalf of a customer (shipper or buyer). The services include transportation, warehousing, packaging, custom clearance, billing as well as tracking and tracing. LSPs have been classified in terms of their alliances with customers, degree of customization, and the knowledge level (Hertz and Alfredsson 2003). Other ways of differentiation would be its geographical base. The tradeoff between economy of scale and customization to the individual customer would determine the services developed by LSPs. To add value to customers, LSPs must possess competencies and scale/scope advantages that are superior to that of customers: co-utilization of resources, creation of specific knowledge, and coordination of activities of a portfolio of customers.

According to Butner and Moore (2006), customers are moving away from piecemeal logistics services, to bigger contracts with LSPs with more customization and flexibility. In response, the LSPs have been evolving to offer knowledge-intensive complex solutions with greater scope. Hertz and Alfredsson (2003) capture this interaction as in Fig. 5.19.

It is apparent that when the need for customization is high, LSPs that are highly accomplished in providing advanced services can satisfy customer needs by creating sophisticated and individualized services. At the other extreme, customers with low customization needs may contract with LSPs that provide simple standard services.

Rabinovich and Knemeyer (2006) suggest a different classification of LSPs using two dimensions: service form (information based, and physical-asset based), and service focus (buyer focused, supplier focused, and delivery focused). This creates six clusters of LSPs, shown in Fig. 5.20.

Physical-asset-based LSPs compete based on transportation assets and/or warehousing assets. The information-based LSPs, on the other hand, create

		Customization	
		Low	High
Process Efficiency	High	**Service Developer** Sophisticated interchangeable modular services	**Customer Developer** Creating knowledge-based individualized services
	Low	**Standard Service** Simple standard services	**Customer Adaptor** Services dedicated to individual customers

Fig. 5.19 Taxonomy of logistics service providers

	Buyer Focused Services	Supplier Focused Services	Delivery Focused
Physical-asset Based	Order-returns	Order consolidation	Picking/Packing
Information Based	Order payment	Order-verification	Freight Forwarding

Fig. 5.20 Focused logistics service providers

competitive advantage through managing information. It follows that LSPs that are primarily physical asset based, will not be that good in providing knowledge intensive customized service; that would fall in the domain of information based LSPs.

The buyer-focused LSPs provide such services to sellers as order payment and returns and exchanges. These LSPs need to be flexible in terms of including services for new products or extensions of existing products. The supplier focused LSPs provide such services as order consolidation and order tracking. Delivery-focused LSPs provide services available through "last mile" carriers, performing functions such as order forwarding. They enable sellers to expand quickly into new geographical markets.

5.9 Restructuring

Several new developments are creating challenges in logistics industry. For example, customers are requiring drop shipment of products to their facilities; retailers are requiring the use of RFID for tracking shipments; and a plethora of new technologies are being used: global positioning system (GPS) by trucking companies, electronic data interchange (EDI) by suppliers, point of sale (POS) by retailers, and smart phones by consumers. Consumers are increasingly concerned about the environmental impact of the products they use, which influences producers' operations. In addition, the unpredictable fuel prices have been causing havoc in the transportation prices. Logistics management would need to restructure in view of the potential of capacity shortages, global expansion, need for agility, security concerns, and environmental concerns.

Whenever fuel prices increase carriers raise prices and cut back on capacity. In addition, shippers are moving goods in small quantities, more frequently, implying LTL shipments. Airlines too have cut more than 25% of capacity in the last decade. Clearly, the ability of individual carriers to overcome these limitations will help to differentiate them from the competition.

Manufacturers are concentrating production and inventory in fewer locations. While generating economies of scale in production, it makes the logistics system more transportation-intensive, lengthening lead-times to customers. In addition, suppliers prefer to ship in full truckloads to minimize the cost of investment in equipment. Economies of scale are also affected by the handling of inventory – shipping pallets is cheaper than shipping cases, and shipping cases is cheaper than shipping individual units.

As industrial activities extend globally, logistics will involve more material and information flows throughout a supply chain from suppliers to customers. This will result in longer transit times and higher cost in logistics management. Individual companies would need to provide adequate transportation, storage, rapid customs clearance, and cost control. The volume of material, information and money flowing across the borders can be problem as well as opportunity. Logistics support is also needed in the global sourcing of components and material and transporting them to assembly centers worldwide. Therefore, logistics needs to be managed as an integrated process that seeks to optimize these flows to include customers, suppliers and manufacturers.

The design and shape of new trade corridors will be strongly affected by the new players actively using them. Logistics service providers should work closely together in order to shape those new transport corridors, i.e. identify future transport corridors, build required infrastructure, and actively contribute to develop services which are optimized for transportation activities in those corridors.

Logistics managers face security concerns from the perspective of complying with new protocols for reducing security threats, as well as in dealing with the impact of an attack that may disrupt operations. Shippers may select security-conscious carriers, using secure ports, and implementing secured packaging.

A growing concern over the environment and energy use also presents a challenge to logistics, as discussed in Chap. 9. Increasingly, consumers are evaluating a product's carbon footprint when making a buying decision that inform consumers how their buying decisions impact the environment. These concerns will clearly have a tremendous influence of the design of future logistics services.

The market for logistics services is currently somewhat fragmented, lacking in transparency and inefficiency in most emerging markets. Online platforms may be an option to better coordinate logistics services. In some countries, these platforms are already functioning successfully. They include relevant information about logistics service providers in the market and serve as freight exchanges and tendering platforms. An online platform which integrates comprehensive information would facilitate the rapid identification of optimal providers for required services. Customers could also tender their needs on the platform and allow logistics service providers to bid and potentially win the contract.

In the past, logistics service providers in emerging countries have strongly focused their activities on urban areas leaving rural areas relatively unexplored. Logistics service providers which extend their activities to rural areas and establish effective distribution networks within them could realize first-mover advantages. By opening up these new markets, logistics service providers could successfully contribute to the design of new supply and procurement markets.

References

AASHTO (2010). Unlocking Freight. http://expandingcapacity.transportation.org/unlocking_freight/images/FreightReportFinal_7710.pdf.

Bartholdi, J., & Gue, K. (2004). The best shape for a crossdock. *Transportation Science, 38*, 235–244.

Brogan, J., Aeppli, A., Beagan, D., Brown, A., Fischer, M., Grenzeback, L., McKenzie, E., Vimmerstedt, L., Vyas, A., & Witzke, E. (2013). Freight transportation modal shares: Scenarios for a low-carbon future. *Transportation Energy Futures Series*. Washington, DC: Cambridge Systematics Inc., and the National Renewable Energy Laboratory U.S. Department of Energy. http://www.nrel.gov/docs/fy13osti/55636.pdf.

Butner, K., & Moore, D. (2006). *Building value in logistics outsourcing: The future of the logistics provider industry*. New York: IBM Business Consulting Services, IBM Institute for Business Value, IBM Global Services.

Campbell, J., & O'Kelly. M. (2012). Twenty-five years of hub location research. *Transportation Science, 46*, 153–169. dx.doi.org/10.1287/trsc.0410.

Chopra, S. (2003). Designing the distribution network in a supply chain. *Transportation Research, Part E, 39*, 123–140.

Clarke, G., & Wright, J. (1964). Scheduling of vehicles from a central depot to a number of delivery points. *Operations Research, 12*, 568–581.

Couriel (2011). Chapter 5: Routing and scheduling. Education and Culture DG, http://www.adam-europe.eu/prj/7095/prj/CourieL_WP2_Chapter5_final.pdf.

Davenport, T. (1993). *Process innovation: Reengineering work through information technology*. Boston: Harvard Business School Press.

Erara, A., Hewitt, M., Savelsbergh, M., & Zhang, Y. (2012). Improved load plan design through integer programming based local search. *Transportation Science*, trsc.1120.0441. http://www2.isye.gatech.edu/~mwps/publications/load_planning_1214.pdf.

Ertek, G. (2012). Crossdocking insights from a third party logistics firm in Turkey. In Ç. Haksoz, S. Seshadri, & A. V. Iyer (Eds.), *Managing supply chains on the silk road: Strategy, performance, and risk*. CRC Press. Taylor and Francis, Boca Raton, FL, USA

Grahovac, J., & Chakravarty, A. (2001). Sharing and lateral transshipment of inventory in a supply chain with expensive low demand items. *Management Science, 47*, 579–594.

Hansen, O., & Cooper, A. (2012). Sustainable energy for all: Opportunities for the transportation and logistics industry. Accenture, http://www.accenture.com/SiteCollectionDocuments/PDF/Accenture-Sustainable-Energy-All-Opportunities-Transportation-Logistics-Industry.pdf.

Hausman, W., Lee, H., & Subramanian, U. (2005). Global logistics indicators, supply chain metrics, and bilateral trade patterns. World Bank eLibrary, http://elibrary.worldbank.org/content/workingpaper/10.1596/1813-9450-3773.

Hays, T., Keskinocak, P., & de Lopez, V. (2004). Strategies and challenges of internet grocery retailing logistics. In J. Geunes, E. Akcal, P. Pardalos, H. Romeijn, & Z. Shen (Eds.), *Applications of supply chain management and E-commerce research in industry*. Dordrecht: Kluwer.

Hertz, S., & Alfredsson, M. (2003). Strategic development of third party logistics providers. *Industrial Marketing Management, 32*, 139–149.

ICF (2002). Economic effects of transportation: The freight story. ICF Consulting Fairfax, Virginia. http://www.ops.fhwa.dot.gov/freight/documents/freightstory_12902.pdf.

Lambert, D., Stock, J., & Ellram, L. (1998). *Fundamentals of logistics management*. McGraw-Hill, Boston, USA.

Lang, G. (2010). Fulfillment systems in multi-channel retailing – customer expectations and economic performance. *8th International Research Conference in Logistics and Supply Chain Management (RIRL)*, Bordeaux, pp. 1–25.

Lockamy, A. (1995). A conceptual framework for assessing strategic packaging decisions. *International Journal of Logistics Management, 6*, 51–60.

Netessine, S., & Rudi, N. (2006). Supply chain choice on the Internet. *Management Science, 52*, 844–864.

OECD (2002). Transport logistics: Shared solutions to common challenges. http://www.internationaltransportforum.org/pub/pdf/02LogisticsE.pdf.

Parakala, K., & Udhas, P. (2012). Hub and spoke operating model, KPMG/NASSCOM, www.kpmg.com/IN/en/IssuesAndInsights/ThoughtLeadership/Hub-and-spoke-model.pdf.

Pyke D., Johnson, E., & Desmond, P. (2001). E-fulfillment: It's harder than it looks. *Supply Chain Management Review, 5*, 26–32

Rabinovich, E., & Knemeyer, M. (2006). Logistics service providers in internet supply chains. *California management Review, 48*, 83–108.

Reddy, J., & Sterling, P. (2009). Packaging value chain optimization. CombineNet and Supply Chain Optimizers, Pittsburgh, http://s3images.coroflot.com/user_files/individual_files/305272_CMHkOK3isqrXuS23XN9K4z7Tl.pdf.

Ricker, F., & Kalakota, R. (1999). Order fulfillment: The hidden key to e-commerce success. *Supply Chain Management Review, 3*, 60–70

Saghir, M. (2004). The concept of packaging logistics. Second World Conference on POM, Cancun, Mexico. http://www.pomsmeetings.org/ConfProceedings/002/POMS_CD/Browse%20This%20CD/PAPERS/002-0283.pdf.

Sople, V. (2007). Material handling equipment: Exploiting productivity potential in supply chain. http://smallb.in/sites/default/files/Material%20Handling%20Equipment.pdf.

Stalk, G., Evans, P., & Shulman, L. (1992). Competing on capabilities: The new rules of corporate strategy. *Harvard Business Review, 70*, 57–69.

Topaloglu, H., & Powell, W. (2007). Incorporating pricing decisions. *Transportation Science, 41*, 281–301.

Wyman, O. (2009). Hub and spoke in insurance: Aggregate and conquer. Oliver Wyman Limited. oliverwyman.com/pdf_files/Oliver_Wyman_Hub_and_Spoke.pdf.

Xu, P., Allgor, R., & Graves, S. (2009). Benefits of reevaluating real-time order fulfillment decisions. *Manufacturing and Service Operations Management, 11*, 340–355.

Transforming the Contemporary Supply Chain

6

6.1 Introduction

In Chap. 1, we discussed a framework for restructuring supply chains that created links between business drivers (such as global marketplaces, product variety, shorter product life cycles, and customer services) and the needed capabilities. This has become even more critical in recent years because the advancement in information technology has enabled companies to improve their supply chain strategies and explore new models for management of supply chain activity. In Chaps. 2, 3, 4, and 5, we discussed the building blocks of a contemporary supply chain. In this chapter we focus on specific transformation approaches for the supply chain as a whole, in view of the business drivers of today such as uncertainties, technology, and globalization. In Chaps. 7, 8, 9, and 10 we would analyze how the modern supply chain would need to evolve in view of the emerging business concerns.

The source of uncertainty may be present in the demand processes, in the supply processes, and in the production processes; as discussed in Chaps. 2, 3, and 4. New technology can help in improving quality, increase processing speed, improve coordination, and create new capabilities through remote control. Globalization presents many contemporary issues with barriers and opportunities, and many emerging issues discussed in Chap. 7. The fulfillment performance, discussed in Chap. 5, can be improved by altering the business model governing the flows of goods and information. While cost can be reduced by increasing efficiency, the question is whether it is possible to so without compromising responsiveness or quality. Value chain segmentation is an approach for modularizing the supply chain that builds on modularized products and processes, discussed in Chap. 3. Thus, there can be multiple approaches for restructuring as shown in Fig. 6.1.

Three dominant themes are discernible in such transformations, irrespective of their drivers: modification of operational control systems, leveraging of emerging technologies, and redesign of the supply chain architecture. While reengineering of

A.K. Chakravarty, *Supply Chain Transformation*, Springer Texts in Business and Economics, DOI 10.1007/978-3-642-41911-9_6,

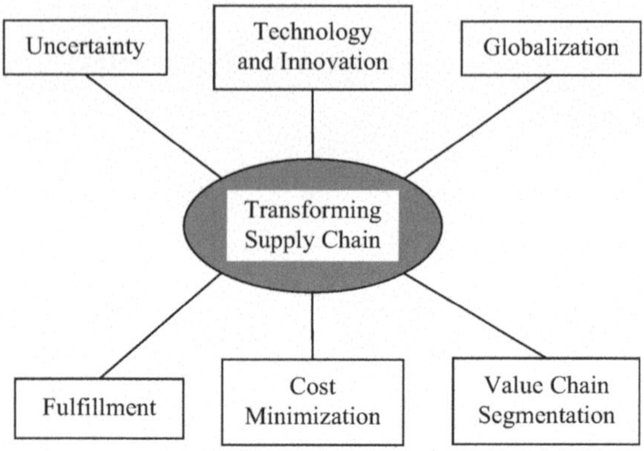

Fig. 6.1 Supply chain transformation

operations is the most common form of transformation, architectural redesign is an undertaking that requires major new financial investment.

For example, to respond to uncertainties in demand a company may simply increase product inventory, modularize the product through a redesign, may make structural changes by investing in flexible processes and equipment, and/or move the order decoupling point upstream or downstream. For rapid fulfillment, the company may modify the order processing system or make investments in shipment-tracking, drop-shipping, and information-sharing. To reduce cost the company may tighten procedures for the reduction of waste, procurement from suppliers, and control of inventory. It may also make structural changes by creating shared services hubs, and restructure the global network to benefit from the country-specific customs duty and tax regimes.

Technologies such as the internet help transform supply chains from a sequential and linear relationship to a network. They enable customers configure their own customized products. Other examples include information sharing for transparency, radio frequency identification (RFID) for tracking shipments, mobile technology for real time decisions, cloud technology for better coordination and reduction of fixed cost, and the Web 2.0 for information exchange. A successful implementation of these technologies requires significant changes throughout the supply chain.

Companies may transform supply chains to benefit from market opportunities in *multiple* countries. However, access to global markets is not achieved merely through designing, manufacturing, and marketing new products. Companies must operate under different environments, constraints, processes, people, and countries. It requires restructuring of relationships with suppliers and customers, access to new technology and its implementation, and a customer focused mindset. For example, companies may split production and sales operations so as to have assembly operations in emerging economies with a subset of components imported from mature economies (discussed in Chap. 7).

Segmentation of the value chain lets companies tailor their supply chain focus to clusters of customers and products, in their portfolio. Customers are served through segment-specific supply chain processes, policies, and operational modes. It calls for a redesign of the supply chain architecture to create clusters of products, customers, and resources, based on shared sets of values. The intent is to design flexible and responsive supply chains that are capable of shifting back and forth between the complexity of customization and the simplicity of standardization.

6.2 Uncertainty

Uncertainty is caused by many factors including increased customer choices, product customization, rapid technological improvements, global competition, and upstream supply fluctuations (Gangadharan 2007). The impact of uncertainty can be minimized by restructuring the supply chain to create special sourcing for demand surge, decoupling order penetration, using flexible processes and capacities, creating information transparency with advanced technology, and adopting "sense and respond". Managing uncertainty in a cost effective manner can lead to improved customer service, lower supply chain costs, and a significant competitive advantage.

6.2.1 Demand-Surge

In some cases, organizations develop two-stream supply chains – one to deliver the base demand, and the other to manage the demand surge. For example, if demand of a company's product varies between 500 and 800 units per month, it would be risk-free to order the first 500 units. An order quantity exceeding 500 will involve both upside and downside risks. The case in point is Griffin Manufacturing in the US, discussed in Chap. 4 (Warburton and Stratton 2002). Griffin's customer (a large retailer) planned to move work to Honduras to take advantage of the low cost of production in that country. While the Honduran supplier's production was geared to large volumes based on economy of scale, it was unable to react quickly to changes in demand. This provided Griffin a niche and it reengineered its process to become an agile supplier. Clearly, splitting of demand created a new competitive environment, and Griffin needed to retool itself to increase its agility.

The retailer needed to restructure the supply chain to manage two channels of procurement. Placement of orders to the Honduran company was greatly simplified with advance orders of quantity Q and long lead times. To manage procurement from Griffin, on the other hand, the retailer needed to track demand D_i per period of product i, for T periods, and place an order for the quantity $Max\left(0, \sum_{i=1,T} (D_i - Q_i)\right)$ with Griffin. As the order quantity varied from period to period, Griffin needed to make arrangements to ensure adequate capacity availability for the retailer's orders.

An added complexity was the management of two unit prices; Griffin charged a premium price, higher than the price paid to the Honduran supplier. Because of the variability in order size, Griffin's decision on production capacity was not risk free. Thus, four decisions needed to be made – Honduran order size Q (by retailer), unit prices (by Griffin and the Honduran supplier and production capacity (by Griffin).

6.2.2 Customer Order Decoupling

The customer order decoupling point separates the demand side of the supply chain (focused on delivery to the end customer) from the supply side (based on procurement and production). The decoupling point can be positioned so as to best suit the need for responding to a volatile demand downstream yet providing level schedule upstream of the decoupling point. The decoupling point approach is essentially based on the concept of postponement.

As an example, consider customers of a supplier of television components, who required next day delivery (Waddington et al. 2002). The supplier used a make-to-stock policy, holding a wide variety of finished goods in inventory. However, it could not react quickly enough if a particular type of product was out of stock. To rectify the problem, the supplier moved the position of the customer order decoupling point (CODP) upstream. Some of the items that were made to stock were now made to order. This necessitated the redesign of some of the manufacturing processes to allow rapid final assembly of items using other generic components held in stock. The items could be assembled, packed and dispatched quickly. Inventory was reduced as a result, dispensing with the need for holding finished goods inventory. The assemble-to-order process made the system responsive, though the delivery lead time increased somewhat.

While moving the CODP upstream from the finished goods warehouse to component manufacturing (as shown in Fig. 6.2) increases responsiveness, the processes lying between the initial and revised CODPs must be restructured. As there will be no inventory of finished goods now, the assembly process will have to be customized to be order-specific. The assembly technology must be flexible to enable assembly of different choice sets of components to match customer orders (assemble-to-order, ATO). Raw material procurement and component manufacturing will continue to operate in a make-to-stock (MTS) mode.

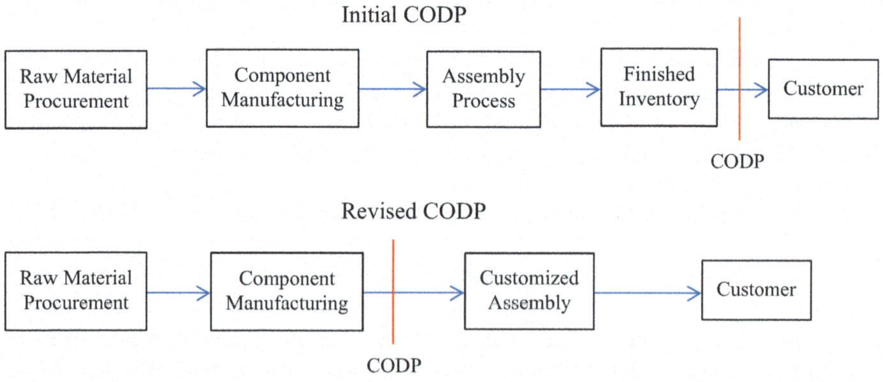

Fig. 6.2 Moving the CODP upstream

Note that with the revised CODP for assemble-to-order (in Fig. 6.2) customers must wait while their order is assembled. To minimize this wait time, the supply chain can be structured for a third option where the finished product is made to stock (MTS) as well as assembled to order (ATO). We call this make to stock and order (MTSO), where the size of inventory is smaller than that in MTS. While the make-to-stock items ensure demand fulfillment for customers who arrive early, MTSO facilitates fulfillment of late orders as well, on an ATO basis. As the demand is not known in advance, a policy that combines MTS of components with ATO can be optimal. To implement this, inventory is allocated to two buckets: bucket 1 with finished products is used to fulfill the demand of early customers, and bucket 2 with the generic components is earmarked for the late customers. The company must decide how much of which products to stock in bucket 1; and what should be the size of inventory in bucket 2.

Consider a scenario with n final products (one component per product type) and demand ξ_i, $i = 1, n$. The supplier builds MTS inventories of Q_i units of finished product i at unit cost c_i, and holds them in bucket 1. The supplier also builds and holds Q units of the "generic" component (unit cost c) in stock in bucket 2. Of this Q units, assume that q_i units are earmarked for conversion to product i $(\sum_{i=1,n} q_i \leq Q)$, at cost r_i per unit, $c \leq c_i \leq c + r_i$. Therefore, the maximum demand of product j that can be satisfied would be $Q_j + \left(Q - \sum_{i \neq j} q_i\right)$. If ξ_i lies in the range $Q_i \leq \xi_i \leq Q_i + q_i$, $Q - \sum_{i=1,n} \min(\xi_i - Q_i, q_i)$ units of the generic product will be left over in inventory. Demand in excess of $Q_i + q_i$ for product i is lost. The supplier needs to determine the optimal mix of inventory Q of the generic components, and MTS quantities Q_i and q_i, as in Chakravarty (2010).

6.2.3 Creating Flexibility in the Supply Chain

Flexibility of different types (volume, variety, expansion, process, program, etc.) has been discussed in the literature (Gerwin 1993). With uncertain demand, the company will primarily be concerned with variability in volume and variety. While volume flexibility requires that the company must possess the capability to increase and decrease production quantities quickly. A large variety implies that the company should be able to switch to different processes rapidly to accommodate a mix of products. We first discuss volume flexibility, followed by mix flexibility for variety. The primary objectives of restructuring the supply chain for volume flexibility include: providing flexibility in manufacturing and/or assembly capacity, structuring contracts with suppliers for flexible procurement volumes, and contracting flexible delivery volumes with distributors.

6.2.3.1 Capacity Sharing

Capacity sharing is one of the ways of increasing the availability of capacity. It enables better coordination of capacity related decisions in a supply chain. It helps moderate the impact of demand uncertainty by offsetting capacity shortage at one location with capacity surplus at another. Competitors such as the international airlines collaborate through code-sharing and alliances. Capacity sharing networks emerge to leverage cost and/or purchasing-power differentials between different regions of the world. Transshipment of finished products between two networked facilities is a form of capacity sharing. The emergence of competing carriers with fast deliveries has made it possible to efficiently move items between facilities.

Rudi et al. (2001) show that companies can increase their profits if transshipment is allowed between them. Grahovac and Chakravarty (2001) consider a supply chain with a distribution center (DC) supplying expensive parts (such as aircraft components) to n retailers. They establish that by permitting transshipment among retailers, the inventory held at the DC decreases while the inventory held by an individual retailer increases. This is an example of free-rider, as the decrease in DC's inventory is not induced by DC's productivity, per se.

If firms do not sell identical products, transshipment would not be very useful for creating flexibility. Exchange of capacity between two networked facilities (firm 1's product can be manufactured in firm 2's facility and vice versa) would be another way of providing flexibility. As an example, Renault's plant in Brazil produces two additional Nissan vehicles, while Nissan's plant in South Africa builds two additional Renault vehicles. A total of 11 vehicles are cross-manufactured (Ghosn 2009). Capacity sharing in network services is implemented through capacity swaps and redundancy agreements (Yankee 2001). Another example is the telecommunication tower, which is shared by multiple network operators (Chakravarty and Werner 2011).

Chakravarty and Zhang (2007) discuss a capacity exchange scenario between two firms and establish how capacity price may be determined, and how a side payment maybe used to coordinate the capacity exchange decisions. They also study a scenario where the firms' capacity investment decisions are made individually and exchange decisions are made as in a centralized system.

6.2.3.2 Building Flexibility

We consider building flexibility in manufacturing and sourcing.

Restructuring manufacturing to include programmable equipment for rapid retooling is one of the ways of supporting product variety (mix flexibility). Mix flexibility may require order-specific choices of suppliers (for procurement) from a large set of suppliers. A job shop, where workers are trained in multiple tasks, is an example of flexible manual facility. Facilities can be organized to be market-specific or centralized.

A flexible manufacturing system, with appropriate processes and equipment, is the key to ensuring product variety. The production processes must be designed to cover a wide range of variations induced by customized demand. Therefore, process design must ensure minimum setup times for product changeover at machines. It must also provide for flexible routing that enables a large number of paths that can be followed from start to finish. Programmability of automation requires flexibility-enhancing processes in terms of sequencing and dispatching decisions.

Flexible sourcing and shipment enhance the ability to offer increased variety in the final product. Sourcing flexibility is a function of the number of available suppliers, the ease of reconfiguring the supply side, and the supplier's production-process flexibility. A larger set of available suppliers increases the buyer's degrees of freedom in choosing suppliers to procure from. Ease of reconfiguring the supply base enables adding new suppliers to the supply base and dropping suppliers that cease to be relevant. Finally, the flexibility of supplier-processes is critical in working with new materials and modified component designs. Furthermore, suppliers must be able to support process standardization for generic components.

Of course, there needs to be a close collaboration with suppliers. However, as collaboration is based on long-term agreements it would have a negative impact on flexibility, because reconfiguring the supply chain is much more difficult when collaboration specifications are binding.

6.2.4 Providing Transparency

In today's fragmented business value network, real-time transparency is a key imperative. Visibility is the ability to know exactly where things are at any time, or where they have been, and why. It is a capability that allows a company to achieve its business goals with event-based information relating to suppliers' production rates, shipment lead times, retail sales, inventory, and customer's demand projections. Visibility can lower inventories, decrease production lead times, increase coordination, facilitate tracking, and decrease fulfillment time. It enables proactive operations, product tracking, on-time delivery, and alerts on product availability and shipments-status.

Del Monte Foods, a large US foods company, has successfully transformed its supply chain (Godfrey et al. 2012). The company created network-wide transparency into demand signals and inventories to implement pull-oriented supply processes. Del Monte's has benefited greatly from service level enhancements, and its order-to-ship lead times dropped by as much as 24 hours for many customers. It enabled a 60-hour order-to-delivery cycle time, which enabled the company to help retailers reduce safety stocks.

In the health care industry the inability to track pharmaceutical products, as it travels from the supplier to the end consumer, can lead to a proliferation of counterfeit drugs. Federal regulations have lately started requiring traceability of drugs: record of ownership, transaction history, packaging configurations, and storage conditions. Visibility is required in terms of product "pedigree" transactions at each stage, recalls, counterfeits, and offshore content. The high cost and time required and internal political struggles are often the biggest barriers to integrating the disparate systems of healthcare IT. The food industry has also started requiring the traceability of fresh foods in the global chain to avoid contamination and the threat of terrorism.

The information that needs to be managed, for providing transparency, extends from the master data on products, bills of material, vendors, materials, and customers, through to transactional data including purchase orders, process plans, delivery notes, and invoices. At its core lies the Master Data Management (MDM) providing the data-reservoir, definitions, and governance needed to manage all transactional data. Companies need to restructure the supply chains so as to adopt Enterprise Application Integration (EAI) software to tie together and capture all relevant data in disparate applications. It is possible to selectively outsource this role to service providers that provide a platform for the managing the visibility of supply chains.

Butner (2007) suggests setting up a virtual command center to enable supply chain visibility over logistics service providers, multiple transportation modes, import and export regulations, and distribution systems. It uses a combination of business processes and technology enablers to effectively monitor and evaluate supply chain activities. It integrates suppliers and logistics partners to provide actionable responses, in response to events and performances. It delivers timely data from existing business applications to stakeholders, providing visibility as well as proactive monitoring and notification. It uses real time information together with business rules to integrate and synchronize actions among partners, and generate alerts and performance indicators. Supply chain partners use data and technology to choose the best option to fill orders, and re-allocate them if stock-outs occur. The virtual command center can bring warehouses online at other locations to respond to peak demand instances, and take them offline when no longer needed. They can redirect in-transit product to the distribution center closest to the demand source, using RFID technology.

6.2.5 Sense and Respond

The focus of sense and response is on achieving real time demand and developing methods to respond quickly so as to fulfill customer needs. Demand sensing is a forecasting method, based on the analysis of real-time information, for generating a demand-forecast. The process relies on developing demand insights based on mathematical analysis of signals, and orchestrating the supply network around these insights (Donnelly 2007).

Demand insights are built using demographic and market data to determine market potential, buying patterns, and market shifts. Companies can store demand insights, and query the data as and when needed. The partners can improve demand-sensing capabilities by increasing the frequency of demand signals, and synchronizing the attribute-based models with market-data. This enables triggering of alerts and exceptions based on shifts in demand insights (Donnelly 2007).

The attribute-based forecasting software includes multiple data sources including shipments and order history, distribution center withdrawals, customer collaboration data, point-of-sales, and marketing data. Configurable business rules enable users to spot trends, identify forecast variances, and respond to other demand stream changes easily and efficiently (Inspirage 2013). The supply chain partners must determine the key attributes that need tracking to improve demand response. The partner must also be knowledgeable about the sources of market data to sense buying behavior, and build a repository for demand signals to model key attributes.

6.3 Crafting Responsiveness in Fulfillment

An order fulfillment process starts with receiving orders from customers and ends with delivering the orders, and the time between the two events – order receiving and order delivery – is the fulfillment response time. While minimization of the fulfillment response time is the primary objective of most customers, it is not so for all customers; customers ordering a book online are willing to wait longer than those that drive to a nearby bookstore. The manufacturer or the supplier, therefore, must first determine the tradeoff between the customer preferences of response time and cost.

Stratex Networks, a leading provider of digital microwave radios, made a strategic decision in 2002 to raise customer service levels by elevating order-delivery performance and reducing order-fulfillment time (Cohen 2006). Stratex was aware that the numerous handoffs inherent in its current business model were a major hindrance to achieving the new objective, which required coordination of order-scheduling, product-configuration, and shipments. The existing structure operated with functional groups for order entry, order management, production, planning, sourcing, and logistics. While the groups interacted frequently, communication about delivery schedules was inconsistent and erroneous, often generating inaccurate order status information. Stratex restructured the operations by creating an order-fulfillment team and moved the logistics related functions to this new group. Links were tightened between customer order management, planning, and procurement by physically moving the groups closer together and the procedures for order confirmation were tightened.

The supply chain can be restructured in different ways to make order fulfillment more responsive. Chopra (2003) suggests the options shown in the first column in Table 6.1. These include: manufacturer's online store delivering customer order, manufacturer using 3PL partners for drop shipping, manufacturer using merge in transit points before delivery, distributor using carriers for delivery from its warehouse, and distributor providing last-mile delivery from its warehouse located near the customer. The capabilities and/or resources necessary for each of the fulfillment options are shown in columns 2–7.

We discuss these fulfillment options below in more detail.

Table 6.1 Fulfillment options

	INV at plant	DIST near customer	Merge point	3PL drop shipping	Carriers	Last mile
Direct delivery	X				X	
Drop ship	X			X		
Merge in transit	X		X		X	
Distributor with carrier		X			X	
Distributor with last mile		X				X

6.3.1 Online Manufacturer Shipping Directly to Customers

As shown in Fig. 6.3, order information flows from the customer to the manufacturer's online store. Inventory is held at the manufacturing plant. The manufacturer ships the products to customers using carriers, as per actual orders. To reduce response time, the manufacturer can contract with carriers with faster modes of transportation. Fulfillment time depends on the distance between the manufacturer's warehouse and the location of the customer. Note that in this model a retailer can also be the immediate customer, and it plays the role of an intermediary. The carrier would now deliver to the retailer, and customers would visit the retail stores to make a purchase. Fulfillment models using a retail store are discussed in Chap. 5. In a variation to this model, Dell uses a merge in transit facility, described later.

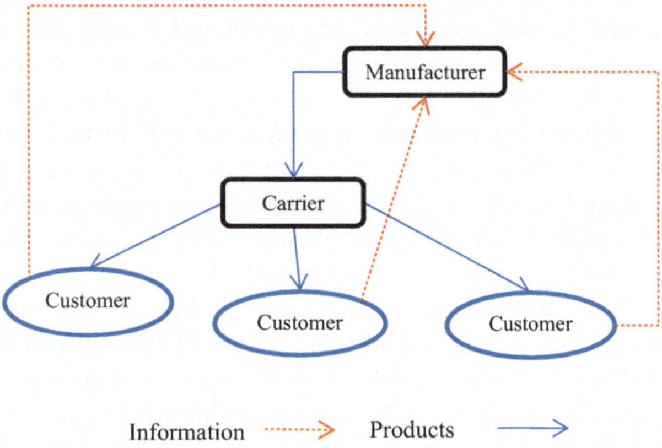

Fig. 6.3 Manufacturer shipping directly to customers

The manufacturer needs to restructure the supply chain by creating an online store, and developing appropriate expertise of managing customer-relations. To reduce the fulfillment time, the manufacture can operate in a VMI mode with visibility of customers' warehouses; the information arrows in Fig. 6.3 will now be reversed. This would require a major restructuring, as the databases of the manufacturer and customers must be linked.

6.3.2 Drop Shipping

As Chopra (2003) points out, online retailers such as eBags, W. W. Grainger, and Nordstrom.com hold little or no inventory, and use drop shippers to deliver goods to the customers, as shown in Fig. 6.4. When a customer orders a product online, or from a mail-order catalog, the retailer transmits the order to the drop shipper, a third-party fulfillment company. That drop shipper packages, labels, and ships the item directly to the customer.

Fig. 6.4 Drop shipping by retailers

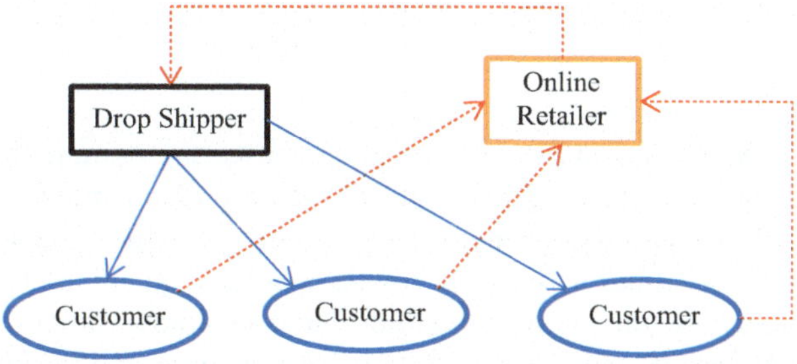

By selecting a drop shipper, closer to customers and with appropriate inventory, the fulfillment time can be reduced significantly. The drop shippers' operations can become complex given that they are required to satisfy different fulfillment times for customers, as specified in orders. If a drop shipper is unable to deliver all items in a customer's order, the order must be split and allocated to multiple drop shippers at multiple locations. This is far from ideal from a customer's perspective, as he/she may receive multiple partial shipments, over time, increasing the complexity of her receiving process.

6.3.3 Merge in Transit

At the merge-in-transit points that operate without inventory, items from multiple manufacturers are consolidated into a single delivery per customer. In assemble-to-order scenarios merge in transit is used for finishing the customized product. To fulfill a customer order – Dell PC with Sony monitor – the carrier would pick up the PC at the Dell factory and the monitor at the Sony factory, and merge the two at a hub before delivering the order. This reduces the cycle times of the consolidated product, as the transportation time (and cost and inventory holding) of the monitor from Sony factory is avoided. This is critical in industries with short product life cycles such as electronics. Information and product flows for the in-transit merge network are as shown in Fig. 6.5.

Fig. 6.5 Merge in transit

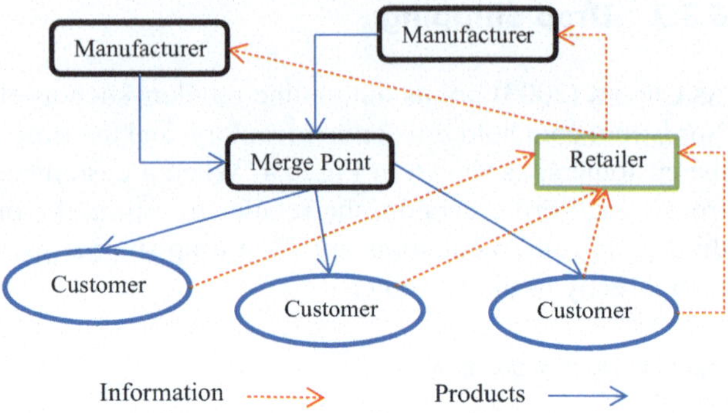

Merge-in-transit fulfillment increases the complexity of material flow control (Ala-Risku et al. 2003), and the supply chain must be restructured accordingly. All items on a customer order must be given individual identification numbers and this information must be made available at all merge points. The problem of matching the items with individual customer orders becomes magnified if the number of items per customer order is large and/or the number of manufacturers involved is large. The database and information system must be restructured to include a relational database and an effective matching algorithm. Tracking orders post merge, however, is simpler as orders are received by customers as a whole. For example, an order with items from five manufacturers would require only one delivery to the customer compared to five, before.

6.3.4 Distributor Warehouse with Carrier Delivery

In this model, a distributor is used with its warehouses located closer to customers, as shown in Fig. 6.6, and make-to-stock inventory is shipped to the distributor warehouse, proactively. Customers place orders with the distributor, who uses carriers to deliver products to customers. Closeness of the warehouses to customers reduces fulfillment time, and the entire order is aggregated at the warehouse before shipping.

Fig. 6.6 Distributor warehouse with carrier delivery

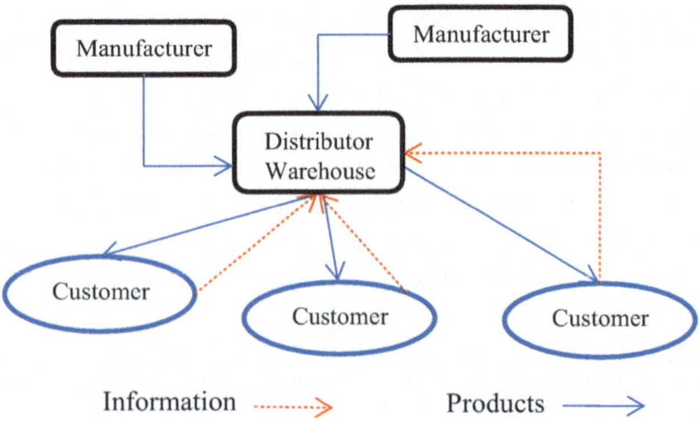

Distributors are used regularly in the beer industry for fulfilling customer orders (Soave 2013). From the time a beer package leaves the brewery to when it arrives at the customer, the distributor is responsible for providing transportation, refrigerated storage, and preservation of perishable malt beverages. Distributors maintain a fleet of trucks and help deliver the customized inventory to a diverse set of customers including restaurants, bars and neighborhood stores.

Distributors such as Grainger process customer orders on the same day and have a sufficient number of warehouses to deliver most orders the next day, using ground transportation (Chopra 2003).

6.3.5 Distributor Warehouse with Last Mile Delivery

The last mile delivery to home or business requires the distributor to design a delivery tour of the delivery-vehicle, satisfying customers' delivery-time requirements, as shown in Fig. 6.7. Vans start from the distributor's warehouse, make deliveries, and return to the warehouse. To facilitate scheduling customers are offered time slots to choose from, which are determined before customer-orders arrive. When delivery routes begin to fill up, the distributor may reduce the number of slots offered and/or change the fees associated with deliveries during each time slot.

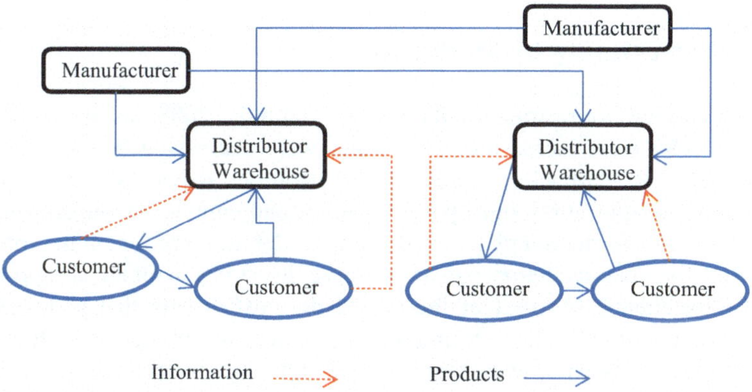

Fig. 6.7 Distributor warehouse with last mile delivery

In setting up their last mile home-delivery operations for fresh food, Peapod determines the number of time slots to offer in a day, the length of time slots, and time slots that would be active in a zip code of a city (Agatz et al. 2007). Peapod uses characteristics such as population density, web penetration, and historical demand data to define service requirements for each zip code. Because delivery trucks may visit several zip codes in the duration of a single time slot (based on customer demand), and the delivery tour may span multiple time slots, assigning active time slots to a zip code cannot be done in isolation. On Peapod's website, individual customers create orders and then select specific time slots for delivery. In order to have a sufficient time for order picking, time slots are closed few hours before the actual delivery.

For a soft drink distributor using last mile delivery, inbound shipments of ingredients and consumables usually come on large trucks but the outbound shipments of finished products to numerous customers are organized using smaller vans. It leads to higher shipment costs for the outbound shipments. In global health care delivery last mile distribution is used to make medicines available at remote locations (USAID 2011), and must address local issues such as infrastructure.

6.4 Supply Chain Cost

Starbucks ran five regional distribution centers in the United States: two were company-owned and the other three were operated by third-party logistics companies (3PL). It also operated two distribution centers in Europe and two in Asia, all of which were managed by 3PLs (Cooke 2010). In addition to coffee, these warehouses also handle other items required by Starbucks' retail outlets, everything from furniture to cappuccino mix.

6.4.1 Restructuring by Starbucks

Starbucks needed to restructure its supply chain in 2008, as the company's operations costs were increasing, while sales were decreasing (Cooke 2010). As much as 70% of Starbucks' supply chain operations expenses were related to outsourcing: transportation, third-party logistics, and contract manufacturing. The supply chain transformation plan included a clear definition of functional roles, and procedures for cost reduction, so as to serve its stores without compromising execution-excellence. As part of the effort, the cost drivers that were inflating prices were identified. The information for restructuring included all existing contracts, the prices paid, and shipping costs. Quantitative models were built to determine what it should cost.

Transportation, distribution, and warehousing made up the bulk of Starbucks' operating expenses because the company ships a large number of different products to its retail centers around the world. To create a single global logistics system for its far-flung supply chain, they built a global map of transportation expenditures that involved collecting all supply chain costs by region and customer. An analysis of those expenditures allowed Starbucks to drop some carriers, retaining only those that provided the best service. The logistics team developed weekly scorecards for monitoring performance of 3PLs and other vendors. It included service metrics, cost metrics, and productivity metrics. Starbucks focuses on high priority performance categories: safety in operations, on-time delivery, order fill rates, and total end-to-end supply chain costs.

6.4.2 Inventory Reduction

Inventory represents unproductive investment, although it may be necessary to hold some inventory to cushion against demand uncertainty. In the supply chain inventory is held at multiple stages, upstream and downstream. Because of the value added, it is more expensive to hold inventory downstream. Customer order decoupling, discussed earlier, is one of the ways of reducing downstream inventory.

6.4.2.1 Multi Echelon Planning

In general, inventory can be positioned at each tier of the supply chain. While it increases responsiveness, it causes decision making to become complex. Such a system with multiple tradeoffs is modeled as multi echelon inventory optimization.

Procter & Gamble achieved considerable success in reducing its inventory investment ($1.5 billion in 2009) by restructuring its supply chain, as Farasyn et al. (2011) have pointed out. They implemented a multi echelon inventory optimization model to minimize inventory costs across the supply chain. The optimal numbers of tiers and links in there structured network were determined by the model. For example, in the supply chain for the "Beauty and Grooming" division of P&G the optimal numbers were 4,000–5,000 stages; 6,000–10,000 links; and approximately 500 finished goods. The optimal decisions were arrived through multidimensional tradeoffs built in the model. Such tradeoffs involved production batch sizes, reaction times, order policies, replenishment timing, facility restrictions, customer orders, transportation-modes, product introductions, promotions, assortment management, and regulatory constraints. In 2011, more than 90% of P&G's business units (about $70 billion in revenues) used the inventory management tool.

6.4.2.2 Replenishments

Inefficient inventory replenishment can be major source of cost in a supply chain. Replenishment management determines optimal order quantities to minimize the combined cost of inventory holding and shortages, for individual suppliers and customers. In a supply chain, additional constraints such as the coordination of flows from vendors to distribution centers (DC) and from DC to stores, moderate replenishment decisions. Moreover, replenishment decisions must incorporate orders from multiple channels such as catalogs, and the internet.

Cabelas is the one of the largest supplier of hunting, fishing, camping and related outdoor merchandise in the US, with a long history in the direct sales channel (Cabelas 2013). Cabelas' inventory management processes were based on the direct sales model. The company subsequently added several retail stores that attracted a large number of customers. Customer preferences varied by geographic region, sporting season and skill level. As a consequence, Cabelas' inventory replenishment operations became complex and unsynchronized with sales. It became hard to keep track of transactions, capacity utilization, new customers, and inventory shortages.

Cabelas restructured its operations by centralizing common functions while still satisfying diverse customer needs through distribution centers (DC). It implemented a new warehouse management system to organize retail replenishments and fulfillments out of the DCs. It also provided the ability to accurately forecast, purchase and replenish inventory in advance of demand. Cabelas could tailor the product-assortments to each channel, using channel-specific attributes and metrics.

Warehouse management was implemented in three DCs, two return centers, and several stores, to integrate inventory with other systems. It was initially focused on the DC-to-store channel and was then integrated into the catalog and internet channels. The replenishment procedure calculates daily orders in response to actual sales and updates inventory positions. Each inventory item is forecast weekly on a rolling horizon basis, and order projections are provided to the DCs and vendors. Additional capabilities include system-generated seasonal profiles, demand alerts, purchase order alerts, and order frequency optimization.

6.4.2.3 ABC Analysis

ABC analysis is a classification method dividing the inventory items into three categories based on unit price and demand: category A comprises the most valuable items and category C the least valuable. Managers can focus on class A items that account for a large proportion of the inventory value. Adapting Pareto's Law "the vital few and trivial many" to inventory control, it is possible to create the pattern: A items accounting for about 80% of the dollar usage; B items accounting for about 15% of the dollar usage, and C items accounting for the remaining 5% of the dollar usage. The steps in the ABC analysis are, (i) determine annual quantity usage of each item, (ii) multiply the annual quantity usage by the unit cost of the item to obtain the annual dollar usage, (iii) arrange the items in the descending value of annual dollar usage, (iv) going down the list determine the cumulative dollar usage value and the total value of all items, (v) express the cumulative values as the percentage of the total value, and (vi) classify items so as to account for the top 80% of value as A, next 15% as B, and the last 5% as C. Class A items are controlled tightly: frequent reviews, demand forecasts, order quantities, safety stocks, and cycle counts. Class B items have similar controls but reviews are less frequent. Class C items are ordered in much larger quantities and have higher safety stocks.

The ABC classification reduces inventory cost, as the class B and C items need not be controlled tightly. To implement the ABC classification, procurement in the supply chain will need restructuring. Because of large size orders, contracts with the suppliers of class C items would be redesigned to include quantity discounts and long lead times. To obtain tighter control on class A items, the relevant suppliers would need close monitoring through information transparency. Databases will have to be linked at the process level to provide the most updated information on supplier's capabilities, and anticipated problems.

6.4.2.4 Inventory Buffer

A large chemical producer selected an inventory optimization application to set adequate safety stock to respond to demand (Aberdeen 2007). The company produced both commodities and specialty products. Their solution was to have a procedure that would automatically calculate the required amount of safety stocks. Although the biggest benefits were in the specialty products, the company treated these items in the same way as the rest of them.

Using data for supply and demand variability, the optimization application computed the optimal stock levels at different stocking points to meet customer service levels. The result has been a significant inventory reduction on the specialty products, while maintaining or improving service levels for all customers. The company also rationalized its shipping locations for improving customer service. It facilitated direct shipments from its plants to reduce transportation costs, without compromising customer shipping needs.

6.4.2.5 Demand Pooling with Vendor Managed Inventory

In a VMI system, the supplier monitors the inventory levels of the retailers, and determines when and how much to deliver to the retailers as well as when and how much to replenish its own inventory at the warehouses. It is known that by employing VMI, one can reduce the operating cost of the supply chain and maintain or improve the service levels desired by customers.

Consider a two-level supply chain comprising a single supplier and multiple retailers (Kang and Kim 2012). The supply chain includes a central distribution center (CDC) and multiple candidate regional warehouses. It is assumed that the supplier is authorized to manage inventory levels of the retailers by a VMI contract. The visibility of the pattern of retailers' demand enables the CDC to pool retailer demands in deciding the inventory levels at its warehouses. As retailers' demands are mostly independent of one another, the variance of demands from a set of retailers is smaller than the sum of the variances of the demands. As the safety stock increases proportionately with demand variance, the safety stock needed for the pooled demands would be lower than the sum of the safety stocks for the individual demands. Therefore, to reduce inventory holding costs, the supplier would aggregate demands from multiple retailers (risk pooling strategy) to reduce the effective total demand variability.

The supplier clearly stands to increase his cost saving from demand pooling by increasing the number of retailers assigned to each warehouse (reduces total inventory in the system). With a dynamic restructuring capability, the supplier would be able to decide how many of its warehouses he should select for use: fewer warehouses if the number of distinct retailer demands decreases. This would ensure allocation of a sufficiently large number of retailer demands to a warehouse. To implement it, the supplier would need a modified system that would enable him to "open and close" warehouses, as needed. Ideally, the system would "pop up" the open warehouses online, and enable the supplier to divert retailer demand to the warehouses in real time.

6.4.3 Shared Services

Shared Services (SS) is the aggregation of similar functions of an organization at a hub to ensure that they are delivered effectively (Frost and Sullivan 2005). Shared Services can be instrumental in driving efficiency, and relieving budget pressures by minimizing service duplication. SS enables companies to eliminate activities that do not add much value, and automate routine processes. The system benefits from standardization, economy of scale, proficiency, and learning. SS leads to cost reduction, high quality service delivery, operational efficiency, waste reduction, and satisfaction of stakeholders. In this business model, the SS hub is run like an independent business unit, delivering services to both internal and external customers at a cost, quality, and timeliness that is competitive with other alternatives the customer might have.

Companies such as IBM and Dell have adopted the SS to keep their supply chain cost low. Supply chain processes that require massive data inputs are ideal candidates for shared services. These include financial services such as credit card processing, billing, transaction management, procurement, revenue management, exception processing, payroll processing, call center services, and employee data records management (Frost and Sullivan 2005). British Telecom (BT) has partnered with Accenture to revamp its global operations through shared services.

Companies may adopt a "hub-and- spoke" model of SS where standardized and data intensive tasks and activities that are not language-dependent are supported from a global hub. Other centers serve as satellites that service localized functions (Accenture 2012). The global hub serves as a vehicle for structured transactional processing, as mentioned earlier. It supports common processes in a standardized environment, while the regional centers deliver functions that are customer facing, requiring interaction with local businesses, customers and vendors in local languages or context. The hub-and-spoke model increases the proximity between the decision makers and the shared services organization.

6.4.4 Managing the Tax Burden

Businesses often launch supply chain restructuring projects that typically do not incorporate the implications of government taxes. But this can mean that the effective costs increase and opportunities are missed, after a restructuring. Taxes have major implications for almost all supply chain activities including business development, R&D, product design, procurement, manufacturing, distribution, logistics, and services (Chakravarty 2005).

A producer driven centralized supply chain often involves a large number of transactions that have tax obligations across many countries. Accounting for all such taxes may lead to a very different structure of the supply chain, and the decisions thereof. The main tax challenges arise from factors that include: buying in many countries with different tax structures; selling to many countries with transfer price decisions; producing goods in many countries with different value-added tax structures; buying or selling goods in consignments that are owned by the sellers until needed by buyers; moving goods across borders for storage, processing, and repair; purchasing and/or supplying services from many countries; becoming the owner of all products from raw material through finished product throughout the whole supply chain.

Consider the implications of value added tax to company Y that restructured its logistics network (Robinson 2012). Company Y purchases material, and it ships manufactured products to destinations all over the world. The group recently reengineered its logistics network in Europe to protect VAT zero-rating for intra-EU shipping. The company had approximately 40 manufacturing plants for finished products, in several countries in Europe. They were responsible for moving products from suppliers' locations to Y's locations. Each supplier shipped products directly to each plant. This resulted in hundreds of suppliers shipping to hundreds of locations. A large number of carriers used various routes and billed the plants for freight in multiple formats. To streamline the flow of material in the network, Company Y restructured the logistics into a hub and spoke model: suppliers stocked material in their warehouses, and then shipped them to the hubs where they were then redistributed with direct routing to destination plants. From a VAT perspective, the modified routing created new services, as the material did not flow directly from the supplier to its final destination. Company Y needed to satisfy a series of conditions to maintain the VAT zero-rating in sales. These requirements have become incorporated in the standard operating procedures for the network.

Many companies have redesigned their supply chain to try to reduce customs duty burden, landed costs and lead times (Robinson 2012). It typically involves trade route design that leverages free trade agreements in creative ways. These strategies can significantly drive down the costs of customs duties and compliance. For example, components with preferential origin status are routed to assembly destinations where (i) preferred-origin is recognized, and (ii) the assembled products, if re-exported, qualify for customs duty suspension. Further efficiencies can be obtained through specific tax-effective models for procurement and manufacturing. For example, purchase commissions may be excluded from the customs value, thereby reducing the amount of import duties to be paid.

6.5 Information Technology

Technology has become the essential tool for supply chain planning and execution. Advanced supply chain technologies enable companies to automate business processes, facilitate real time execution, and support visibility. However, the supply chain can become difficult to manage if critical business processes are not restructured adequately for deploying the technology. As supply chain processes are refined and new technology is introduced, the supply chain structure must be realigned to support these initiatives. The technologies that can make significant difference to supply chain performance include RFID, information sharing, mobile devices, cloud computing, and Web 2.0.

6.5.1 Radio Frequency ID

RFID refers to a set of technologies that are used to identify and transmit information from tagged objects using radio waves. It comprises a tag embedded with a silicon chip, and an antenna that enables communication with a remote reader (Thangamuthu 2008). The tags that are attached to cases, pallets, or other units transmit radio frequency signals when they are present in the "read zone" of a reader. The reader decodes the signal, and this information is transferred to the database for storage and analysis. The host computer system runs the software that converts data that is collected from the tags into useful information. Unlike bar codes, RFID readers can simultaneously scan many tagged items, irrespective of their location. In the context of a supply chain, RFID has been found useful in manufacturing (detect defective parts), warehousing (inventory and fleet management), and retailing (stocking level, speedy checkout).

Business processes must be reengineered to improve and automate them and to integrate the RFID data into SCM, ERP, and other enterprise applications such as new work methods and performance measurements (Thangamuthu 2008). The supply chain would also need to redesign the engineering operations. For example, many companies currently manage both the storage and retrieval tasks in a warehouse by a centralized computer using static information. With RFID, storage locations within a warehouse can be determined in real time by assessing the tradeoffs between storage and retrieval times, and the available space. The typical warehouse design will change substantially to implement such tradeoffs (McFarlane and Sheffi 2003).

Similarly, a typical packaging operation can be made very responsive by redesigning it so as to work with RFID. RFID tags are attached to empty boxes to be filled. This tag is linked to a customer order, characteristics of which may still be unknown. The tag is read at the time the empty box enters the loading station and at that point the automated packing system interprets the order "recipe" on the RFID tag, and proceeds to locate the required set of items to be loaded into the box (Hodges et al. 2002).

Use of RFID in the reverse supply chain would also require restructuring. RFID helps in keeping product information updated so that information about the item is known throughout their lifecycle. For example, with these tags, the recycling center can detect the parts inside a discarded appliance and the pattern of their usage. The pattern of usage of the appliance can aid in deciding which parts to recycle, which can be used in remanufacturing, and which should be discarded and where.

Hanebeck (2004) discusses a RFID induced restructuring in a warehouse of a cornflakes manufacturer. Tags were attached to pallets and RFID antennae were fixed on every dock door to monitor all receiving and shipping activities within the warehouse. As soon as a pallet enters the warehouse the attached tag is read. As the RFID-enhanced warehouse management system is aware of the cartons (ID) on a pallet, the system reconciles receipts against Advanced Shipping Notice (ASN) documents, purchase orders, and invoices. Discrepancies, if any, are notified to the relevant parties in real-time. Picking and shipping transactions are triggered by an enterprise system and are communicated to material handlers who then execute pick and ship orders.

If there is a flaw in one of the ingredients, manufacturing can quickly determine which production batches were affected by matching the RFID tags with the flawed ingredient list, and it can then produce a list of IDs that need to be recalled. For each box with the flawed ingredient, the respective recipients can be determined. The company can contact each recipient and recall the product.

6.5.2 Mobile Technology

Mobile technology can streamline operations, reduce costs, increase worker productivity, and enable real-time response. Technology impacts mobility by enhancing workflow, improving access to knowledge, and increasing the speed of transactions. It causes changes in the systems and infrastructure of the supply chain, with implications for networks, devices, databases and applications. Companies must restructure for a new system of task management involving supply chain partners.

With the remote task processing capability of the mobile technology, the centralized task management can assign, modify, optimize, and redistribute tasks to available resources (ATT 2009). For example, the field sales personnel can enter a new customer demand using a mobile device, which also identifies its location. With this input, the company's software can quickly determine transshipment opportunities from nearby depots, using the device's location; and commit to a delivery date. Such mobile workflows help task completion time. In addition, the real-time monitoring reduces downtime, as decisions can be made quickly as and when situations arise. The real time decision making improves asset utilization, and enables dynamic control of inventory.

One of the most effective applications of mobile technology is the tracking of shipments on their way to customers. In manufacturing, the same technology using wireless sensors can integrate scheduling, parts control, inventory management, equipment maintenance, and monitoring of key performance indicators to support continuous process improvement (ATT 2009).

To deliver real-time information in the mobile supply chain, the network must also be enhanced in different ways that include access, performance, intelligence, and security. Mobile networks should be able to provide "anytime anywhere" broadband access and provide a strong performance in terms of availability and speed. It should incorporate intelligence for seamless handoff between different technologies. Such networks must also provide end-to-end monitoring and a flexible, adaptable security policy (ATT 2009). Database must be restructured to incorporate more dynamic features. Real-time data processing and monitoring lets databases such as CRM return data on-demand, and incorporate alerting systems that distribute data for rapid response. Applications must be modified to provide a high degree of interoperability in interacting with ordering, supply chain and customer systems.

6.5.3 Information Sharing

In a typical supply chain, decisions at each stage are based on information received from its immediate suppliers and customers (decentralized information). Thus, each stage makes locally optimal decisions. In such a system, customer order information can become distorted, becoming lumpy as it moves upstream. This is also known as the Bullwhip effect. As mentioned earlier, Lee et al. (1997) mathematically established the bullwhip effect in a decentralized supply chain with no information sharing.

Note that a supply chain partner must determine what types of data it should share; it may be risky given that its partner could also be working with others, exogenous to the supply chain. Because the characteristics of information flows are determined by the content of information being shared, there is a need to explore content-specific information sharing. Li and Shaw (2001) have identified four types of data that can be shared: (i) no information sharing, (ii) sharing inventory status, (iv) sharing shipment information, and (iv) sharing demand information.

In scenario (i), decision makers at each tier of the supply chain build forecasts, based only on the orders received from the tier immediate downstream; they are not aware of the status of its tiers further downstream. In scenario (ii), each tier of the supply chain shares information about its inventory and actual demand with its supplier, something quite common in the grocery and fashion retailing industry, and in vendor managed inventory. Scenario (iii) assumes that each tier is aware of its downstream customer's shipment data. HP and IBM, for example, request sell-through data on withdrawn stocks from their resellers' central warehouse (Li and Shaw 2001). Scenario (iv) assumes that real-time demand information is transmitted from the end-consumer back through every tier in the supply chain. Sharing of POS data, and collaborative planning and forecasting are examples of demand sharing.

Li and Shaw (2001) established that compared with inventory sharing, demand sharing incurs lower inventory cost when the variability of the end demand is high. However, compared with "no information sharing", sharing of inventory and shipment information improve fill rate and decrease inventory.

Liu and Kumar (2003) describe a hub-and-spoke information sharing architecture, where a central hub communicates with all partners. The hub aggregates, stores, and maintains information about each partner, and makes decisions that are communicated to individual partners (HP's hub and spoke architecture is described in Chap. 3). The e-hub-based architecture possesses "plug-and-play" capability. With only a few integration points between applications, they can be used as needed. Collaborative planning, forecasting, and replenishment (CFPR) is based on the idea of such a centralized hub. Standardization plays an important role because all partners must use common standards for information description, storage, and exchange.

6.5.4 Cloud Technology

Cloud technology provides a common platform where companies can be interconnected on-demand through information exchange, as shown in Fig. 6.8.

A supply chain partner can connect effortlessly, using a web browser, to new product and service offerings, customer preferences, distribution channels, and other business capabilities (Aivazidou et al. 2012). Customers and suppliers can be on the same page while accessing cloud-based applications; software and data can be stored at a remote location or in-house. The cloud environment, therefore, is very conducive to supply chain coordination.

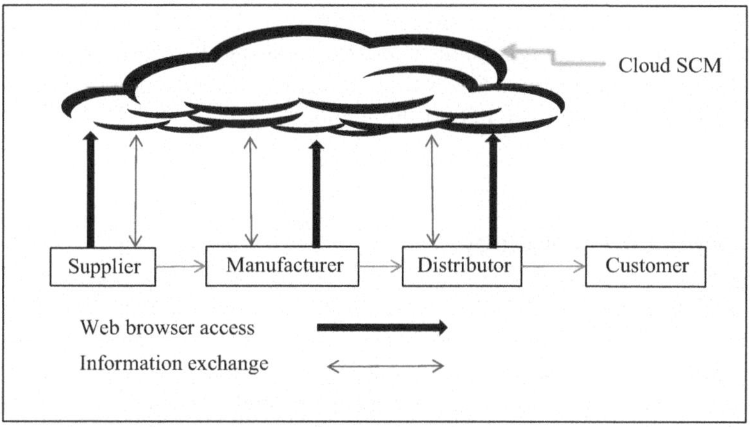

Fig. 6.8 Cloud supply chain (Source: Aivazidou et al. 2012)

Cloud computing provides three types of services: infrastructure, platform, and software. Infrastructure service focuses on providing the "backbone" resources such as network memory and storage capacity. The platform includes operating system, data base, and programming language execution, where users may develop and deploy applications. Application providers install and operate application-software in the cloud, which can be accessed by all registered users. While the cloud users cannot manipulate the cloud infrastructure and platform, they have control over applications. Schramm et al. (2011) describe some of the supply chain processes already available on the cloud. These include procurement, logistics, spare parts management, and reverse-logistics, among others.

The cloud-based platforms provide databases, which may be used for procurement and order fulfillment. For example, it may contain information about suppliers; supplier's product list, prices, and availability. Companies can use this information to select suppliers that match the buyer's needs, according to specifications and time schedules. It may also provide databases of integrated product catalogs, discussed in Chap. 5. Moreover, cloud-based tools can help companies in developing supply contracts.

The cloud can be used for providing logistics support to supply chain partners. It does so through several software applications such as inventory replenishment, order processing, shipment tracking, fleet management, and route planning. It provides visibility; multiple supply chain partners can access the same applications and update their status. Other processes that have been hosted efficiently in a single cloud-based platform include warranty validation, returns processing, spare parts inventory, and distribution.

Pfizer, a global pharmaceuticals company, used the cloud platform to manage the company's global network (Chandar 2012). Pfizer operates in two different markets: a high-valued patented-drug market that relies on rapid response, and a generic-drug market that is focused on inventory optimization and efficiency. The move to the cloud enabled Pfizer to quickly adapt to ever changing customer needs. It enabled Pfizer to communicate effectively with all its partners in drug development, and also obtain visibility into the total supply chain. The transition to cloud solution has transformed Pfizer from a state of "zero shipment traceability" to be accountable for more than 40,000 shipments.

6.5.5 Web 2.0 Technology

Web 2.0 is a social web, with people connecting with other people to share information (Barlow 2012). It supports a socially connected community and builds on characteristics such as openness, participation, and collaboration. Services, technologies, and applications, associated with Web 2.0 include blogs, wikis, podcasts, RSS feeds, social networking, forums, multimedia sharing services, tagging and social bookmarking, text messaging, and instant messaging.

The types of supply chain activities that benefit from Web 2.0 include marketing and advertising; collaborating and strengthening relationships with customers and suppliers; information and knowledge transfer; and networking for research. These activities may be classified as either internal such as communications within and across departments, or external such as communicating with customers, suppliers, and partners.

Web 2.0 can foster internal collaboration among employees through training, discourse, and knowledge management (Mamgai and Jolly 2009). Osborne, an electrical equipment manufacturer, implemented an internal Wiki based solution to avoid duplication of effort and to improve decision making processes within the organization (Kenney 2008).

Companies in a supply chain may use Web 2.0 for customer service and for acquiring new customers. Nokia, for instance, hosts a social networking site where users are free to express their views on Nokia products (Nokia 2013). This information is then used by the company in developing their next generation products. General Motors has developed and built a social networking platform where suppliers and company representatives interact frequently through Wikis to address pressing supply issues (Saran 2007).

6.6 Global Value Chain

Over the past two decades, manufacturing and service activities in emerging countries have increased substantially. The increase in foreign direct investment (FDI) has caused an expansion of global trade, and has extended the reach of global supply chains to many products (computers, office equipment, telecommunications, video and audio equipment, semiconductors, and apparel). In addition with the rise of new entrants, the global supply chain has become very fragmented, which has made it possible to partition the value chain into a number of segments that can be located in different countries. Therefore, the global value chain, spanning multiple countries, must be carefully analyzed and restructured to be able to satisfy all global stakeholders (discussed in Chap. 7).

6.6.1 Value Chain

There are three major value chain categories: producer-driven, buyer-driven, and multi-polar (APEC 2012). The producer-driven chain is the one where a large multinational manufacturer plays the central role in coordinating a geographically distributed network of suppliers. This company generally retains control of product design, innovation, production, distribution, and access to customers. This type of chain tends to be a characteristic of capital-intensive and technology-intensive industries such as automobile, aviation, and communication. It is typical of how companies such as Toyota, Boeing, and General Electric (GE) manage their particular networks of suppliers and buyers. This usually gives them considerable influence over other suppliers in the network.

The buyer-driven chain, on the other hand, is a relatively more recent development in international production where large retailers and brand names such as Walmart and Carrefour play the role as supply chain orchestrators. They source from a decentralized network of independent suppliers, while retaining control over product definition, process specifications, and standards. Some retailers may engage directly with increasingly powerful first-tier suppliers such as Li & Fung of Hong Kong, who help organize all upstream production-related activities on their behalf. To compete in a new global environment, first-tier suppliers need to impose stringent standards and use fewer preferred manufacturers. The buyer-driven chain tends to be a characteristic of labor-intensive, consumer goods industries, such as apparel, footwear, agriculture, and consumer electronics. The participation requirements such as technical capability and sophistication are usually less stringent than in the producer-driven model. This creates opportunities for producers from emerging economies, including small and medium enterprises, to enter the global supply chain (discussed in Chap. 7). Enterprises move to higher value added activities as they enhance their technical capability and sophistication.

The multi-polar value chain is a less common. It is characterized by multiple power centers in different parts of the value chain (APEC 2012). There is no overall dominant lead firm with the power to determine the ultimate shape of final products, but some do exert control over certain key activities throughout the chain. For example, Intel, Microsoft and Fujitsu, are lead firms in their own production chains within the personal computer value chain. However, a specific personal computer marketed by Fujitsu may reflect a mix of strategies: Microsoft's software strategy, Intel's strategy for semiconductors, and Fujitsu's marketing strategy.

6.6.2 Outsourcing

The development of global trading has led to the increase in buying and selling of components, more so than the finished products. Companies have been importing parts, assembling them into products, and re exporting them (known as vertical specialization). This has now morphed into global outsourcing (details in Chap. 4); companies such as Apple, Hewlett Packard, and Nokia have outsourced most of their manufacturing (Sturgeon 2002). Companies in the auto industry have also been outsourcing modules and major systems for vehicles in large quantities. In both cases, this has led to increasingly globalized supply systems able to provide parts and components to major customers at many points around the world.

In the apparel industry, there are now brand name designers (such as Liz Claiborne, Zara, Hugo Boss, and Armani), who design and market the products that are manufactured elsewhere (in multiple countries). The designers and large retailers determine product characteristics including design features, process specifications, quality targets, compliance (with labor and environmental standards), and time schedules. Thus, coordination without a full ownership of the supply chain has become an important feature of global transactions. In this model, the suppliers do not have to be concerned about product design and the idiosyncratic market requirements of importing countries. These issues are addressed by the buyers (retailers). On the other hand, specialization in only the production activities in the value chain may leave the supplier with a limited understanding of market requirements and few opportunities to develop capabilities in the high-value functions of design and marketing.

6.6.3 Upstream/Downstream Restructuring

From a supplier's perspective, access to global markets is not achieved merely through design, manufacture, and marketing of new products. The need to operate under different environments, constraints, processes, people, and countries, provides the opportunity to a prospective supplier to upgrade its capabilities. For such enterprises the task is to insert themselves into the wider network. It takes restructuring of relationships with other suppliers and customers; accessing new technology, skills, competence, and support services (needed in a global network); and a customer focused mindset.

As buyers and sellers are linked together in a network, restructuring the seller's upstream operations would also imply a restructuring of the buyer's downstream operations.

6.6.3.1 Supplier's Perspective

Four stages in restructuring can be identified: produce to specification, develop internal design capability for customers' products, offer a product line to customers to choose from, and develop into a marketing and sales organization similar to that of the indigenous firms with which it would then compete. It follows that the supplier can be represented by a node in the supply chain while it is in stage 1. In stage 2, it would control its suppliers upstream. In stage 3, it manages a few channels downstream, and in stage 4 it controls a significant portion of the supply chain with its own marketing and distribution capabilities.

In a global network, suppliers in stage 1 are exposed to new products with intricate specifications, tight schedules and stringent quality standards. They are expected to meet requirements that may not apply in their respective domestic markets. This creates a gap between the capabilities required for the domestic market and those required for the global market. New business opportunities and higher margins often act as motivators for many to upgrade their capabilities to bridge this gap. They gradually excel in manufacturing, but they do not control procurement and logistics; the buyers usually manage these tasks.

Another complication for these suppliers is that they typically operate in two markets – domestic and global – with different expectations and incentive structures. For example, a shoe manufacturer operating in Brazil and Italy needs to satisfy two sets of customers. While Italian customers highly value innovative shoe designs, they are typically less sensitive to price in comparison to the Brazilian customers.

Once a supplier has attained the global production standard, he may look to broaden his scope in stage 2. With a good knowledge of his production processes and his second tier suppliers, he is in a strong position to assess and monitor the upstream suppliers. He would need a new capability of managing relationships with its suppliers: sense emerging problems in supplier's operations, supplier's ability to develop new processes as needed, and their possible over-commitment of capacity. The tier 1 supplier needs to have perfect visibility of his suppliers. These tier 1 suppliers evolve into lead suppliers for the buyers.

In stage 3, the lead suppliers learn to better leverage commonalities in resources. For example, instead of producing a certain product (say A) for a customer, they may explore producing product B as well, if B requires many of the same components that A does. In fact, they may produce several products based on commonality of resources, and ask customers (including the original customer) to choose from them. Clearly, the new capability they must possess in this stage is product design and development. A good example is Flextronics that transitioned from being contract manufacturer of mobile phones, to design and develop its own mobile phones that were far superior to the design of the original customer (Huckman and Pisano 2006).

In stage 4, the supplier transitions to an OEM with new capabilities of marketing, distribution, and access to consumers. It begins to sell directly to consumers. Observe that in stages 1–3, the supplier worked with product standards *provided to it*; the margins were low but it did not have to be concerned whether the product would sell (low risk). In stage 4, on the other hand, it needs to master the global marketing and distribution channels, with their inherent uncertainties. Therefore, in addition to the capabilities of international marketing, it must be able to manage risk satisfactorily.

6.6.3.2 Buyer's Perspective

As the supplier X evolves from stage 1 into stage 4, the buyer must also evolve. Essentially, the supply chain evolves through decentralized restructuring. In stage zero, prior to stage 1, the OEM performs the task of supplier X, and operates a producer driven supply chain. In stage 1, as supplier X emerges, the OEM modifies the supply chain so as to incorporate supplier X as an outsourcing partner.

First, the OEM transmits information about the task-bundle that X is supposed to perform. It includes engineering drawings; detailed process specifications of the task-bundle; specification of the inputs (material, components, services, and know-how); information on all equipment needed to perform the tasks; and manuals for training the workers. Second, the OEM must ensure that all inputs are procured and delivered to supplier X, as and when needed. Third, it must provide logistics for transporting supplier X's output (components and subassemblies) to the OEM's facility for further processing. Fourth, it must reengineer information visibility to monitor supplier's performance and capacity. Finally, it must structure a mutually agreed contract with X specifying delivery schedules, and payments.

Clearly, the OEM needs to develop new managerial capabilities for ensuring availability of inputs for the supplier X and transportation of outputs from the supplier, performance evaluation, and payment management. These capabilities replace some of the OEM's current capabilities in a producer driven supply chain.

In stage 2 as supplier X becomes the integrator of its upstream supply chain, it takes over the function of managing the upstream, from the OEM. The OEM is still responsible for providing the specifications for the task bundle to supplier X, and arranging the logistics for receiving X's outputs. As the OEM relinquishes additional responsibilities to supplier X, it can better focus on its core activities such as customer management. The decrease in OEM's overhead cost leads to reduction in its breakeven quantity. It now becomes profitable for the OEM to serve smaller niche markets. In essence, the supply chain becomes segmented between upstream of X and downstream of X.

When supplier X transitions to stage 3, the OEM permits X to deviate a little and recalibrate the output-specifications it provides to X. As a result, the OEM does not need to invest in a lot of research for creating specifications; it is now the responsibility of supplier X. As supplier X builds a few finished products, the OEM chooses the products that are closest to what the OEM's customers prefer. The OEM may also buy the intermediate products and services from X, assemble the final product, and sell them to its customers. The OEM retains all customer facing tasks.

With supplier X in stage 4, the OEM has no obligation to purchase X's outputs (products or services). It totally relinquishes the responsibility of providing specifications and quality standards to X. In essence, the OEM would now be a buyer of finished products from X, and sell them to its customers. It leverages its knowledge of customers to create a profit margin. Global retailers such as Wal-Mart fall into this category.

6.6.4 Integrators in Supply Chain

Bliss and Haddock (2008) use the term "integrator" to imply buying components from supplier X while also selling the finished product in X's home market. For example, such integrators possess the capabilities of sourcing, manufacturing, and selling, in addition to innovation, in the same country (such as China); these are integrated, optimized, and leveraged globally. Bliss and Haddock (2008) describe the operations of Hansgrohe, a German plumbing products company, as an example. Hansgrohe supports sales of imported products in several cities in China and, at the same time, continues production in China (other products) at low cost for export to other countries, so as to support its global operations network.

Wal-Mart, in 2008 sourced goods worth US$30 billion from China (Bliss and Haddock 2008). Wal-Mart is now rapidly adding new stores for sales in China. Home Depot acquired HomeWay in China to grow Home Depot's sales in China, although it has been sourcing large volumes of products from China.

A recent addition to the growing list of global supply chain integrators in China is Goodyear. It now produces some of its advanced tires in China for the domestic and overseas markets while leveraging China as an increasingly important market for exports. In an effort to accelerate growth in China, Goodyear has built a franchise network selling tires and other products and services related to aftermarket car repairs (Bliss and Haddock (2008). However, although the company had become a role model in establishing dual-mode operations, it had not redesigned its logistics and supply chain operations to support the dual-operating mode. This has resulted in problems: overall lead times were up to 22 weeks, which was a major problem in an industry with an average product life cycle of 45 weeks. The company had high inventory levels; it also had frequent out-of-stock and high obsolescence cost (Bliss and Haddock (2008). Lastly, it had very high transportation costs, due to the need to use air transportation for boxed shipments as well as many expedited air shipments in order to meet lead-time requirements.

6.7 Segmentation

A supply chains is usually structured as a linear vertical chain, or an integrated chain. The vertical supply chains can be designed for specific products or customers, leading to a large number of supply chains. This would require operating a large number of independent supply chains, with implied redundancies. The integrated supply chain, on the other hand, requires all products and programs to make use of common channels of supply and demand so as to share resources. The segmented supply chain combines the best of both: it supports the specific needs of products or customer groups, while reducing complexities and redundancies.

Supply chain segmentation allows for grouping of products, customer, and resources by shared sets of values. Each product-group can profitably meet the needs of a specific customer group on its own terms. This can lead to increased customer satisfaction, while at the same time increasing operational efficiency and profitability. Instead of taking a "one-size-fits-all" approach, different supply chains cater to different customer groups. While each segment operates like a distinct supply chain, meeting its objectives through tailored operations, the segments are integrated and coordinated to enable sharing of resources such as warehouses, information systems, and logistics.

In the grocery industry, for example, supply chains handle thousands of products with a wide range of characteristics: some require refrigeration; some have a short shelf life; some are bulky; some are expensive; some have high demand; and some have lumpy demand due to promotions (Allain et al. 2010). Therefore, the grocery stores may segment their supply chains based on commonality of attributes: type of produce, dry goods, frozen, and seasonal items. Stocking and replenishment procedures may be distinct for each segment, but resources can be shared across them. For example, delivery to remote stores may combine frozen, produce, and dry goods on the same truck. That is, segments are tailored to be customer and/or product specific, while optimizing the use of resources.

6.7.1 Segmentation Process

Allain et al. (2010) suggest four steps in segmenting a supply chain, while emphasizing that the process requires creative thinking in order to match the right solution with the right situation. These include: identifying goals and scope; creation of segments by analyzing data (customer demand, product characteristics, supply sources, customer and storage locations, routing resources, and inventory); developing operating procedures for each segment; and framing an implementation strategy (financing, change management, performance metrics, etc.) for each segment.

Examples of possible goals of segmentation would be (a) merging of two or more vertical supply chains, such as production and delivery of refrigerators and dish washers into a single supply chain; (b) expand existing supply chain for mobile-phone services to new customer groups; (c) segmenting customers geographically as demographics change; and (d) segmenting customers by their preference for customization and order fulfillment.

The most common factors used in supply chain segmentation are listed in Table 6.2.

Table 6.2 Factors for segmentation

Factor	Characteristics
Product	High/low volume, low/high margin
	Product variants, pricing
	Life cycle, handling characteristics, shelf life, size and weight
Customers	Degree of comfort with replenishment uncertainty
	Instant fulfillment or assemble to order
	Service expectations
Technology	Technologies provide specific advantages in production and execution and in creating synergies
Market	Geopolitics, market volatility
	Channel idiosyncrasies, seasonal/constant demand
Sourcing	Quality, cost, availability, lead-time, flexibility
Commercial environment	Infra-structure, transportation modes, customs duties, legislation

Note that companies may choose to create segments based on some subset of these factors. Becks (2012) discusses a scenario of segments being created based on customer preferences. If a customer has a high preference for certain attribute such as customization, he/she is given a high score on a 0–100 scale. Consider the preference scores for five customers shown in Table 6.3.

Table 6.3 Preference scores

	Low cost	Customization	Responsiveness	On time delivery	Perfect fulfillment
Customer 1	20	100	40	60	80
Customer 2	100	10	80	60	30
Customer 3	10	70	30	30	50
Customer 4	90	20	70	70	30
Customer 5	50	40	50	60	50

Observe that the pattern of preference scores for customers 1 is similar to that of customer 3; and the pattern of customer 2 is similar to that of customer 4. Customer 5, on the other hand, stands alone. Hence, customers 1 and 3 will be placed in segment #1, customers 2 and 4 in segment #2, and customer 5 will be placed in segment#3.

6.7.2 Applications

Becks (2011) discusses the application of supply chain segmentation in the computer industry. The company builds laptops, desktop computers, and smartphones. The company was regionally based. It restructured its fulfillment strategy to meet the needs of the mature markets in North America and Europe, emerging markets in

Asia, and additional customers in Latin America. It needed to satisfy the unique needs of three categories of customers, and redesign the logistics networks to deliver products into these three regions.

The company identified two supply chain segments: production and distribution of make-to-stock products, and production and distribution of assemble-to-order products. The first segment was designed for high volume products with low cost and price sensitive customers with transaction focus; it emphasized high forecast accuracy that allowed the company to leverage its production capacity. The second segment involved a larger product portfolio with high demand variability and short lead times; product costs were high. Customers in this segment tended to be more relationship focused. This supply chain segment needed to be agile and responsive. The company decided to centralize the bulk of its procurement, manufacturing, and logistics functions in order to drive cost reductions, resource sharing, and serviceability. Both segments served all three markets (North America, emerging economies in Asia, and markets in Latin America).

To create operating procedures, the company set up new operational teams based on skills and experience for each market type. It then implemented process flow mapping to enable monitoring and adjustments for continuous process improvement. Customer satisfaction improved across all three markets and the number of exceptions dropped dramatically in both segments. It decreased the need for trouble shooting, and reduced the overall supply chain risk.

Malik et al. (2011) describe a segmentation approach based on demand characteristics. The first supply chain segment, based in China, was dedicated to products with high and stable demand. The second segment based in North America was responsible for three other demand scenarios: high demand and high volatility (based in the US), low demand and high volatility (based in the US), and low demand and low volatility (divided between the US and Mexico). In addition to reallocating the production resources, the company needed to alter its information and planning processes significantly. For the most volatile items (now produced in the US), the company produced directly to customer orders instead of forecasts, and created an infrastructure for rapid delivery. For low volume and stable demand products, assigned to US and Mexico, the company created a very simplified forecasting process.

Bitran et al. (2006) suggest segmentation based on hierarchical decomposition of the supply chain based on functions. For example, all the tasks required to complete a major sub-assembly of a product could be grouped as a segment with a single coordinator. The authors call such coordinators mini-maestros, and describe how the Hong Kong based Li and Fung have been using such segments to manage their global supply chain of suppliers, for many years.

The supply chain of Boeing 787, discussed in Chap. 3, also resembles a hierarchical decomposition. It has approximately 50 tier 1 partners (Tang and Zimmerman 2009). These strategic partners (known as integrators) assemble different parts and subsystems produced by a subset of tier-2 suppliers. In turn, each tier 2 supplier assembles parts and subsystems produced by a subset of tier-3 suppliers, and so on. Thus, each tier 1 partner can be thought of managing a segment of the supply chain; a tier 2 supplier manages a smaller segment nested within a larger segment. The authors point out that Toyota has been using supply chain segments similar to that of Boeing 787 to develop new cars with shorter development cycle times and lower development costs (Tang 1999).

6.7.3 Case Study

Allain et al. (2010) describe a case study of segmenting the supply chain of a public sector health care system. Two criteria (demand and geography) were chosen to create three supply chain segments: distribution to hospitals, high demand products to all locations, low demand products to all locations. It was also decided that they would outsource both transportation (to private carriers) and distribution centers.

6.7.3.1 Segment 1: Distribution to Hospitals

This segment serviced a large number of items delivered only to hospitals. Hospitals mostly located in urban settings are easily accessible, making resupply from warehouses very efficient. Fill rate and cost per unit were identified as the most relevant performance indicators. Operating procedures were developed to resupply the hospitals bi-weekly, reducing the need for large storage space. Temperature-sensitive products were transported and stored in special equipment. Most products were supplied by the public sector third party logistics (3PL) providers, who were also responsible for designing an optimizing the routing plan. Products procured from the private sector were delivered directly to the hospitals with specified delivery dates.

6.7.3.2 Segment 2: High Demand Products to All Service Locations

This segment was created because 30 out of the 70 products represented more than 75% of the demand at the service locations. They also represented the bulk of transportation and storage requirements. Fill rate and response time were identified as the most relevant performance indicators. Operating procedures were developed to facilitate resupply every 2 months. To keep buffer stock to a minimum the products obtained from the public sector were stored at the zonal DCs. 3PLs were used to design an optimized routing plan from the central level to the zonal DCs and from the zonal DCs to the service locations. Products obtained from the private sector were delivered by specified delivery dates.

6.7.3.3 Segment 3: Low Demand Products to All Service Locations

This segment was created to ensure that the service locations were not denied of low-demand products, if they were needed. Because they are slow moving, it was decided to resupply them once a year. The performance indicator chosen was efficiency. This implies that the buffer stock would be larger than the buffer stock for the products in segment 2. Products in segment 3 would follow the same model as that in segment 2, except that they would be ordered and delivered once a year instead of every 2 months.

To manage the processes and technologies associated with the operating procedures of segments, Becks (2011) suggests architecture of a Business-network. A Business Network is cloud-based (discussed earlier), and offers the software, content, and services needed to manage processes of multiple trading partners across multiple tiers of the supply chain. Specifically, a Business Network enables process integration, visibility, and collaboration across a community of trading partners.

The network will be supported by a technology platform that can execute the operating procedures of different segments. All trading partners in a segment would be accessible in real time, irrespective of individual partner protocols and formats. The platform would possess the capability to configure supply chains for any combination of attributes of products, customers, service preferences, and fulfillment needs.

References

Aberdeen. (2007). *Supply chain cost-cutting strategies*. http://www.areadevelopment.com/article_pdf/id94571_Supply_Chain_Cost-Cutting_Strategies.pdf

Accenture. (2012). *Shared services location strategy*. http://www.accenture.com/SiteCollectionDocuments/PDF/Accenture-Shared-Services-Location-Strategy-Asia-Pacific-Perspectives.pdf

Agatz, N., Campbell, A., Fleischmann, M., & Savelsbergh, M. (2007). *Challenges and opportunities in attended home delivery*. Working Paper, ISYE, Georgia Tech. http://www2.isye.gatech.edu/~mwps/publications/modelCD26April07.pdf

Aivazidou, E., Antoniou, A., Arvanitopoulos, K., & Toka, A. (2012). Using cloud computing in supply chain management: Third-party logistics on the cloud. In *2nd International Conference on Supply Chains*. http://www.logistics.teithe.gr/icsc2012/fullabstracts/session_3/3_5_ICSC_12_AIVAZIDOU.pdf

Ala-Risku, T., Kärkkäinen, M., & Holmström, J. (2003). Evaluating the applicability of merge-in-transit. *International Journal of Logistics Management, 14*, 67–82.

Allain, L., Goentzel, J., Bates, J., & Durgavich, J. (2010). *Reengineering public health supply chains for improved performance: Guide for applying supply chain segmentation framework*. Arlington: USAID | DELIVER PROJECT, Task Order 1.

APEC. (2012). *Concepts and trends in global supply, global value and global production chains*. http://publications.apec.org/file-download.php?id=1283&filename=2012_psu_GSCs_GVCs_GPCs.pdf

ATT. (2009). *The mobile enterprise: Moving to the next generation*. http://www.att.com/Common/about_us/files/pdf/WP-MobEnt_18448_v1_5-28-09.pdf

Barlow, A. (2012). *Web technologies and supply chains*. Glasgow Caledonian University, Scotland. http://cdn.intechopen.com/pdfs/18521/InTech-Web_technologies_and_supply_chains.pdf

Becks. (2011). *Supply chain segmentation: The next step in supply chain excellence*. E2open. http://info.e2open.com/rs/e2open/images/WP_SC_Segmentation.pdf

Becks. (2012). *Supply chain segmentation 2.0: What's next?*. E2open. http://info.e2open.com/rs/e2open/images/White%2520Paper_SC%2520Segmentation%2520Part%2520II.pdf?mkt_tok=3RkMMJWWfF9wsRonva%2FBZKXonjHpfsX56eUtW6C3lMI%2F0ER3fOvrPUfGjI4HScF0dvycMRAVFZl5nQkKFfCQcg%3D%3D

Bitran, G., Gurumurthi, S., & Sam, S. (2006). *Emerging trends in supply chain governance*. MIT Sloan School of Management, Working Report. http://mitsloan.mit.edu/pdf/ups.pdf

Bliss, C., & Haddock, R. (2008). *Integrating China into your global supply chain*. Booze & Company. http://www.boozallen.com/media/file/Integrating_China_into_Your_Global_Supply_Chain.pdf

Butner, K. (2007). *Blueprint for supply chain visibility*. Somers: IBM Global Services. http://www-935.ibm.com/services/uk/bcs/pdf/blueprint_for_supply_chain_visibility.pdf

Cabelas. (2013). *Cabela's outfits the outdoor lifestyle with multichannel supply chain solutions*. Manhattan Associates. www.manh.com/library/MANH-Cabelas_CaseStudy.pdf

Chakravarty, A. (2005). Global plant capacity and product allocation with pricing. *European Journal of Operational Research, 165*, 157–181.

Chakravarty, A. (2010). *Market driven supply chains*. Boston: NOW Publishers.

Chakravarty, A., & Werner, A. (2011). Telecom service provider portal: Revenue sharing and outsourcing. *European Journal of Operational Research, 215*, 289–300.

Chakravarty, A., & Zhang, J. (2007). Lateral capacity exchange and its impact on capacity investment decisions. *Naval Research Logistics, 54*, 632–644.

Chandar, S. (2012). *Global supply chain network in the cloud*. http://cmuscm.blogspot.com/2012/10/global-supply-chain-network-in-cloud.html

Chopra, S. (2003). Designing the distribution network in a supply chain. *Transportation Research Part E: Logistics and Transportation Review, 39*(2), 123–140.

Cohen, S. (2006). *The new supply chain organization*. PRTM. http://www.google.com/url?sa=t&rct=j&q=&esrc=s&frm=1&source=web&cd=3&cad=rja&sqi=2&ved=0CDwQFjAC&url=http%3A%2F%2Fwww.gsb.stanford.edu%2Fsites%2Fdefault%2Ffiles%2Fdocuments%2FPRTM_The_New_Supply_Chain_Org.pdf&ei=OchLUeDZMoGPiALh6oDoBg&usg=AFQjCNE6OEMHXJU91JtSHcCI4irs5FEiNQ&sig2=NwwhtZEM2xEwKc9BEzHoBA

Cooke, J. (2010). *From bean to cup: How starbucks transformed its supply chain*. Supply Chain Quarterly. http://www.supplychainquarterly.com/topics/Procurement/scq201004starbucks/

Donnelley, R. (2007). *The role of product completion in a demand-driven supply network*. Chicago. http://www.werc.org/assets/1/workflow_staging/Publications/673. PDF

Farasyn, I., Humair, S., Kahn, J., Neale, J., Rosen, O., Ruark, J., Tarlton, W., Velde, W., Wegryn, G., & Willems, S. (2011). Inventory optimization at Procter & Gamble: Achieving real benefits through user adoption of inventory tools. *Interfaces, 41*, 66–78.

Frost and Sullivan. (2005). *Shared Services and Outsourcing (SSO) hub potential analysis*. http://www.portalcentre.com/codenavia/portals/msc/images/pdf/awards_accolades/frost_sullivan_sso.pdf

Gangadharan, R. (2007). *Supply chain strategies to manage volatile demand*. Supply and Demand Management. http://www.sdcexec.com/article/10289792/supply-chain-strategies-to-manage-volatile-demand

Gerwin, D. (1993). Manufacturing flexibility: A strategic perspective. *Management Science, 39*, 395–410.

Ghosn. (2009). *Ten years into the alliance: Renault and Nissan are taking cooperation to a higher level*. http://www.ameinfo.com/198894.html

Godfrey, G., Frey, B., Naik, S., & Evers, A. (2012). *The consumer-centric supply chain: Achieving high performance through dynamic, demand-driven execution*. Accenture. http://www.accenture.com/SiteCollectionDocuments/PDF/Accenture-The-Consumer-Centric-Supply-Chain.pdf

Grahovac, J., & Chakravarty, A. (2001). Sharing and lateral transshipment of inventory in a supply chain with expensive, low demand items. *Management Science, 47*, 579–594.

Hanebeck, H. (2004). *Processes management and RFID*. Globe Ranger. http://www.globeranger.com/staging/pdfs/futureoftheedge/ProcessManagementandRFID.pdf

Hodges, S., Thorne, A., Garcia, A., Chirn, J., Harrison, M., & McFarlane, D. (2002). *Auto-ID based control demonstration phase 1: Pick and place packing with conventional control*. Auto ID Center White Paper, CAM-AUTOID-WH06. http://research.microsoft.com/pubs/132557/cam-autoid-wh006.pdf

Huckman, R., & Pisano, G. (2006). *Flextronics International*. 9-604-063. Harvard Business School, Boston.

Inspirage. (2013). *Demantra demand management*. http://www.inspirage.com/applications/oracle-value-chain-planning/demantra-demand-management/

Kang, J., & Kim, Y. (2012). Inventory control in a two level supply chain with risk pooling effect. *International Journal of Production Economics, 135*, 116–124.

Kenney, B. (2008). *Web 2.0 tools transform Osborne's management strategies*. Industry Week, July. http://www.industryweek.com/articles/web_2-0_toolstransform_osbornes_management_strategies_16504.aspx

Lee, H., Padmanabhan, P., & Whang, S. (1997). Information distortion in a supply chain: The bullwhip effect. *Management Science, 43*, 546–558.

Li, J., & Shaw, M. (2001). The effects of information sharing strategies on supply chain performance. In *Proceedings of 8th ECIS*. http://citeseerx.ist.psu.edu/viewdoc/summary?doi=10.1.1.144.4916

Liu, E., & Kumar, A. (2003). *Leveraging information sharing to increase supply chain configurability*. http://test.scripts.psu.edu/users/a/x/axk41/icis03_reprint.pdf

Malik, Y., Niemeyer, A., & Ruwadi, B. (2011, January). Building the supply chain of the future. *McKinsley Quarterly, 1*, 1–10. http://www.mckinsey.com/insights/operations/building_the_supply_chain_of_the_future

Mamgai, A., & Jolly, S. (2009). Web 2.0: Reshaping organization strategy in the flat world. *SETLabs Briefings, 7*, 79–86. http://www.infosys.com/infosys-labs/publications/Documents/reshaping-organization-strategy.pdf

McFarlane, D., & Sheffi, Y. (2003). *The impact of automatic identification on supply chain operations*. Auto ID Center-Cambridge Laboratory, Cambridge University. http://web.mit.edu/sheffi/www/documents/TheImpactofAutomaticIdentificationonSCOperations.pdf

Nokia. (2013). *Wiki*. http://wiki.forum.nokia.com/index.php/Wiki_Home

Robinson, P. (2012). *Managing indirect taxes in the supply chain*. Ernst & Young. http://www.ey.com/Publication/vwLUAssets/Managing-indirect-taxes-in-the-supply-chain/$FILE/Managing-indirect-taxes.pdf

Rudi, N., Kapur, S., & Pyke, D. (2001). A two-location inventory model with transshipment and local decision making. *Management Science, 47*(12), 1668–1680.

Saran, C. (2007). General motors oils supply chain with web 2.0 lessons. *Computer Weekly*, 12 Nov, http://www.computerweekly.com/news/2240083863/General-Motors-oils-supply-chain-with-Web-20-lessons

Schramm, T., Nogueira, S., & Jones, D. (2011). Cloud computing and supply chain: A natural fit for the future. *Logistics Management*. http://www.logisticsmgmt.com/article/cloud_computing_and_supply_chain_a_natural_fit_for_the_future

Soave. (2013). *America's beer distributors deliver choice and value, one beer at a time*. http://www.abdi.org/public/documents/Soave-07-03-08.pdf

Sturgeon, T. (2002). Modular production networks: a new American model of industrial organization. *Industrial and Corporate Change, 11*, 451–495.

Tang, C. (1999). The supplier relationship map. *International Journal of Logistics, 2*, 39–56.

Tang, C., & Zimmerman, J. (2009). Managing new product development and supply chain risks: The Boeing 787 case. *Supply Chain Forum, 10*, 74–86.

Thangamuthu, K. (2008). *Issues with supply chain and RFID in the retail industry*. Middleton: SAS Institute. http://support.sas.com/rnd/papers/sgf2008/RFID.pdf.

USAID. (2011). *Using last mile distribution to increase access to health commodities*. Arlington: USAID | DELIVER PROJECT, John Snow, Inc. http://villagereach.org/vrsite/wp-content/uploads/2011/12/UsinLastMileDist.pdf

Waddington, T., Childerhouse, P., & Towill, D. (2002). *Engineer your supply chain to cope with demand uncertainties*. Control. http://www.littoralis.info/iom/assets/200112-14.pdf

Warburton, R., & Stratton, R. (2002). Questioning the relentless shift to offshore manufacturing. *Supply Chain Management, 7*, 101–108.

Yankee Group. (2001). *Coopetition: Intelig and Electronet swap infrastructure*. The Yankee Group Research Report, January 26.

Part III

Evolving with Emerging Business Paradigms

The emergence of novel business paradigms – non applicability of the traditional laws of supply and demand, dominance of negative externality effects, and anomalies of high growth rate coexisting with high supply side uncertainty – must be recognized in transforming supply chains. Healthcare delivery and humanitarian relief, for instance, do not follow known supply/demand relationships; the negative externality effects are increasing sustainability concerns; and emerging economies, with dysfunctional business infrastructure, must manage high growth rates.

Incorporating Emerging Economies in the Supply Chain

7

7.1 Introduction

The ripples from the rapid pace of change in the emerging economies (EE) are being felt worldwide. These EE countries include Brazil, Russia, India and China (BRIC), and CIVETS (Columbia, Indonesia, Vietnam, Egypt, Turkey, and South Africa). In the next 15 years, 85% of the world's population, and 57% of the nearly one billion households with earnings exceeding USD 20,000 a year will live in the EE. Seven EE countries – China, India, Brazil, Mexico, Russia, Turkey, and Indonesia – are expected to contribute about 45% of global GDP *growth* in the coming decade. This economic scenario not only provides opportunities for the multinationals, it also creates a sense of urgency. For example, as much as 35% of Unilever's current sales come from emerging markets, GE expects 60% of its growth to come from emerging economies, and P&G allocates 30% of its R&D budget of $6.5 billion to these countries. Moreover, Coca Cola and Pepsi are now earning 20% or more of their revenues from the emerging markets. Emerging economies are expected to represent a large share of the growth in mature industries; automobiles that have hitherto been the engine of growth in mature economies (ME) such as US, Europe, and Japan have now set up major operations in China, India, and Brazil.

That notwithstanding, global companies do face a complex set of challenges in penetrating these markets. These challenges range from an unstable economic environment and poor infrastructure to lower price expectations in EE. The new challenges in EE are summarized in Table 7.1.

All these factors directly impinge on the design and execution of supply chains. For example, customers in EE seek products with features that may be different from those of the customers in ME and, in addition, EE customers usually expect lower price points. Clearly, identifying the country-specific product features would be the first step in satisfying international demand. Coca-Cola, for instance, uses cane sugar as sweetener in some countries and corn syrup in others. McDonalds offers vegetarian burgers in India, and KFC offers chicken in India that is spicy.

A.K. Chakravarty, *Supply Chain Transformation*, Springer Texts in Business and Economics, DOI 10.1007/978-3-642-41911-9_7,

Table. 7.1 Challenges in emerging economies

Supply chain performance	Challenges in emerging economies
Rapid response	Erratic demand forecast, inferior infrastructure, logistics congestion, unreliable data, informal systems
Asset utilization	Unpredictable economic system, complex taxation, lack of transparency, inefficient facilities
Efficiency	Regulations, bureaucracy, inferior infrastructure, fragmented markets

The country-specific variations clearly add complexity across products and marketing channels as well as the supporting supply chain functions – sourcing, manufacturing, distributing, and retailing.

An increasing number of multinationals are recognizing that they are not tapping into global growth opportunities with sufficient speed and scale. Therefore, formulating a strategy that incorporates the EE has become the top priority for many. As the EE becomes larger, diverse and competitive multinationals must develop approaches for dealing with variations within countries, and concentrate resources on the most promising regions of the world. Traditional aggregated approaches might not work, as they fail to account for the variability and rapid changes within these economies.

The extended global supply chain is exposed to a higher level of risk from severe disruptions in both supply and demand. The disruption in air travel due to the volcanic eruption in Iceland in 2010, which spewed ash across Europe, drove many supply chain experts in companies to juggle production and shipments worldwide to avoid interruptions in production (Lohr 2011). Apple buys almost a third of its flash memory from Toshiba, with the rest coming mainly from South Korea. As the lead time between chip orders and delivery can exceed several months, Apple is exposed to considerable procurement-risk. It mitigates the risk by holding adequate inventories.

The changes in supply chains would clearly necessitate going beyond just tweaking the strategies framed primarily for the ME. At the same time, it is not feasible to manage multiple country-specific supply chains. The answer may lie in agility; a supply chain that can quickly adapt to different country-specific environments, or a meta-supply-chain for coordinating several country-specific supply chains. In the next two sections the supply chain environments specific to ME and EE are described.

7.2 Mature Economies

We describe the common traits of supply chains in ME – customization, processes, logistics, infrastructure, and customer-service – to contrast them from those in EE.

7.2.1 Customization

With economic maturity in ME, customers take it as given that their basic needs would be satisfied. Therefore, they tend to be more selective and are often willing to pay a premium for convenience. In a tepid economy, however, acquisition of customers has become very important and competitive, in terms of increasing the value to a customer. In addition because of the high wages in ME, production costs have increased, increasing the cost of buying goods. On the positive side, reliable supply chains in ME enable quicker response, flexibility, and a low risk. Companies need to analyze the economics of this tradeoff. A high customization requirement suggests that upscale products (time sensitive) should be produced at local facilities that can react quickly to changes in demand. US manufacturers use local facilities for producing products that are complex and are subject to high demand volatility. They also use facilities in a low cost country for products with stable demand. As discussed earlier, while it costs more to manufacture in ME, the US customers are typically willing to pay premium prices for high-end products and for convenience. In a similar vein, Mattel produces miniature cars in China that has low unit production and shipping (to US) costs; it produces the larger Fisher Price cars (high unit shipping cost) in Mexico, which is closer to the markets in US.

7.2.2 Process Maturity

With mature processes it is possible to derive efficiencies from automation and capital intensive processes. KW (2011) describes how two companies in Europe (unnamed) drastically reduced inventory and shortages by using a common warehouse with shared delivery routes to retailers. Sharing of the warehouse was enabled by investment in warehouse automation and a well-structured distribution system that only existed in ME. Both the handling and transportation costs were reduced as they used a large number of common drop-off points to deliver their products.

Use of information and communication technology has enabled the design and operations of manufacturing systems to evolve. Planning tools such as enterprise resource planning (ERP) help to streamline internal processes. Companies have painstakingly implemented CRM (customer relationship management) and ERP along with computerized forecasting. Together, these systems can quickly determine what products customers want and how to locate and speedily deliver specific quantities at specific times to specific customers. Advanced Planning Systems (APS) have mainly led to the centralization of warehousing facilities and distribution centers. Smooth running of application processes such as ERP, CRM, and APS requires a rigorous discipline in reporting, tracking, and expediting tasks. Workers must be trained to desist from ad hoc decisions and actions that are exogenous to the processes in the system.

The report on US Advanced Manufacturing (PCAST 2011) states that US workers and systems are very competitive in sophisticated manufacturing of products using processes derived from scientific discoveries and technological innovations. The report sees great potentials for technologies such as Nano-scale Carbon Materials (graphene, buckeyballs, and carbon nanotubes), next-generation optoelectronics (technology for converting light into electrons, and vice versa), and flexible electronics (semiconductor crystals for infection sensing bandage etc.).

7.2.3 Logistics and Infrastructure

Most countries in ME have perfected the implementation and deployment of efficient logistics infrastructure. For example, the US accounts for about 25% ($900 billion) of the $3.5 trillion of annual revenue that the transportation sector generates worldwide. On an average per-capita basis, the estimates indicate that approximately 68 tons of freight (135,338 lb) valued at $45,324 were transported 15,310 ton-miles for every American resident in 2002 (DOT 2006). In 2002, more than three-quarters of the weight of all shipments and over half the value moved in local and short haul shipments within 250 miles from origin. However, long haul shipments carried 82% of the ton-miles.

As shown in Fig. 7.1, note that the logistics cost as a percentage of product-value is significantly low in the US in comparison to the emerging economies such as Brazil. In addition, most countries in ME possess very efficient retail industries and reliable distribution systems that make use of the logistics infrastructure.

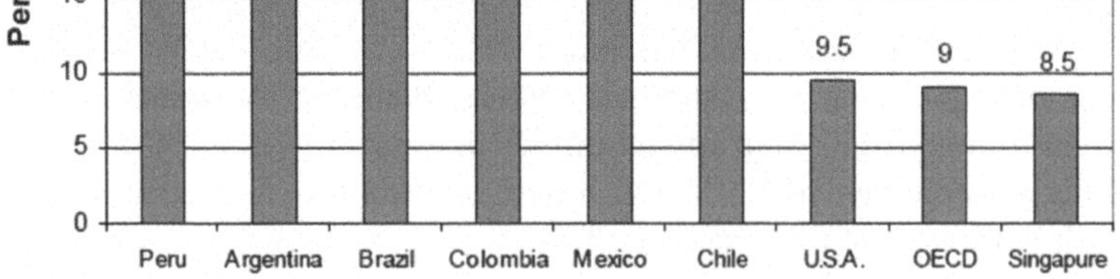

Fig. 7.1 Logistics cost as a percent of product value (Source: Gonzalez et al. 2007; Guasch and Kogan 2006)

A manufacturer or a distributor may choose among three alternatives to transport its materials. The company may operate a private fleet of owned or rented vehicles. It may use a carrier for transporting materials through direct shipments regulated by a contract. The company can also use a carrier that leverages common resources (vehicles, crews, terminals) to fulfill several clients' transportation needs. Major retailers such as Wal-Mart manage high volumes of goods. Huge loading bays and elevated platforms are configured so that forklifts can load pallets of merchandise onto tractor-trailers that take the goods directly to stores with large retail spaces.

In the US while some long-standing freight trends persist, new ones are emerging. Trucking is the mode most used for shipping, and its market share continues to increase. Air freight and express delivery are growing rapidly. An increasing number of customers appear to prefer faster delivery of products that are of high-value per unit weight. It is changing the industry structure, creating new alliances among shippers, carriers, and logistics providers.

The market economy sometimes may work to the detriment of efficient logistics services. For instance in some regions of a country logistics providers may be denied land claims, as they may not generate sufficient employment per square feet (OECD 2002). Other regions may prefer logistics providers that together form clusters to create logistics nodes. This often leads to sub-optimal supply chains and inadequate accessibility of the main infrastructure for industry and business. Netherlands plans to solve this problem by integrating its logistics and freight traffic flows from its industrial parks (OECD 2002). They must, however, account for different levels of policy making to obtain efficient freight distribution networks.

Switzerland's spends heavily on R&D and can quickly commercialize its research into products that also leads to a high rate of patenting. Switzerland's public institutions are among the most effective and transparent in the world. These include an independent judiciary, strong rule of law, and a highly accountable public sector. Its competitiveness is enhanced by excellent infrastructure and a well-functioning market.

7.2.4 Customer Service

Companies in ME excel in customer service as one of the ways of retaining or increasing market share. Advanced business logistics permit companies to send and receive invoices electronically, order components in advance, share demand forecasts with its suppliers and customers and pay retailers to stock certain goods and carry inventory. Collaboration with customers, by communicating more closely, is another approach. This allows better monitoring of demand signals and assessment of customer's purchase habits. Procter & Gamble maintains a sizable presence near Wal-Mart's headquarters. Collaboration with Wal-Mart enables P&G to increase or decrease the supply of products as and when needed.

UPS uses many technology tools to enhance customer service. For example, Enhanced Tracking enables customers track packages. Shipping & Handling enables buyers to compare total shipping and handling charges for different service levels. Time-in-Transit provides the buyer with the time-in-transit in business days for ground shipments between any two postal codes. Service Mapping generates a color-coded map displaying UPS ground transit time for any ZIP code. Finally Service Mapping is a valuable tool for just-in-time inventory planning.

7.3 Emerging Economies

We describe the common traits of supply chains in EE – country-specific preferences, informal logistics infrastructure, labor intensive systems, and distribution systems.

7.3.1 Country-Specific Preferences

In emerging markets such as China, India, and Brazil the purchasing power is rising rapidly. However, it is far from certain that Chinese and Indian consumers' preferences will converge to those of the ME. The distinctive local consumption patterns are driven by cultural norms. Customers in an emerging economy are more likely to prefer functional products to innovative high-end products. As discussed in Chap. 2, understanding the unique needs of the consumers is extremely important as it can reveal new market opportunities. As Blanco (2009) points out, Nokia's design of handsets for the Indian market included a flashlight to overcome frequent power outages and a dust-cover to protect against a dusty environment. Similarly, for the Chinese rural customer Nokia provided Chinese characters, as customers were not comfortable with Roman characters.

In an emerging economy with a low purchasing power relative to ME, it is necessary to modify the product for affordability. Motorola created an emerging-market-handset program to develop low cost handsets (less than $30). P&G introduced a simple unisex diaper in Brazil that is easy to produce. It sold for half the price of their existing product, and enabled P&G to grow its diaper sales in Brazil significantly (Blanco 2009).

7.3.2 Informal Logistics Infrastructure

Companies in ME mostly make decisions based on the premise that the logistics and business infrastructure would continue to function in a reliable way. In contrast, reliable infrastructure in the EE is a luxury, and not something companies can count on. A large number of service providers do not often play by the same set of rules that work in ME, and they are utterly uncoordinated. In addition, because of a fragmented market there are not many logistics providers available for transporting raw materials and finished products end-to-end. As a consequence, delivery delays are quite common, and placing advance-orders does not always secure the order. Very often supplies are bought (cash on delivery) from delivery trucks that work on the same principal that the ice cream vans use in a mature economy.

In Brazil, up to 15% of all apparel is sold door-to-door in poorer neighborhoods. Monitoring of delivery trucks while on the move is far from perfect. Lack of planning, lack of reliable data, and lack of information visibility, cause delivery trucks to travel mostly empty on their return trips. The system also suffers from a lack of 3PL service providers and their expertise in organizing the fulfillment process.

To succeed, multinational companies must work around infrastructural shortcomings. They need to develop strategies for EE that are different from those they use at home, and must find novel ways of implementing them. At the same time, such strategies need to be synergistic with one another, and they must leverage the company's current capabilities.

7.3.3 Labor-Intensive System

Both manufacturing and order fulfillment systems are usually very labor intensive. Unlike in an automated system, workers in a manual system may drift from what they are instructed to do. While it creates flexibility in the system, it can become chaotic; as it is hard to coordinate a large number of workers distributed throughout the supply chain. Rigorous worker training can, of course, help minimize this problem; but in the absence of an installed base of expertise, training is usually limited to the basics. As a consequence, most companies plan to manufacture only functional products in the EE. While some countries in the EE have begun to build innovative products such as advanced information technology, the numbers are relatively low. Khanna et al. (2005) summarize the labor practices in EE as shown in Table 7.2.

Table 7.2 Labor markets in EE (Source: Khanna et al. 2005)

U.S./EU	Brazil	Russia	India	China
Level of unionization varies among countries. Industrial actions take place often in Europe, but are rare in the United States	Trade unions are strong and pragmatic, which means that companies can sign agreements with them	Trade unions are present but their influence is declining except in certain sectors, such as mining and railroads	Trade union movement, with political links, is active and volatile, although it is becoming less important	Workers can join the government controlled trade unions. There have been recent strikes at Hong Kong owned manufacturing facilities

7.3.4 Distribution System

Customers in EE tend to buy in smaller quantities because of income and space limitations. Typical homes are small and so are retail stores and transportation vehicles. Small mom-and-pop stores cater to this need by delivering small quantities of products to a large number of locations. Upstream, the supply chain comprises a chain of wholesalers. The wholesalers as well as their delivery trucks become progressively small downstream and, in some cases, the vehicle for final delivery may well be a bicycle. Therefore in a country such as India, large multinational companies have a huge number of delivery points compared to that in an ME.

Companies operating in the rural markets in EE face unique challenges because of the low population density, and a poorly developed transportation infrastructure. In India 85% of villages have less than 5,000 people in them on average, implying that 612 Million people in India live in low-density areas (Neuwirth 2012). Companies may be forced to manage thousands of sales outlets to satisfy demand, increasing inventory and transportation costs.

For fast moving consumer goods, the products may be sold through village-retailers. To get products to these retailers, the company may use a hub-and-spoke model in which the company delivers products to a central distribution point and then independent entrepreneurs purchase the products and deliver them to the retailers. For consumer durables, demand should be aggregated by population centers to decrease inventory and transportation costs.

Operating a cascading distribution network in a rural setting can also be challenging because of infrastructure limitations. In addition, fragmentation of distribution services exacerbates coordination and accountability problems. Because of these issues many companies "piggyback" on distribution networks of other established companies in the region (Terpstra and Yu 2001). Companies may also use local non-profit organizations such as women's organizations and farmer federations. Unilever uses female teams to sell its cosmetic products in small packages in Indian villages, and it employs a motorcycle fleet to reach Indonesian villages (Baillie 2007).

7.4 Integrated Supply Chains

It follows from the foregoing discussion that the strategy for supply chains designed exclusively for producing and selling functional products in EE, would be different from that of the supply chains designed exclusively for producing and selling innovative products in ME. Therefore, the supply chains that incorporate both the EE and ME would have some unique features. We elaborate this in Fig. 7.2, by modifying Lee (2002) framework. The four resulting supply chain types: responsive, trade-driven, risk-hedging, and export-driven, are discussed below.

	Demand Volatility	
	Low (Functional products)	**High** (Innovative products)
Mature economy Stable supply process Stable trading process	*Trade Driven SC* **EE** to **ME**	*Responsive SC* **ME** to **ME**
Emerging economy Evolving supply processes Evolving trading process	*Risk hedging SC* **EE** to **EE**	*Export Driven SC* **ME** to **EE**

Fig. 7.2 Supply chain models

7.4.1 Responsive Supply Chain

For producing and selling innovative products in the ME such as those in high-fashion and computer industries, the supply chain must be responsive. This is so, for two reasons: in the ME logistics and business processes are sophisticated enough for sustaining a responsive supply chain, and customers are willing and able to pay for customized services. Note that the companies in ME need to offer a variety of products and make them available to customers quickly. Thus, to enhance the responsiveness the supply chain may use agile manufacturing, postponement, distribution centers, VMI, and information transparency.

Agile manufacturing is one way of creating responsiveness, as it enables rapid changeover from one product to another, and can vary production volumes frequently. However, it needs to be supported by a well trained work force and an adaptable system, which is more likely to be achievable in ME. Because of high wage cost in ME and high cost of sophisticated technology, the implementation of agile technology can be expensive. If the customers in ME (with high purchasing power) are willing to pay premium prices, investment in agile technology in ME can be justified.

Postponement of production and procurement of components for delayed differentiation of the final product is another way of achieving responsiveness. Postponement may take different forms. In MEs, logistics companies such as UPS and FedEx are now capable of orchestrating the assembly and delivery operations on behalf of the manufacturer. This makes it possible to complete the assembly of the final product at a facility close to the customer for quick delivery (Chap. 2). Another option would be to extend suppliers' ownership of inventory for a longer duration: suppliers holding inventory of finished goods at their own facilities, and delivering them to retailers as consignments. Clearly, suppliers would be justified in charging a premium price. For new products with unknown demand (as in the high-fashion industry) it would be prudent to build a small size facility first and add capacity if and when demand picked up. As the failure rate of the innovative product is high, it would help if the facility possessed built-in flexibility to switch between products.

Customer services must be refined for responsiveness in supply chains; it should be geared to working as a part of the customer's team to understand what problems the customer is facing and then to come up with innovative solutions. The company can then figure out the role it would like to play in this emerging solution in terms of products and services. Building such collaborative relationship requires trust building and communication savvy – information technology with personal reach.

Companies may also use some of the following approaches (discussed in Chap. 6): use drop-shipping where contract manufacturers directly ship the finished products to customers, and use 3PL services with merge-in-transit capability. The distribution centers may also be owned and operated by 3rd party service providers. Vendor managed inventory (VMI) with its proximity to the customer, would be very effective from distribution centers. To implement merge in transit, the logistics partner would pick up the major components of an order (such as the CPU and monitor) from different contract manufacturers and "assembles" them while in transit. Assembly may mean sorting the components according to customer preferences as per order transmitted by the retailer.

7.4.2 Trade Driven Supply Chain

As the MEs may not to be competitive in producing functional products, most of the needs would be satisfied through imports. We call this the trade driven supply chain, shown in Fig. 7.2. For example in the apparel industry, products that are not fashion-sensitive are produced in the EE, usually in Asia, and imported in large volumes by retailers such as Wal-Mart. Manufacturers in EE obtain large production contracts with lead times of several months and therefore their exposure to demand side risk is negligible. However, they are exposed to a considerable supply side risk, discussed later.

The shift from producing to importing functional products is altering how competitiveness of the firm is measured. The ability to analyze and manage the total delivered cost (cost of ownership) from supplier to buyer is assuming greater importance than it did before. One way is to break down the international and domestic logistics costs by product and delivery route so as to optimally allocate customs-related costs such as tariffs. This may become complicated if trade agreements were to be figured in, as discussed in Chap. 6. The delivery cost must also account for inventory carrying costs. In addition, companies must look for the most effective combination of carriers, routes and freight forwarders to satisfy their delivery needs at a minimum cost.

7.4.3 Risk Hedging Supply Chain

The consumption pattern in EE countries is primarily focused at functional products; demand is growing rapidly because of the increasing purchasing power in countries with large populations. This implies that the supply chains for production and consumption of functional products mostly remain in the EE. As the supply side risk in the EE is high, we call this the risk hedging supply chain (Fig. 7.2), as in Lee (2002). The highest concentration of activities in EE falls in this category, as companies must produce functional products for home consumption and a smaller quantity for export to MEs. Note that the instability in the supply side processes may limit the exploitation of arbitrage opportunities in resource-cost.

Because of low costs in component manufacturing and assembly, most companies prefer to locate these two major processes in the EE. However, to hedge against supply side risk, they may choose to import some of the components from ME at higher cost (Blanco 2009). By assembling the imported components in EE and exporting the finished products to the ME, the manufacturer can take advantage of economic incentives such as the duty-drawback and local-content rules in EE. On the other hand, if the company chooses to use locally produced components, it would need to carry a large inventory of components to limit its exposure to supply side risk.

The second concern is providing sufficient capacity with an increasing trend in demand in the home country (EE). One option for the company would be to expand its production facility. The company must decide between expanding each department in the facility (cutting, drilling, inspection etc.) and creating new production departments in a new facility by replicating all existing departments there.

The above two concerns (supply risk, and the increasing trend in local demand) can be in conflict and there lies a paradox. The paradox is an offshoot of the Hayes et al. (2004) observation that the manufacturing processes become stable by the time the demand of a product matures to high volumes. However, because of the high growth in the EE and the global trade links between the ME and EE, we are witnessing an anomaly in that *large volumes* must be produced in the EE with an *unstable* supply process. The combination of uncertainty and an increasing demand suggests that the size of the buffer and imports of components from ME must keep increasing as well. The perfect solution would be to peg the size of imports to the status of supply – increase import size only if the supply decreased or is delayed. The problem is that the imports do not materialize instantly and therefore additional buffer is needed to provide an adequate cover for the shipping lead time. To overcome this problem companies may reserve capacity with component manufacturers in the ME. The advance commitment carries little risk, given an increasing demand. The fact that customers in the EE are far less sensitive to delays in fulfillment, may counteract the need for super-size inventory buffers.

7.4.4 Export Driven Supply Chain

The market for innovative products produced in ME and sold in EE, though not zero, is very small. We do not consider this further.

7.4.5 Reverse Factoring

Global outsourcing of manufacturing has eroded the assets of companies in ME, and reduced the availability of working capital. As a consequence buyers in ME are demanding extended payment terms from suppliers in EE, and delaying inventory ownership where possible. In response, suppliers are seeking funds at low interest to cover their procurement and production. This is not available easily because of the unstructured capital markets in EE, and therefore suppliers incur a high cost in borrowing. This cost-shifting to suppliers adds to the financial instability of the supply base. To alleviate the stress on suppliers, 3rd party funding agents, known as factors, are making up the gap in working capital by financing supplier's receivables and *debiting the buyers*. A full discussion is provided in Sect. 7.6.

7.5 Business Models of Supply Chain Integration

The crucial question companies need to answer is how to restructure the supply chain so as to benefit from market opportunities in *multiple* countries. For example, they could choose to manufacture and distribute innovative products exclusively in a few MEs, by building expertise in quick response and in handling volatile demand. They could charge a premium price but they would also need to absorb the high cost of labor and material. As an alternative, they could choose to manufacture functional products in the EE and distribute them in both the EE and ME. This reduces cost because of the scale economy of large volume production in EE. As a third option they could have operations in both ME and EE. That is, build and sell innovative products in ME and build functional products in EE with sales in both ME and EE. The fourth option would be to split their production and sales operations so as to be have assembly operations in EE but have some of the components imported from ME, and vice versa. The fifth option would build a resilient supply network so that if the conditions in one country worsened they could move their operations to another country without any glitches. Clearly, the global supply-production-distribution network must draw upon the strengths of all geographical regions. We discuss these supply-production-distribution options next.

7.5.1 Managing Functional Products

As explained with respect to Fig. 7.2, countries in ME are losing ground in producing functional products to the EE, especially in food, shoes, toys, and clothing (daily-wear). This is due to a combination of factors related to the products and countries in ME and EE. Companies in the EE have acquired the expertise to manufacture functional products at low cost, not just because the costs of labor and material in the EE are low. The managerial expertise required is not sophisticated, as functional products have stable demand that is easy to forecast. As a consequence, as shown in Fig. 7.3, an increasing number of companies are locating production of functional products in the EE, and distributing them to customers in both ME and EE. Nokia uses this production model in India and China. Although the profit margins are low, the large production volumes generate scale economy advantage. The evolving and unstable supply processes, however, may undo some of this advantage. While companies in ME make money, the shift in production from ME to EE can be controversial because of large job losses in the ME.

The decisions a typical company in ME needs to make include procurement of components, assembly of these components to configure products, distribution of products to markets in the EE, and shipment of products to the markets in ME. As explained earlier the uncertainty on the procurement side will be significant, and there will be some uncertainty in demand, though not large. Note that retailers in ME may overcome the supply uncertainties (in EE) by using the EE facilities only

Fig. 7.3 Functional products

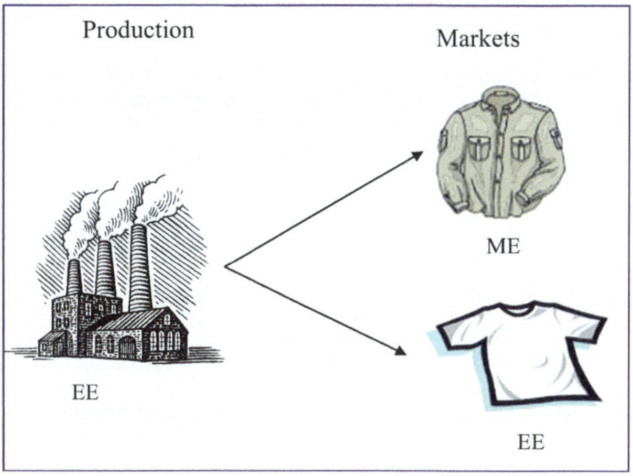

for assembling products for sale in ME. The company develops suppliers in EE to have their facilities and products tailored for export, and not for local consumption.

To elaborate the business model, consider a simple case where a company produces and sells a single product *entirely* in EE. Assume that the per unit cost of manufacturing the product (including procurement of components) and shipping it to the market is c, and unit sales price is p. The company needs to decide how many units to produce and ship (Q), given that there is a huge uncertainty in the supply of components (S) and some uncertainty in demand (D). We may capture the supply uncertainty by denoting the actual delivery from suppliers for an order of Q as γQ, where γ is a random variable between 0 and 1. It is clear that the manufacturer's expected sales will equal the minimum of supply and demand, expressed as $E_\gamma\ E_D\ Min(\gamma Q, D)$. As the manufacturer only pays for the amount supplied, γQ, his expected profit will be $p\ E_\gamma\ E_D\ Min(\gamma Q, D) - c\ E_\gamma\ \gamma Q$. He will choose order quantity Q to maximize his expected profit. Note that the *expression* for optimal value of the order quantity is independent of the probability distribution functions of D and γ, the actual value will be specific to these distributions (Rekik et al. 2010).

Example. Consider a company that faces four possible values of demand (discrete): 200, 400, 600, and 800; and three possible values of γ: 0.2. 0.5, and 0.8. It can be verified that for an order size of 900, the possible values of realized sales quantities, corresponding to the values of demand and γ, will be as shown in Fig. 7.4.

Assuming the demand probabilities as 0.3, 0.4, 0.2, and 0.1; probabilities of γ as 0.5, 0.3, and 0.2; the expected sales will be (180)(.5)(.3) + (180)(.5)(.4) + (180)(.5)(.2) + (180)(.5)(.1) + (200)(.3)(.3) + (400)(.3)(.4) + (450)(.3)(.2) + (450)(.3)(.1) + (200)(.2)(.3) + (400)(.2)(.4) + (600)(.2)(.2) + (720)(.2)(.1) = 27 + 36 + 18 + 9 + 18 + 48 + 27 + 13.5 + 12 + 32 + 24 + 14.4 = 278.6.

		Demand			
		200 prob = 0.3	400 prob = 0.4	600 prob = 0.2	800 prob = 0.1
Quantity Delivered by Supplier	180 prob = 0.5	180	180	180	180
	450 prob = 0.3	200	400	450	450
	720 prob = 0.2	200	400	600	720

Fig. 7.4 Realized sales quantities for order size of 900 units

Assuming unit sales price of \$20, expected revenue will be \$5,578. As the company only pays for the quantities actually delivered, we compute the procurement cost based on expected number of units delivered, which will be $\{(180)(0.5) + (450)(0.3) + (720)(0.2)\} = 366$. Assuming a unit purchase price of \$4 the expected cost of purchase will be = \$1,476. Thus, for an order quantity of 900 units, the company's profit will be 5,578 – 1,476 = \$4,102. The company can do similar computations with different order sizes Q, and select the value that maximizes its profit.

In many cases it may not be easy to determine the distribution function of γ, whereas, the distribution of available supply S, like demand, may be easier to determine (S is independent of Q). If this is the case, the company would need to investigate two supply scenarios ($S \geq Q$ and $S \leq Q$) and two demand scenarios ($D \geq Q$ and $D \leq Q$) generating four supply chain scenarios, shown in Fig. 7.5. For each of the four scenarios, the unit profits and the likelihood of the scenarios are listed, from which the expected profit can be determined. It is assumed that if $S < Q$,

	Demand exceeds production $D \geq Q$	Production exceeds demand $Q \geq D$
Supply exceeds production $S \geq Q$	Sales = Q Revenue = pQ Profit = $(p-c)Q$ Prob = $P(D \geq Q).P(S \geq Q)$	Sales = D Revenue = pD Profit = $pD - cQ$ Prob = $P(D \leq Q).P(S \geq Q)$
Production exceeds supply $Q \geq S$	Sales = S Revenue = pS Profit = $pS - c_2Q$ Prob = $P(D \geq Q).P(S \leq Q)$	Sales = Min (S , D) Revenue = $p \underset{D,S}{E} Min(D,S)$ Profit = $\underset{D,S}{E} Min(D,S) - c_2Q$ Prob = $P(D \leq Q).P(S \leq Q)$

Fig. 7.5 Supply chain scenarios

the company makes up the difference of $Q - S$ from a different source at a higher cost ($c_2 \geq c$). On the other hand, if $S > Q$, the company only pays for the amount Q at unit cost c. The supplier can either take back the amount $S - Q$ or leave it at the company's warehouse and get paid only when used.

For the general case with sales in both EE and ME, we denote the shipped quantities as Q_E and Q_M. The unit cost of production and shipping would now be denoted as c_E and c_M, respectively. Similarly, unit sales price will be p_E and p_M. Without loss of generality, we assume $p_M - c_M \geq p_E - c_E$. We will now have 12 supply chain scenarios to consider as shown in Appendix 1.

7.5.2 Importing Components from ME

Because of the high supply uncertainty in EE, a fraction of the components may be produced in the ME, and others procured in EE. They are then assembled in the local facility in EE. The finished functional products are sold in EE as well as ME, as shown in Fig. 7.6.

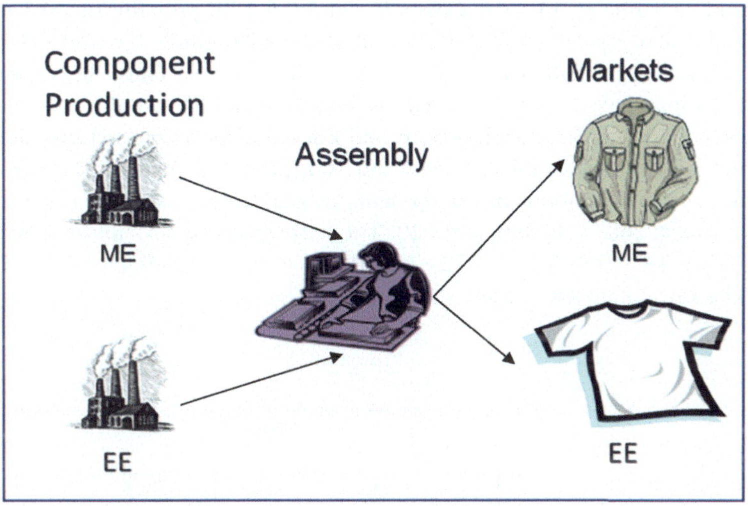

Fig. 7.6 Importing components

This business model allows for a faster market reaction, and mediation of supply risk in EE. Dell implemented this framework in Brazil in 1999, while Coca-Cola has used it for their bottling model. The company can claim duty draw-back on the imported components that are assembled in EE and the products exported back to MEs.

GE Healthcare makes parts for its diagnostic machines in China, Hungary, and Mexico and develops the software for those machines in India (Khanna et al. 2005).

The company created this system when it realized that the market for diagnostic machines was small in most emerging countries, and used the Indian facility as a global sourcing base. GE Healthcare has learned to use its operations in all emerging countries as part of a system that allowed the company to produce equipment cheaply for the world market.

To illustrate how the profits may be computed, we consider a simple case where finished products are sold only in the EE. Thus, consideration of duty draw-back and demand in ME are not included in this analysis. Note that if the total quantity produced is Q, a fraction of it will be produced using components procured from EE $(= q_E)$. The remainder is produced using components procured from ME $(= q_M)$. Assuming we can specify the distributions of γ_E and γ_M, the total amount of components delivered for assembling Q would equal $\gamma_E q_E + \gamma_M q_M = Q$; Q is clearly a random variable. The manufacturer's expected sales will be $E_{\gamma_M} E_{\gamma_E} E_D [Min(\gamma_E q_E + \gamma_M q_M, D)]$ and, therefore, his expected profit is expressed as $p E_{\gamma_M} E_{\gamma_E} E_D [Min(\gamma_E q_E + \gamma_M q_M, D)] - \{c_E q_E E_{\gamma_M}(\gamma_E) + c_M q_M E_{\gamma_E}(\gamma_M)\}$, where q_E and q_M are decision variables.

Example. Consider a variation of the example analyzed earlier. The demand distribution and unit sales price are as in previous example but the data on supply is different. Assume γ_E to have values 0.2 and 0.6 with probabilities 0.8 and 0.2 respectively; and γ_M values of 0.4 and 0.8 with probabilities 0.6 and 0.4 respectively. Assume $q_E = 500$, $q_M = 400$, $c_E = 4$, and $c_M = 7$. Thus the quantity of components purchased and delivered by suppliers in EE would vary between 100 (prob = 0.8) and 300 (prob = 0.2); and components purchased and delivered by suppliers from ME would vary between 160 (prob = 0.6) and 320 (prob = 0.4). Therefore, the total amount of components available for assembly can take four possible values: 260, 420, 460, and 620. The corresponding probabilities would be (0.8)(0.6) = 0.48, (0.8)(0.4) = 0.32, (0.2)(0.6) = .012, (0.2)(0.4) = 0.08, respectively. The sales quantities are as shown in Fig. 7.7.

		Demand			
		200	400	600	800
Delivered Quantity	260	200	260	260	260
	420	200	400	420	420
	460	200	400	460	460
	620	200	400	600	620

Fig. 7.7 Realized sales quantities

As before, we compute expected sales as (200)(0.48)(0.3) + (260)(0.48)(0.4) + (260)(0.48)(0.2) + (260)(0.48)(0.1) + (200)(0.32)(0.3) + (400)(0.32)(0.4) + (420)(0.32)(0.2) + (420)(0.32)(0.1) + (200)(0.12)(0.3) + (400)(0.12)(0.4) + (460)(0.12)(0.2) + (460)(0.12)(0.1) + (200)(0.08)(0.3) + (400)(0.08)(0.4) + (600)(0.08)(0.2) + (620)(0.08)(0.1) = 302. The expected revenue will be $6,040.

The expected cost $c_E q_E E_{\gamma_M} \gamma_E + c_M q_M E_{\gamma_E} \gamma_M$ = (2,000){(0.2)(0.8) + (0.6)(0.2)} + (2,800){(0.4)(0.6) + (0.8)(0.4)} = 2,128. Hence, the profit will be equal to $3,912. Computing the best purchase quantities will involve a 2-dimensional search in q_E and q_M.

Note that the profits for examples corresponding to Figs. 7.4 and 7.7 are not comparable because the assumed values of γ and unit purchase costs are not equivalent. In general, it is possible to determine the conditions in γ and c that would make the profit in one scenario (Fig. 7.4) exceed the profit in the other (Fig. 7.7). Analysis like this would clearly help establish when the risk of supply uncertainty in EE outweighs the cost of imported components from ME.

For the case when distribution of γ is hard to determine we would have two demand scenarios, ($D \geq q_E + q_M$, and $D \leq q_E + q_M$) and four supply scenarios ($S_E \geq q_E$ & $S_M \geq q_M$; $S_E \leq q_E$ & $S_M \leq q_M$; and $S_E \geq q_E$ & $S_M \leq q_M$; and $S_E \leq q_E$ & $S_M \geq q_M$). This would generate 8 supply chain scenarios. In the general case when the product is sold in both EE and ME, the shipment quantities will need to be broken down as Q_E and Q_M, so that the quantity of product available in ME for sale would equal $\gamma_E q_E + \gamma_M q_M - Q_E$. As a consequence, the demand side will have four scenarios, and there will be 16 scenarios for the supply chain. The decision variables are Q_E, Q_M, q_E and q_M.

7.5.3 Distribution Risk in EE

Because of the distribution channel complexities and lack of local knowledge, companies may prefer to use general distributors. As mentioned earlier, the distributors in the rural EE can be the village retailer, who owns the inventory. The downsides of using such distributors are that they have limited reach and tend to concentrate only on fast moving products. As a consequence, the distributor's objectives may not be aligned with the long term interests of the company. Other alternatives would be companies forming joint ventures, acquiring existing networks, and creating own distribution networks. For example, Home Depot acquired HomeWay to reach its customers in China, while Lenovo of China built its distribution network for computers from scratch which now includes over 5,000 stores (Blanco 2009). It would be interesting to study whether it would be in Lenovo's interest to share its network with competitors such as Dell and, if so, how. The presence of Dell can invigorate the market for computers increasing Lenovo's sales. Even if its sales decrease Lenovo could more than recover its sales losses by charging Dell for use of the network, through an appropriate fee structure.

A typical supply chain for functional products produced in EE and sold in both EE and ME is shown in Fig. 7.8. Because of the distribution uncertainties, it is sold in EE through a distributor.

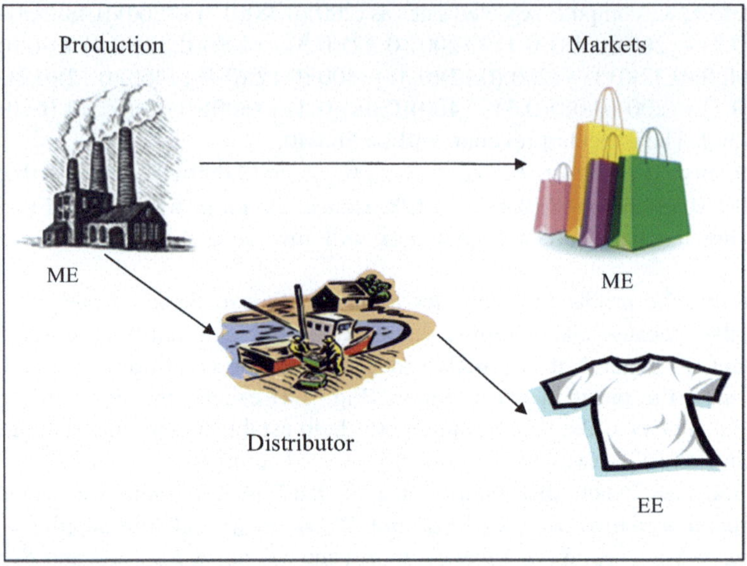

Fig. 7.8 Distribution in EE

Consider a simple case of a company producing a functional product in EE and selling it *only* in EE through a distributor (Rekik et al. 2010). Because of distribution uncertainties, we assume that out of Q units shipped only βQ units arrive at the market. As there is also a supply side uncertainty in EE modeled as γ in Fig. 7.4, we may express the manufacturer's expected profit as $p\, E_\beta\; E_\gamma\; E_D\, Min(\beta\gamma Q, D) - c\, E_\gamma\, \gamma Q$.

Example. We augment the example shown in Fig. 7.4 by including distribution effectiveness β. Assume two values of β (0.3, 0.8) with probabilities (0.7, 0.3). It follows that there will be six possible values of the product $\gamma\beta$ (0.06, 0.16, 0.15, 0.4, 0.24, 0.64), given that γ: 0.2. 0.5, and 0.8. The corresponding values of quantities shipped to customers will be (54, 144, 135, 360, 216, 576) with probabilities (0.35, 0.15, 0.21, 0.09, 0.14, 0.06). Fig. 7.4 is modified as shown in Fig. 7.9.

Expected sales quantity = (54)(0.35)(0.3) + (54)(0.35)(0.4) + (54)(0.35)(0.2) + (54)(0.35)(0.1) + (144)(0.15)(0.3) + (144)(0.15)(0.4) + (144)(0.15)(0.2) + (144)(0.15)(0.1) + (135)(0.21)(0.3) + (135)(0.21)(0.4) + (135)(0.21)(0.2) + (135)(0.21)(0.1) + (200)(0.09)(0.3) + (360)(0.09)(0.4) + (360)(0.09)(0.2) + (360)(0.09)(0.1) + (200)(0.14)(0.3) + (216)(0.14)(0.4) + (216)(0.14)(0.2) + (216)(0.14)(0.1) + (200)(0.06)(0.3) + (400)(0.06)(0.4) + (576)(0.06)(0.2) + (576)(0.06)(0.1) = 150.07

Assuming unit sales price of \$20, expected revenue = \$3,001.4

Expected quantities procured = {(180)(0.5) + (450)(0.3) + (720)(0.2)} = 366. Assuming a unit purchase price of \$4 the expected cost of purchase will be = \$1,476.

Expected number of units shipped = (54)(0.35) + (144)(0.15) + (135)(0.21) + (360)(0.09) + (216)(0.14) + (576)(0.06) = 166. Assuming a shipping and handling cost of \$2 per unit shipped, distribution cost = \$232.

Therefore, expected profit = 3,001.4 – 1,476 – 232 = \$1,293.4. In this example the distribution uncertainty reduces the company's profit by (4,102 – 1,293.4)/ 4,102 = 68.5%.

		Demand			
		200 prob = 0.3	400 prob = 0.4	600 prob = 0.2	800 prob = 0.1
Shipped Quantity	54 prob = 0.35	54	54	54	54
	144 prob = 0.15	144	144	144	144
	135 prob = 0.21	135	135	135	135
	360 prob = 0.09	200	360	360	360
	216 prob = 0.14	200	216	216	216
	576 prob = 0.06	200	400	576	576

Fig. 7.9 Realized sales quantities

7.5.4 Innovative Product with High Demand Volatility

Innovative products are primarily sold in the ME. It typically comprises innovative components (produced in ME) and simple components (produced in EE), as shown in Fig. 7.10.

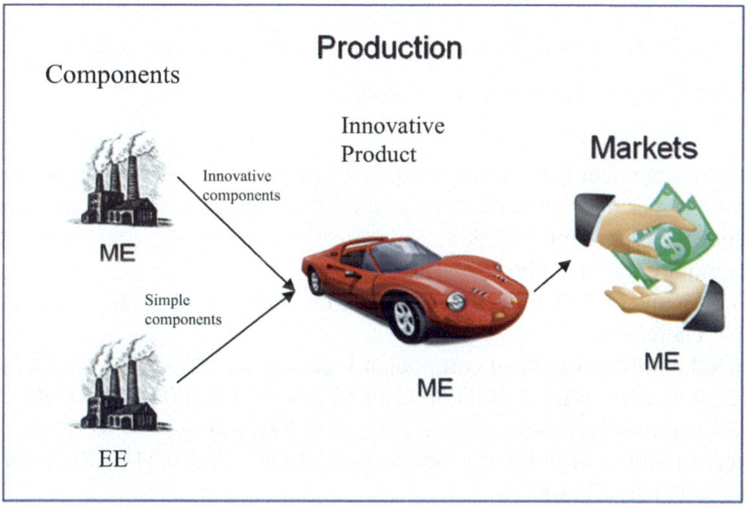

Fig. 7.10 Produce only innovative product

Assume that the company in ME sells the innovative product that has a random demand of ξ. Also assume that the company assembles one innovative component and one simple component to produced one unit of the final product. It procures q_E units of simple components from EE, and q_M units of innovative component from ME. The actual delivered quantities are $\gamma_E q_E$ and $\gamma_M q_M$, where γ_E and γ_M are random variables, not exceeding one, that capture procurement uncertainties in the EE and ME respectively. It follows that the number of units of the final product that would be available for sale in ME, $Q = Min(\gamma_E q_E, \gamma_M q_M)$. A complete analysis of this business model is provided in Appendix 1.

Example. Consider a company that orders components from EE and ME with $q_E = 1,000$ and $q_M = 600$. Assume two possible values of γ_E with associated probabilities in parenthesis as 0.3, (0.8) and 0.6, (0.2), and two possible values of γ_M with associated probabilities in parenthesis as 0.6, (0.3) and 0.8, (0.7). Therefore the possible values of finished product Q that can be produced and its associated probabilities will be as shown in Fig. 7.11.

	$q_E = 300$ probability = 0.8	$q_E = 600$ probability = 0.2
$q_M = 360$ probability = 0.3	$Q = 300$ probability = 0.24	$Q = 360$ probability = 0.06
$q_M = 480$ probability = 0.7	$Q = 300$ probability = 0.56	$Q = 480$ probability = 0.14

Fig. 7.11 Available supply of innovative products

Assume component E's procurement cost \$10 per unit, component M's procurement cost \$40 per unit, and assembly cost \$20 per product. Assume sales price of the product \$200 per unit.

Procurement and assembly

The company in ME will pay only for the number of units of components delivered. Therefore,

Expected number of units of component E procured = 300(0.8) + 600(0.2) = 360

Expected number of units of component M procured = 360(0.3) + 480(0.7) = 444

Component expected procurement cost = 10(360) + 40(444) = \$21,360

Expected number of units of products assembled = 300(0.24) + 300(0.56) + 360 (0.06) + 480(0.14) = 328.8

Note that because of the mismatch between supply and demand, the expected number of units of products assembled is less than the expected number of units of component E or component M procured.

Therefore, the expected assembly cost = 20(328.8) = \$6,576.

		Demand				
		200 prob = 0.1	300 prob = 0.2	400 prob = 0.4	500 prob = 0.2	600 prob = 0.1
Supply	300 prob = 0.24	200 prob = 0.024	300 prob = 0.048	300 prob = 0.096	300 prob = 0.048	300 prob = 0.024
	300 prob = 0.56	200 prob = 0.056	300 prob = 0.112	300 prob = 0.224	300 prob = 0.112	300 prob = 0.056
	360 prob = 0.06	200 prob = 0.006	300 prob = 0.012	360 prob = 0.024	360 prob = 0.012	360 prob = 0.006
	480 prob = 0.14	200 prob = 0.014	300 prob = 0.028	400 prob = 0.056	480 prob = 0.028	480 prob = 0.014

Fig. 7.12 Sales quantities of the assembled product

Sales and Revenue

Assume demand and associated probabilities to be 200 (0.1), 300 (0.2), 400 (0.4), 500 (0.2), and 600 (0.1). The possible values of sales will be as shown in Fig. 7.12.

Expected sales units = 200(0.024 + 0.056 + 0.006 + 0.014) + 300(0.048 + 0.112 + 0.012 + 0.028 + 0.096 + 0.224 + 0.048 + 0.112 + 0.024 + 0.056) + 360(0.024 + 0.012 + 0.006) + 400(0.056) + 480(0.028 + 0.014) = 305.68

Expected revenue = \$200(305.68) = \$61,136.

Expected profit = 61,136 − 21,360 − 6,576 = \$33,200

The company will choose the values of q_E and q_M that maximize its expected profit.

For the general case, the expressions for expected profit and cost are shown in Appendix 1.

7.5.5 Splitting Production

Malik et al. (2011) describe a US-based manufacturer who was losing ground to competitors, as it had sent production to EE (China). The increasingly volatile customer demand was eroding the company's profit in its supply chain. The company decided to split its supply chain into distinct groups. For high-volume functional products with stable demand, the company kept the sourcing and production in EE. Products with volatile demand and those with low demand (irrespective of volatility) were assigned to ME for production. Note that low-volume products needed to be switched over from one to another frequently, which was possible only with a flexible production system. Products with volatile demand require production quantities to be variable and therefore production capacity must be adjusted frequently through outsourcing, overtime, and additional shifts. Assigning these products to the manufacturers in ME made economic sense because these companies had invested in flexible production technology and an efficient warehousing network. They could quickly adjust production capacity, and could provide quicker customer fulfillment. Some of these products required unique manufacturing processes in which the skilled ME workforce excelled. For the most volatile products, the products were made to order in ME. The company produced functional products in EE, based on forecast. Together, these changes helped the company reduce its sourcing and manufacturing complexity, and decreased its cost of goods sold by about 15%. It also improved its service levels and shortened lead times to 3 days from 10.

Instead of splitting production by product type, one may split demand of a product by its volatile and stable components, as shown in Fig. 7.13.

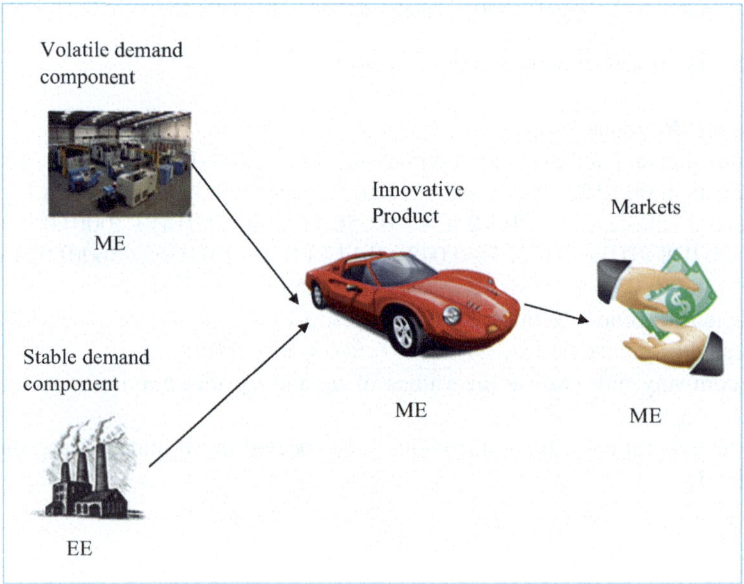

Fig. 7.13 Splitting into volatile and stable demand components

In general, we can restate demand splitting as allocating production of a product to two countries, one in EE and the other in ME. In Fig. 7.14, the variation of demand in time of a product in ME is shown. Observe that the demand ξ has a "floor" value of K units. Therefore if the demand pattern holds up, K units can be contracted out for production risk-free.

As the stable component of demand of K units can be assigned to EE ex ante, a sufficiently long lead time will be available for the supplier to manufacture and ship the product from EE to arrive at ME before they are needed. Because of the low unit cost in EE, retailers in ME will be justified in taking a calculated risk by ordering a quantity $Q \geq K$ from supplier 1 in EE. After the demand ξ is known, retailer would order $\xi - Q$ units from a supplier 2 in ME, with production capacity R. The excess demand $(\xi - Q) - R$ is lost. While supplier 1 is a low cost mass-producer (unit price w_1), supplier 2 must be agile as it would need to absorb a highly variable demand exceeding Q, and will be justified in charging a higher unit

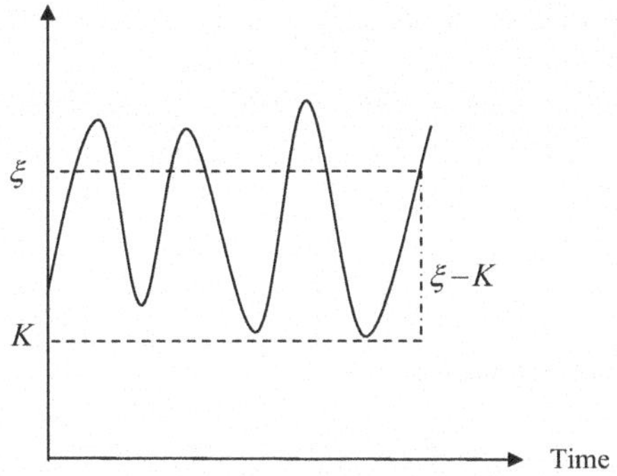

Fig. 7.14 Splitting demand between two suppliers

price w_2 ($w_2 \geq w_1$). Note that $Q = 0$, implies that the product is produced entirely in ME. A very high value of Q, on the other hand implies a very low value of R, as the quantity assigned to supplier 2 will be very small. Note that these two extreme cases are equivalent to the model in Malik et al. (2011) of allocating all or none of the demand of a product to a single supplier.

The retailer maximizes her expected profit from sales in ME, by choosing an appropriate value of Q, and supplier 1 maximizes his profit by choosing an appropriate value of w_1. Finally, supplier 2 maximizes profit with respect to w_2 and R. Note that the retailer transfers her demand risk to supplier 2, who will therefore be justified in charging a premium price. The details of this model can be found in Appendix 2, where the optimal values of Q, w_1, w_2, and R are established.

Example. Consider the decisions in a company where demand and corresponding probabilities (in parenthesis) are 1,000 (0.1), 1,500 (0.2), 2,000 (0.3), 2,500 (0.2), 3,000 (0.1), 3,500 (0.1). The unit costs (capacity + production + shipment) of suppliers 1 and 2 are assumed to be \$5 and \$8, respectively; and the retailer's unit sales price is \$20. We would like to compute the expected profits of the three parties when the retailer orders $Q = 1,500$ units from supplier 1, and supplier 2's production capacity is $R = 1,000$ units. The wholesale prices of the two suppliers are $w_1 = \$7$ and $w_2 = \$15$, respectively.

Retailer's expected profit

$$= \left| \begin{array}{l} \{1,000(0.1) + 1,500(0.2) + 2,000(0.3) + 2,500(0.2) + 2,500(0.1 + 0.1)\}20 \\ -\{500(0.3) + 1,000(0.2) + 1,000(0.1 + 0.1)\}w_2 - 1,500w_1 \end{array} \right.$$

which can be simplified to profit $= 40,000 - 550w_2 - 1,500w_1 = 40,000 - 6,900 - 10,500 = \$19,600$.

Supplier 1's expected profit $= 1,500w_1 - 1,500c_1 = 1,500(7 - 5) = \$3,000$

Supplier 2's expected profit $= 550w_2 - 1,000(8) = 9,900 - 8,000 = \$1,900$

Note that while supplier 2 receives orders that vary in size from 0 to 1,000, it only delivers 550 units on average.

The values of Q, R, w_1, and w_2 chosen above may not be optimal. The optimal values can be determined by solving the four nonlinear simultaneous equations, shown in Appendix 2.

7.5.6 Resilient Supply Network

In addition to the supply side risk in EE, other sources of considerable risks exist that have major supply chain implications for the long term. These include the evolving terms of global trade, uncertain capital flows, rising labor cost, and currency revaluations. Because of such risks, some organizations are building a diverse and resilient portfolio of supply chain assets as hedges against uncertainty.

Malik et al. (2011) describe a company with a manufacturing hub in China. The company was concerned at its reliance on the Chinese hub because business conditions in China could change rapidly. The company examined how its cost structure might change for a range of scenarios of global business conditions including wages, currency rates, commodity prices, and logistics costs. The company determined that the long term risks associated with wage inflation and currency rate fluctuations were too high so that Mexico became a preferable alternative. As a hedge, the company built a supplier base in Mexico with the objective of switching operations from China to Mexico quickly if it became necessary.

As another example, Malik et al. (2011) discuss a company that is examining how low-cost countries might help it hedge against a range of labor cost, tariff, tax, and exchange rate scenarios. The company is also factoring in unexpected supply disruptions, including fires, earthquakes, and labor-related strifes.

An American industrial manufacturer chose to broaden its footprint in Brazil and Mexico to hedge against swings in foreign-exchange rates. In particular, the company invested in spare capacity to make several innovative, high-end components that it had formerly produced only in Europe and the United States because of the advanced machining and engineering requirement. The investment is helping the company hedge against currency fluctuations by quickly transferring production of the components across its global network to match economic conditions.

7.6 Global Supply Chain Finance

Global supply chain finance (GSCF) enables retailers and suppliers in different countries reduce their cost of capital through financial cooperation (Lamoureux and Evans 2011). It is used frequently in EE where most companies do not have readily available sources of financing and incur a high cost of capital because of it. In addition, firms in ME may want to delay payment to firms in EE for multiple reasons: tightening of global credit markets that limit cash flows, delays in verifying the quality of purchase, and the intricacies of refunds (for returns) from firms in EE. According to Demica (2007), 73% of large corporations looked to extend payment terms with their suppliers in 2007. Suppliers often depend on financing companies (and customers in ME) to provide them with bridging finance, usually in the form of cash advances that are settled when the customers make payment.

GSCF formalizes financing of suppliers through 3rd party finance companies or banks, also known as factors. It facilitates risk-reduction and fast delivery by increasing the visibility of cash flows in the supply chain. It employs a technology platform that links logistics and tracking of goods to the decisions of finance providers. This enables cash to be injected into the supply chain with a minimum risk.

Heller Financial and Wal-Mart have a GSCF arrangement in Mexico (Klapper and Vittas 2003). Wal-Mart offers its Mexican suppliers the option of receiving immediate payment of 80% of the sale through Heller Financial. The remaining 20%, less interest and service charges, is paid by Heller when it receives payment from Wal-Mart. The sellers can receive short term financing because they would be borrowing on Wal-Mart's credit risk (Wal-Mart owes money to the suppliers). Wal-Mart gains by cutting its transaction cost by paying a single bill to Heller rather than to a large number of suppliers. Moreover, Heller gains from the interest and fees charged to the suppliers, and because it can reduce its cost of capital based on Wal-Mart's accounts receivable, which has a credit rating equal to that of Wal-Mart.

7.6.1 GSCF Process

The process flow related to GSCF is shown in Fig. 7.15. The buyer places an order using a technology platform, and the supplier submits an invoice to the system and requests a credit grant from the bank that is also linked to the platform. The bank pays the supplier a percentage of the invoices less interest, and holds the rest in reserve. When the contract matures the buyer makes a payment to the bank, and the bank pays the amount held in reserve to the supplier. The finance risk of the bank is transferred from the "bank and supplier" to the "bank and buyer", because approval of the invoice by the buyer forms the basis for the bank's decision to grant credit.

In essence, the seller's accounts receivable are purchased by the bank at a discount. The bank advances less than 100% of the face value of the receivable even though it takes ownership of the entire receivable. The difference between this advance amount and the invoice amount creates a reserve, and may be used to cover any deficiencies in the payment of the invoice. When the invoice is paid in full by the buyer, the reserve amount is remitted to the seller. The credit rating of the buyer, which usually exceeds that of the seller, is often used to determine the finance-rate offered. The seller's cost of capital is thus reduced.

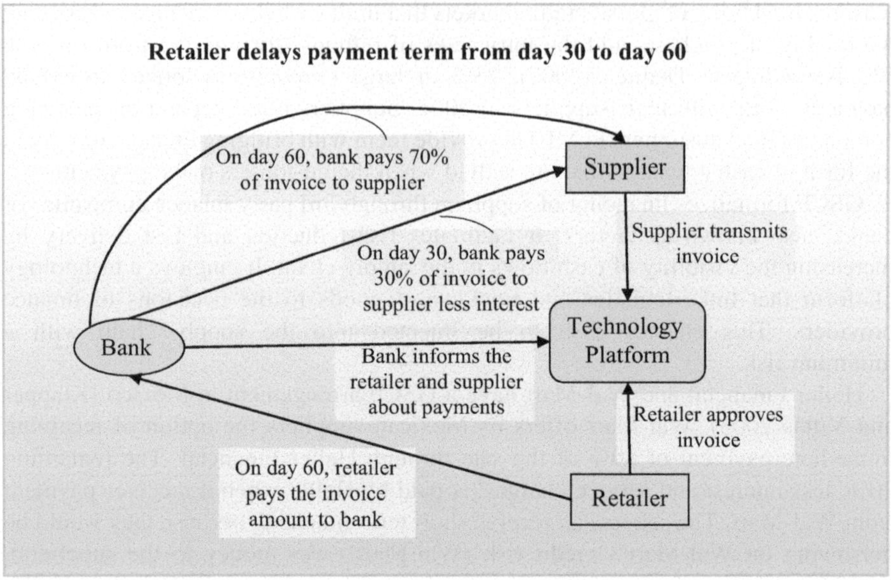

Fig. 7.15 Global supply chain finance

The buyer benefits because from an increased stability of the supply chain due to the indirect support of the supplier by the buyer. It reduces the risks of supplier's bankruptcy and its failure to deliver. Supplier benefits from lower finance charges because of the better credit rating of the buyer. In addition, the technology platform generates benefits for all parties; it can handle a large number of orders in a short time creating scale economy, and it increases information transparency leading to better coordination and quick response. The bank gains because it is not required to carry high capital reserves as per solvency requirements of the central bank; buyer's higher credit rating permits the loans to be classified differently. In addition, setting up GSCF service creates enhanced opportunities for banks to offer additional supply chain financing products to same clients (cross-selling).

As the GSCF is a win-win-win for the buyer, supplier, and financier, it is expected to evolve further to support financial services that are triggered by events at every stage of a physical supply chain. For instance, banks and finance companies may use *dynamic pricing* in purchasing outstanding invoices. They will be able to dynamically lower the interest rates on loans outstanding, as the goods get closer to their destination; the technology platform can provide the tracking information captured in the order processing system. Thus, there will be a convergence of cash management and trade-finance, as procurement, invoicing, and financing are integrated within a single platform.

According to Hurtrez and Salvadori (2010), GSCF has the potential of unlocking as much as $500 billion of liquidity by accelerating the order-to-cash cycle for suppliers and extending the procurement-to-pay cycles for buyers. For this to happen, however, technology platforms that facilitate global exchange of data among trading partners must work on a common set of standards. The authors suggest that SWIFT can be a potential standard as it aims to integrate purchase orders with financial instruments.

7.6.2 Supply Chain Equilibrium with Financing

Consider a chain comprising a retailer in ME, a manufacturer in EE, and a bank in ME. Demand for the product is ξ, where ξ is a random variable. Clearly, there are different ways of managing the demand risk: (i) only retailer bears the risk, (ii) only bank bears the risk, (iii) only supplier bears the risk, and (iv) the three parties share the risk. We consider a simple case where only the supplier bears the risk. This implies that the retailer pays the bank an amount equal to the minimum of her sales revenue and invoice value. In turn, the bank pays the supplier what it receives from the retailer less the service charges.

On day 1, retailer orders a quantity Q of the product at wholesale unit price w with payment due on day 60. She sells the product at unit price p. As shown in Fig. 7.15, on day 30 the bank pays the supplier 30% of the invoice value. On day 60, retailer pays the invoice amount wQ to the bank, only if her revenue $p\xi$ exceeds wQ. Otherwise, she only pays the amount $p\xi$. Therefore, retailer's expected payment to the bank is expressed as $E_\xi Min(p\xi, wQ)$. As the retailer can only sell an amount that does not exceed either supply or demand, her expected revenue will be $pE_\xi\{Min(Q,\xi)\}$. Hence, retailer's expected profit will be $pE_\xi\{Min(Q,\xi)\} - E_\xi Min(p\xi, wQ)$.

The bank pays a fraction α of the invoice value, less interest charge, to the manufacturer on day 30. The bank pays the remainder $wQ(1-\alpha)$ on day 60 to the supplier, if $p\xi > wQ$; otherwise, it pays $p\xi - \alpha wQ$ provided it is nonnegative, that is, it pays $Max(p\xi - \alpha wQ, 0)$. Assuming an annual interest rate of r ($r < 1$), the bank would pay an amount $wQ\alpha\left(1-\frac{r}{12}\right)$ to supplier on day 30, where $r/12$ is the interest rate for 1 month (30 days). The bank's expected payment to the supplier on day 60 would equal $wQ(1-\alpha)P(p\xi \geq wQ) + E_{p\xi \leq wQ} Max(p\xi - \alpha wQ, 0)$, where $P(p\xi \geq wQ)$ implies probability of revenue exceeding wQ. As it receives the amount $E_\xi Min(p\xi, wQ)$ from the retailer, the bank's expected profit would be $E_\xi Min(p\xi, wQ) - wQ\alpha\left(1-\frac{r}{12}\right) - wQ(1-\alpha)P(p\xi \geq wQ) - E_{p\xi \leq wQ} Max(p\xi - \alpha wQ, 0)$.

The manufacturer incurs a cost cQ. On day 30, he receives from the bank an amount αwQ less interest charge for 30 days. If $p\xi > wQ$ the bank pays $wQ(1-\alpha)$ to the supplier on day 60 and, if not, it pays $p\xi - \alpha wQ$ provided it is nonnegative. It follows from the discussion above that the expected value of the payment received by the supplier would equal the amount $wQ\alpha\left(1-\frac{r}{12}\right) + wQ(1-\alpha) P(p\xi \geq wQ) + E_{p\xi \leq wQ} Max(p\xi - \alpha wQ, 0)$.

The sequence of decisions is as follows. For a given w, the retailer determines her reaction function by maximizing her expected profit in Q. The manufacturer's reaction function is then obtained by maximizing his expected profit in w, subject to the retailer's reaction function. Finally, the bank maximizes its expected profit with respect to the bank rate r, subject to the reaction functions of the retailer (in Q) and manufacturer (in w). Note that the bank's optimal expected profit can be negative. The profit functions of the retailer, bank, and manufacturer are shown in Appendix 3.

Example. Assume that demand of the product equals 200, 400, 600, 800, and 1,000 with associated probabilities 0.1, 0.2, 0.4, 0.2, 0.1. Unit sales price is \$20, supplier's wholesale unit price is \$10 and his unit manufacturing cost is \$5. Assume order quantity of 500. Assume that the bank pays 30% of the invoice value on day 30 and the balance on day 60. The bank charges interest at 12% per annum. Clearly, the invoice value will be \$5,000, and the potential revenue of the retailer would vary between \$4,000 and \$20,000.

Retailer's expected revenue = 20{200(0.1) + 400(0.2) + 500(0.7)} = \$9,000

Note that the retailer's revenue will be below the invoice value of \$5,000 only when demand equals 200 (with probability 0.1). Therefore retailer will pay \$5,000 with probability 0.9 and \$4,000 with probability 0.1, the expected payment being \$4,900. Therefore,

Retailer's expected profit = 9,000 – 4,900 = \$4,100

The bank receives a payment of \$4,900 from the retailer on day 60. On day 30, it pays the supplier 30% of \$5,000 (=\$1,500) less interest charge of 1,500(0.12/12) = \$150. Therefore payment on day 30 equals 1,500 – 150 = \$1,350. On day 60 the bank pays the supplier the balance of the invoice value of \$3,500 if it receives full payment from the retailer; the probability of which is 0.9. If not (with probability 0.1), the bank pays the amount received from retailer (=\$4,000) less 30% of invoice value (=\$1,500) =4,000 – 1,500 = \$2,500. Therefore,

Bank's expected payment to supplier = 1,350 + (3,500)(0.9) + (2,500)(0.1) = \$4,750

Bank's expected profit = 4,900 – 4,750 = \$150, and return on investment = 10% per annum. This is a good financing deal for the bank, as its cost of capital is expected to be lower.

The supplier's expected receivable from the bank is \$4,750 and his manufacturing cost is \$2,500. Therefore, supplier's expected profit is \$2,250.

The bank takes no risk in this example, as its initial payment of \$1,500 is fully covered by the retailer's payment even if her revenue hits a low point of \$4,000. However, under certain conditions the bank's profit can be negative. Consider the scenario with the retailer ordering 600 units (instead of 500) at a wholesale price \$16 (instead of \$10) and the bank paying 70% of invoice on day 30. Now the initial payment to supplier will be \$6,720. As the two low revenue points of the retailer (\$4,000 and \$6,000) fall below the initial payment, the bank will be exposed to a risk of 30% of losing money.

Note that because of the bank-charges, the supplier's profit is lower than what it could be. To recoup some of his loss, the supplier may increase his wholesale price and, in response, the retailer may decrease her order quantity. As a consequence, profits of both the retailer and the bank would decrease. As is well known in game-theory, any party moving away from the equilibrium point will be worse off.

Appendix 1

The company's revenue with Q units of the assembled product can be expressed as $\pi = p\left\{\int_{\xi\le Q} \xi f(\xi)d\xi + Q\int_{\xi\ge Q} f(\xi)d\xi\right\}$, where $Q = Min(\gamma_E q_E, \gamma_M q_M)$, and p is the unit sales price. Note that $Q = \gamma_E q_E$ if $\gamma_E \le \gamma_M q_M / q_E$, and $Q = \gamma_M q_M$ if $\gamma_M \le \gamma_E q_E / q_M$. Therefore, with $g_E(\cdot)$ and $g_M(\cdot)$ as the pdf of γ_E and γ_M respectively, the expected revenue is expressed as

$$\text{Rev} = p\left[\underset{\gamma_M}{E}\ \underset{\gamma_E\le\frac{\gamma_M q_M}{q_E}}{E}\ \underset{\xi}{E}Min\{\gamma_E q_E, \xi\} + \underset{\gamma_E}{E}\ \underset{\gamma_M\le\frac{\gamma_E q_E}{q_M}}{E}\ \underset{\xi}{E}Min\{\gamma_M q_M, \xi\}\right]$$

Assuming unit procurement costs of c_E and c_M, and unit assembly cost of c_A, the component procurement cost is expressed as $c_E q_E E_{\gamma_E}\gamma_E + c_M q_M E_{\gamma_M}\gamma_M$. Next, note that the expected number of units assembled is expressed as $q_E\left\{E_{\gamma_M}\left(E_{\gamma_E\le\frac{\gamma_M q_M}{q_E}}\gamma_E\right)\right\} + q_M\left\{E_{\gamma_E}\left(E_{\gamma_M\le\frac{\gamma_E q_E}{q_M}}\gamma_M\right)\right\}$. Therefore, the cost of assembly would be $c_A\left[q_E\left\{E_{\gamma_M}\left(E_{\gamma_E\le\frac{\gamma_M q_M}{q_E}}\gamma_E\right)\right\} + q_M\left\{E_{\gamma_E}\left(E_{\gamma_M\le\frac{\gamma_E q_E}{q_M}}\gamma_M\right)\right\}\right]$.

Therefore, the expected profit would be

$$\pi = \left| \begin{array}{l} p\left[\underset{\gamma_M}{E}\ \underset{\gamma_E\le\frac{\gamma_M q_M}{q_E}}{E}\ \underset{\xi}{E}Min\{\gamma_E q_E,\xi\} + \underset{\gamma_E}{E}\ \underset{\gamma_M\le\frac{\gamma_E q_E}{q_M}}{E}\ \underset{\xi}{E}Min\{\gamma_M q_M,\xi\}\right] \\ -\left\{c_E q_E \underset{\gamma_E}{E}\gamma_E + c_M q_M \underset{\gamma_M}{E}\gamma_M\right\} - c_A\left[q_E\left\{\underset{\gamma_M}{E}\left(\underset{\gamma_E\le\frac{\gamma_M q_M}{q_E}}{E}\gamma_E\right)\right\} + q_M\left\{\underset{\gamma_E}{E}\left(\underset{\gamma_M\le\frac{\gamma_E q_E}{q_M}}{E}\gamma_M\right)\right\}\right] \end{array} \right.$$

Appendix 2

For given w_1, w_2, and R, retailer determines her reaction function (Stackelberg game) by maximizing her profit in Q. Supplier 1 determines w_1, and supplier 2 determines w_2 and R by maximizing their respective profit functions, subject to the retailer's reaction function.

The retailer's expected profit is expressed as

$$\pi = p\left\{ \int_{\xi \le Q+R} \xi f(\xi) + (Q+R) \int_{\xi \ge Q+R} f(\xi) \right\} - w_2\left\{ \int_{Q \le \xi \le Q+R} (\xi - Q) f(\xi) + R \int_{\xi \ge Q+R} f(\xi) \right\} - w_1 Q\ .$$

From the first order condition in Q, we express retailer's decision as $(p - w_2)F(Q+R) + w_2 F(Q) = p - w_1$.

Supplier 1's profit is expressed as $\pi_1 = (w_1 - c_1)Q$. From first order condition in w_1, given the retailer's reaction function $(p - w_2)F(Q+R) + w_2F(Q) = p - w_1$, we have $w_1 = w_2 Q f(Q) + (p - w_2) Q f(Q+R) + c_1$.

Supplier 2's profit is written as $\pi_2 = w_2\left\{ \int_{Q \le \xi \le Q+R} (\xi - Q) f(\xi) + R \int_{\xi \ge Q+R} f(\xi) \right\}$ $-c_2 R$, so that from the first order condition in w_2, given the buyer's reaction function, is expressed as

$$\frac{w_2}{(p - w_2) f(Q+R) + w_2 f(Q)} = \frac{\int_{Q \le \xi \le Q+R} (\xi - Q) f(\xi) + R \int_{\xi \ge Q+R} f(\xi)}{\{F(Q+R) - F(Q)\}^2}\ .$$

Similarly, the first order condition in R leads to,

$$\frac{w_2 f(Q) F(Q+R) + (p - w_2) f(Q+R) F(Q)}{(p - w_2) f(Q+R) + w_2 f(Q)} = 1 - \frac{c_2}{w_2}\ .$$

Appendix 3

Assume $F(\xi)$ to be the continuous distribution of demand ξ.

Retailer's Profit

As the retailer can only sell an amount that does not exceed either supply or demand, her expected revenue will be $p\underset{\xi}{E}\{Min(Q,\xi)\}$. Retailer's expected payment to the bank is expressed as $\underset{\xi}{E}Min(p\xi,wQ)$. Hence, retailer's expected profit will be $p\underset{\xi}{E}\{Min(Q,\xi)\} - \underset{\xi}{E}Min(p\xi,wQ)$.

Bank's Profit

Amount received from the retailer $= \int\limits_{\xi \le \frac{wQ}{p}} p\xi f(\xi)d\xi + wQ \int\limits_{\xi \ge \frac{wQ}{p}} f(\xi)d\xi$

Payment to the supplier on day 30 $= \alpha wQ\left(1 - \frac{r}{12}\right)$

Payment to the supplier on day 60 $= wQ(1-\alpha) \int\limits_{\xi \ge \frac{wQ}{p}} f(\xi)d\xi + \int\limits_{\xi \le \frac{wQ}{p}} Max(p\xi - \alpha wQ, 0)f(\xi)d\xi$

Therefore, the Bank's expected profit is written as

$$\pi_B = \left| \begin{array}{l} \left\{ \int\limits_{\xi \le \frac{wQ}{p}} p\xi f(\xi)d\xi + wQ \int\limits_{\xi \ge \frac{wQ}{p}} f(\xi)d\xi \right\} \\ -\left\{ \alpha wQ\left(1 - \frac{r}{12}\right) + wQ(1-\alpha)\int\limits_{\xi \ge \frac{wQ}{p}} f(\xi)d\xi + \int\limits_{\xi \le \frac{wQ}{p}} Max(p\xi - \alpha wQ, 0)f(\xi)d\xi \right\} \end{array} \right.$$

Supplier's Profit

Manufacturer's profit would be equal to the amount received from the bank less manufacturing cost. Therefore the supplier's expected profit is expressed as

$$\pi_M = \left\{ \alpha w Q(1-r) + wQ(1-\alpha) \int\limits_{\xi \geq \frac{w}{p}Q} f(\xi) d\xi + \int\limits_{\xi \leq \frac{w}{p}Q} (p\xi - \alpha w Q)^+ f(\xi) d\xi \right\} - cQ$$

References

Baillie, D. (2007). *Unilever in India*, South Asia Unilever, Mumbai. http://www.unilever.com/images/ir_Unilever-in-India_tcm13-113961.pdf

Blanco, E. (2009). *Winning in emerging markets: Five key supply chain capabilities*. Research paper from the Center for Transportation & Logistics, MIT. http://ctl.mit.edu/research

Demica (2007). *Steady supply*, 8, Demica research series. pp. 1–7. http://www.demica.com/images/PDFs/supply_chain_finance_steady_supply.pdf

DOT (2006). *Freight in America: A new national picture*. Bureau of Transportation Statistics. http://www.bts.gov/publications/freight_in_america/pdf/entire.pdf

Gonzalez, J., Guasch, J., & Serebrisky, T. (2007). *Latin America: Addressing high logistics costs and poor infrastructure for merchandise transportation and trade facilitation*. The World Bank. http://www.iadb.org/res/ConsultaSanJose/files/Infrastructure_Guasch_SP_Final.pdf

Guasch, J., & Kogan, J. (2006). Inventories and logistic costs in developing countries: Levels and determinants – A red flag for competitiveness and growth. *Revista de la Competencia y e la Propiedad Intelectual*, año 1, n.o 1, Lima, Perú. Policy Research Working Paper 2552. Washington DC: World Bank. http://www.iadb.org/res/ConsultaSanJose/files/Infrastructure_Guasch_SP_Final.pdf

Hayes, R., Pisano, G., Upton, D., & Wheelwright, S. (2004). *Operations, strategy, and technology: Pursuing the competitive edge*. New York: Wiley.

Hurtrez, N., & Salvadori, M. (2010). Supply chain finance: From myth to reality. *McKinsey Analysis*, October, pp. 22–28. http://www.mckinsey.com/clientservice/Financial_Services/Knowledge_Highlights/Recent_Reports/~/media/Reports/Financial_Services/MoP9_Supply%20chain%20finance.ashx

Khanna, T., Palepu, K., & Sinha, J. (2005). Strategies that fit emerging markets. *Harvard Business Review, 83*, 63–76.

Klapper, L., & Vittas, D. (2003). *The use of reverse factoring*. Washington, DC: World Bank.

KW. (2011). *Winning in two worlds: Supply chain flexibility*. Knowledge@Wharton. http://knowledge.wharton.upenn.edu/article.cfm?articleid=2683

Lamoureux, J., & Evans, T. (2011). *Supply chain finance: A new means to support the competitiveness and resilience of global value chains*. Export Development, Canada. http://www.international.gc.ca/economist-economiste/assets/pdfs/research/TPR_2011_GVC/12_Lamoureux_and_Evans_e_FINAL.pdf

Lee, H. (2002). Aligning supply chain strategies with product uncertainties. *California Management Review, 44*, 105–116.

Lohr, S. (2011, March 16). Stress test for the global supply chain. *The New York Times, Business Day*.

Malik, Y., Niemeyer, A., & Ruwadi, B. (2011). Building the supply chain of the future. *McKinsey Quarterly, 1*, 1–10. January. http://www.mckinsey.com/insights/operations/building_the_supply_chain_of_the_future

Neuwirth, B. (2012). *Marketing channel strategies in rural emerging markets: Unlocking business potential*. Chicago: Northwestern University.

OECD. (2002). Transport logistics: Shared solutions to common challenges. Paris: OECD. http://www.internationaltransportforum.org/pub/pdf/02LogisticsE.pdf

PCAST. (2011). *Report to the president on ensuring American leadership in advanced manufacturing*. Executive office of the president president's council of advisors on science and technology. http://www.whitehouse.gov/sites/default/files/microsites/ostp/pcast-advanced-manufacturing-june2011.pdf

Rekik, Y., Sahin, E., & Dallery, Y. (2010). *A comprehensive analysis of the newsvendor model with unreliable supply*. Working paper, Laboratoire Genie Industriel, Ecole Centrale Paris, France.

Terpstra, V., & Yu, C. (2001). Piggybacking: A quick road to internationalization. *International Marketing Review, 7*, 52–63.

8 Humanitarian Relief Chain

8.1 Introduction

Onset of a major disaster requires immediate effort to save human lives. The International Strategy for Disaster Reduction (ISDR) estimates that the global cost of natural disasters would exceed $300 billion annually by the year 2050 (European Commission 2003). The most deadly disasters since 1900 are shown in Fig. 8.1.

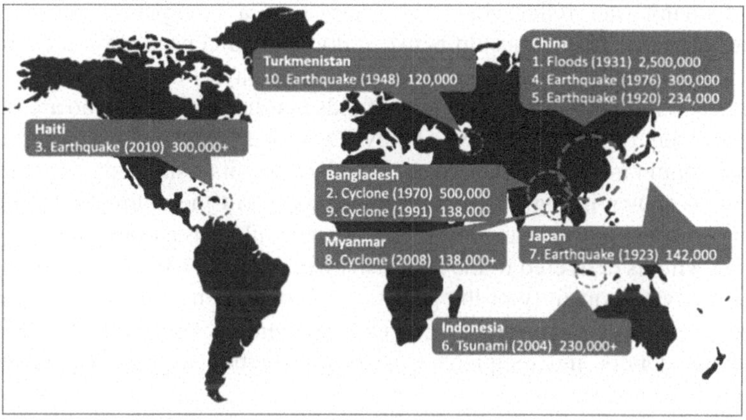

Fig. 8.1 Most deadly natural disasters since 1900 (Source: Bruno et al. 2011)

As the disasters are not frequent, it is not cost effective for individual countries or organizations to hold excessive amounts of resources in reserve for relief effort. Organizations often pool their resources to respond to major disasters. Another factor with major implications is that organizations do not possess same capabilities: while some have funds, others have expertise in logistics, medical care, aid dispensing, evacuation, and rehabilitation. Therefore, multiple organizations comprising the UN, independent relief organizations, charities, national governments, military, and private sector companies are involved in humanitarian relief, as shown in

A.K. Chakravarty, *Supply Chain Transformation*, Springer Texts in Business and Economics, DOI 10.1007/978-3-642-41911-9_8,

Fig. 8.2. The pooled resources such as money, expertise, equipment, supplies, and transportation vehicles, are channeled to specific disaster sites. Obviously, this effort must be coordinated for it to function smoothly. The cluster of entities can be looked upon as a network (suppliers, service providers, and "customers"), and is better known as the "humanitarian supply chain".

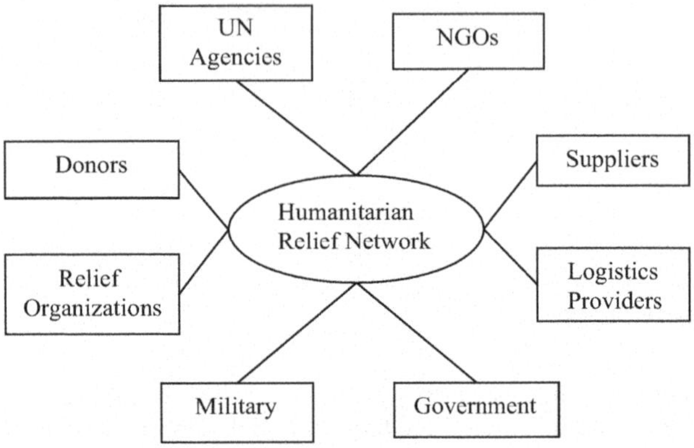

Fig. 8.2 Humanitarian relief network

The humanitarian relief chain, in contrast to a commercial supply chain, embodies an *indirect* relationship between donors and consumers – the consumer (aid recipient) neither places purchase orders nor pays for the delivered service (Binder and Witte 2007). Donors provide funds to relief providers who are typically non-government organizations (NGOs). NGOs procure supplies from international and local suppliers, and deliver them to aid recipients using 3rd party logistics providers, as shown in Fig. 8.3. An important aspect of the indirect relationship is that while the donors (public and private) finance all transactions, they have little control on what is delivered to the recipients who, in turn, have no role in monitoring the quality and quantity of the goods. As a consequence, the NGOs have little incentive to deliver good-quality services efficiently. The indirect relationship between the donors and recipients creates information asymmetry, causing the supply chain to underperform.

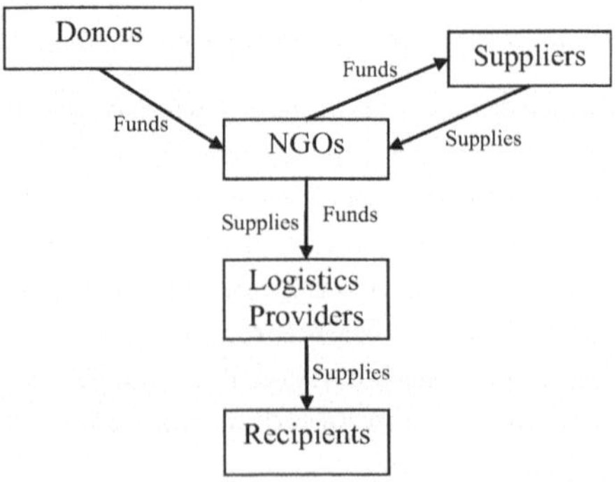

Fig. 8.3 Flows in humanitarian relief chain

Note that the relief chain must possess certain capabilities such as, managing funds, procurement and delivery of supplies, disaster mitigation, preparation, evacuation, and managing a set of disparate organizations. These are distinct activities with diverse goals and objectives, relationships, and implementation practices. For example, the disaster fund management process must begin well before a disaster strikes. It must identify the donors and match them with relief providers, and ensure efficient allocation and channeling of funds. The activities – mitigation, preparation, and evacuation – help in shaping the demand for supplies. Support activities include protection of facilities and supply routes, raising awareness, mobilization of transportation vehicles, and management of traffic flow. Logistics activities include prepositioning of inventory at strategic locations, setting up warehouses, rapid delivery of supplies if needed, managing a mix of transportation modes, transporting equipment and workers, repairing communication and transportation infrastructure, devising procurement contracts, and using 3rd party service providers if feasible. Most such logistics efforts must be coordinated with NGOs, government and international agencies, and the military.

8.2 Relief Chain Capabilities

The external drivers of a relief chain are markedly different from those of a commercial supply chain, as they are primarily defined by social values. The possible external drivers for a relief chain are shown in Fig. 8.4.

Fig. 8.4 External drivers

Social value, accruing from saving lives or minimizing suffering, is the prime driver of humanitarian relief. While lives saved immediately after a disaster generates the largest social value, lives saved days after the disaster continue to generate value. However, the social value diminishes if the victims suffer pain and distress waiting for relief; the decay in unit social value increases in waiting time. Clearly relief supplies, arriving after the surviving victims are dead, generate no social value.

While faster delivery saves more lives by decreasing the wait time for relief, it can be expensive given that faster transportation modes are priced higher. Similarly, sending larger quantities of relief supplies in real time can save more lives. However, the *expected social surplus* may decrease if large inventories remain unused because investment in inventory cannot be fully recovered. At locations that are vulnerable to disasters, the likelihood of having unused inventory is low. Hence, prepositioning of inventory assumes added urgency in south Florida, which is often in the path of hurricanes.

Possible consequences of a disaster and losses, both tangible and intangible, are listed in Table 8.1. Many proactive measures can be instituted to minimize the tangible losses because they are quantifiable. Intangible losses such as grief and suffering, on the other hand, are not directly measurable; they are managed through recovery programs after the disaster. The effects may linger for years, however.

Table 8.1 Disaster damage

Damage type	Tangible losses	Intangible losses
Death	Number of productive people	Grief for the dead, and anxiety for the surviving
Injury	Medical cost, loss of productive days	Cost of suffering
Physical damage	Replacement and repair cost	Cultural losses
Cost of relief	Cost of preparation and mobilization	Stress of relief work
Social disruption	Temporary housing, transportation	Social contacts and cohesion
Environmental impact	Clean up and repair cost	Health risks

As mentioned earlier, protection, preparation, and evacuation are the means of shaping demand, as they reduce demand for relief supplies. Evacuation can create a positive or negative social value depending on whether the site from which evacuation takes place is actually hit by the disaster. For example, when people are evacuated from a region in anticipation of a disaster and the disaster does not materialize, a negative value is created because of the inconvenience of evacuation.

The internal drivers, in contrast, attempt to maximize the effectiveness of the relief effort. In a typical humanitarian relief scenario, the prevailing chaotic conditions increase the likelihood of resources-wastage, and therefore the role of the internal drivers cannot be overstated. Typical internal drivers are shown in Fig. 8.5.

In Fig. 8.5, the three major logistics functions are procurement, warehousing, and distribution of supplies, discussed in Chap. 5. Efficiency in the allocation and monitoring of funds is another critical internal driver, it being the fuel for the relief-engine. Note that the effectiveness of relief effort also depends on making sure that supplies are not misused, whether they are in warehouses or distributed to the

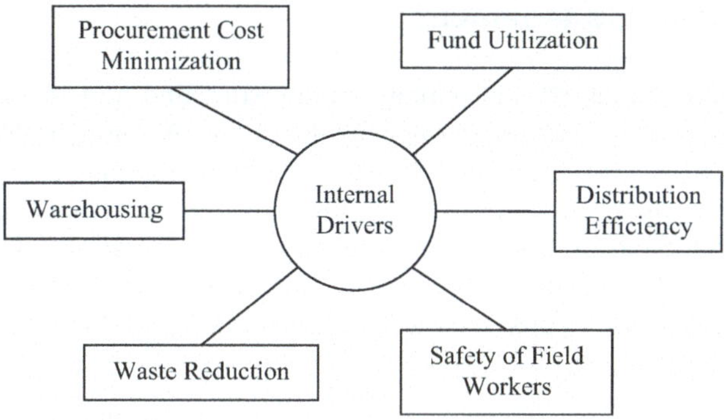

Fig. 8.5 Internal drivers

victims. Safety of field workers and relief volunteers is critical, as no relief effort would succeed otherwise.

Next in Fig. 8.6, we illustrate the capabilities that the humanitarian relief chain should possess, using the framework developed in Chap. 1. Note that the cluster of capabilities, labeled as "demand management" refers to *external drivers*, whereas the other three clusters labeled as "fund management", "logistics", and "collaboration" relate to *internal drivers*.

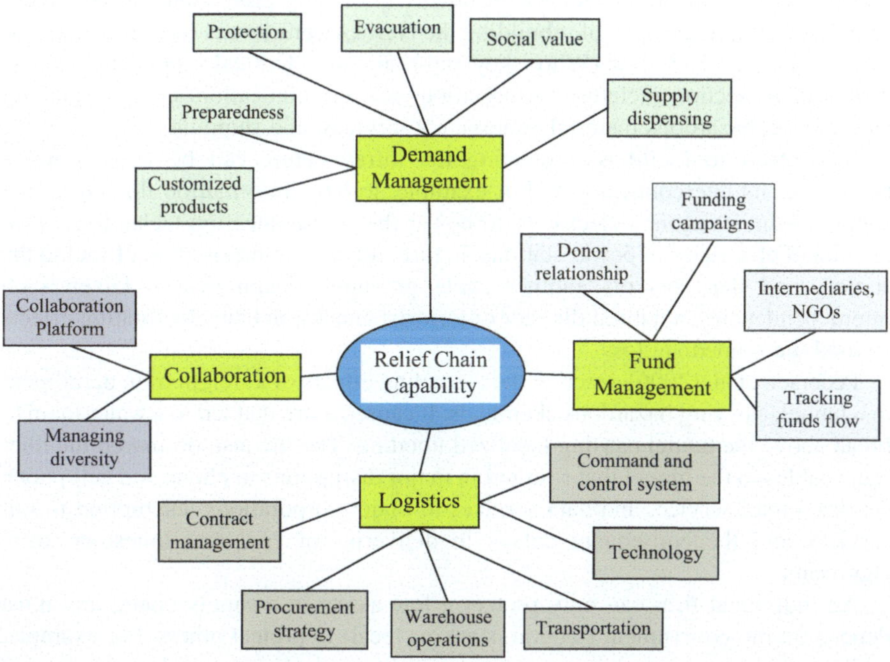

Fig. 8.6 Relief chain capabilities

8.3 Demand Management

While both the commercial and humanitarian chains find best ways of satisfying demand, they shape the demand in opposite directions. A commercial supply chain attempts to increase customer demand, while the humanitarian relief chain aims at decreasing the demand of supplies. Note that proactive measures such as investment in the protection of property and life, advance preparation for disasters, and evacuation can help reduce the negative impact of a disaster. The protection and preparedness functions clearly require new capabilities, not found in a contemporary supply chain.

8.3.1 Protection

Protection minimizes the need for disaster relief. It mitigates the overall risk to supply chain assets, networks, and processes. It may include a wide range of activities, such as building resilience, incorporating redundancy in the system, designing facilities that are resistant to hazards, promoting workforce security, and implementing cyber security. Some examples are fortification of production and service facilities (as in terrorism), strengthening of levees for flood control (as in hurricanes), discouraging construction of multi storey buildings (as in earthquakes), monitoring violations against building codes, and tightening regulations to fend off financial melt-downs. Providing protection can be expensive. For example, the cost of rebuilding the New Orleans levees (post Katrina) was estimated to be $10 billion (Whoriskey and Hsu 2006). Critical supply chain assets that need protection include transportation, telecommunication, energy, banking and finance, hazardous material, emergency services, and shipping.

In contrast to facilities, protecting the infrastructure can be very complex because of its interconnectivity. For example, to have information flowing in the supply chain, it requires electricity to power the communication technology. Production of electricity in power stations, in turn, requires transportation of fuel to the station, and that sets off another cycle of supply chain events. Given such interdependencies, a natural disaster or terrorist attack can have debilitating effects beyond the immediate loss.

Pedersen et al. (2006) describe the cascading effects of a freight train derailment in a tunnel, carrying hazardous chemicals. It caused a fire that led to a water main to break above the tunnel causing localized flooding. The fire also destroyed the fiber optic cables in the tunnel that resulted in major disruptions to phone and cell phone service, email service, and data services to major corporations. Disruption to rail services and the consequent delays in deliveries of coal and limestone were significant.

An individual firm can only protect a few assets in a supply chain, and must depend on the government (federal, state, or local) to protect others. For example, while a firm can protect its production facility, warehouses, and transportation vehicles; it cannot be expected to protect the shared assets such as the power supply grid, and the telecommunication network. In its 2020 Disaster Prevention Plan, the Vietnamese government proposes to spend $18 billion – around $13 billion for structural measures such as building reservoirs, dams and dykes; and $5 billion for non-structural measures (Bruno et al. 2011).

A firm's investment in protection reflects risk/reward tradeoff that is based on the cost of protecting an asset, the potential loss if the asset is damaged, and the vulnerability of the asset. While the firm must assess the cost of protection and potential loss, the government can help with information on vulnerability and risk assessment. For example, companies look to the National Hurricane Center (NHS) as a source of information on natural hazard indications, warnings, and threat assessments. The firm can then figure out how much to invest in protecting certain assets. Models can help in quantifying the benefits in terms of reduction in lost lives, social disruption, and financial loss (Chakravarty 2011). A cost-benefit analysis of different options can identify the best set of options.

The procedures and protocols that help in asset protection include measures such as stipulating standards for products and processes for companies to follow, providing scenario-specific guidelines for physical structures, facilitating exchange of best practices, and implementing methodologies for assessing disaster-vulnerability. The structure of a facility can be designed to resist the impact of disasters, depending on the type of hazard, the impact site, type of construction, and performance requirements (WBDG 2012).

The key strategy to protecting a building from high winds caused by tornados, hurricanes, and gust fronts is to design the structure to withstand the expected lateral and uplift forces. For example, roofs and gables must be braced with straps and secured to the foundation, and doors and windows must be protected with shutters. Protection measures against earthquake include "shear wall" structures, braced and moment-resisting frames, diaphragms, and dampers for energy dissipation. According to research reports (Chandler 2010), new construction materials based on substances such as silk, can enhance material performance in high stress situations, because silk has outstanding extensibility and strength.

Flood mitigation techniques include elevating the building, dry flood-proofing of buildings to prevent water entry, wet flood-proofing to enhance the resistance to water damage; and floodwalls to keep water away from the building. The Netherlands is at the forefront of protecting people from floods through a program called "delta works", which guards the estuaries from storm surges by enabling a collection of locks, sluices, channels, bridges, slides, dams, dikes, storm surge barriers and gates to work together (Wonders 2013). Protected by the delta works, the land between Amsterdam and Rotterdam is now heavily industrialized and provides most of the country's supply chain activity.

8.3.2 Preparedness

Disaster preparedness aims at raising awareness so that people may take appropriate precautions, and heed government alerts that may include evacuation.

Governments and organizations help by developing a number of measures such as emergency-warning protocols, drills to test disaster response, first-aid centers near disaster sites, storage of water and food, evacuation plans, and a reliable communication system. Volunteers must be trained in treating diseases, preventing escalation of devastations, and use of equipment. Essential service providers such as the police, fire services, medical services, and the military, must be kept in a state of readiness. All essential services personnel and volunteers must be well aware of the emergency command structure.

Raising awareness of the disaster damage is critical. It can be achieved through the use of media, education, and social committees. The media can help build awareness through campaigns on the risks of catastrophes, and benefits of protection and preparedness. Schools may teach appropriate response behavior. For example, minor adaptations of behavior such as the tendency of avoidance and denial can significantly decrease the damage to society. Many relief agencies possess considerable expertise and capacity to help in countries, especially with high casualty rates, through training courses (Bruno et al. 2011).

8.3.2.1 Financial Preparedness

The goal is to increase risk sharing among multiple agencies to alleviate the risk on the society and governments. There are four methods of risk transfer: traditional insurance, micro insurance that protects low-income regions from *focused* risks, catastrophe bonds, and catastrophe pools where premium is based on actuarial peril-index and payout is within 15 days of an event (Bruno et al. 2011). There are innovative insurance solutions that allow a *finite* cover of *large* risks that cannot be absorbed by traditional indemnity covers. Neighboring countries may engage in pooling of commodities and equipment (pumps, building materials, mobile hospitals etc.) to avoid price-spikes. The international pools are based on peril-index models so that the likelihood of two regions needing the equipment simultaneously is low.

8.3.3 Evacuation

Evacuation implies moving people out of the harm's way to safe locations. It can be disruptive for the evacuees and expensive for the state. It has the potential to create anxiety in the community, and threatens safety and security of assets. But it saves lives and minimizes the need for a very expensive post disaster relief effort. Dependent on the nature of the threat, an alternative could be "shelter in place" (Queensland Government 2010).

Evacuation plans must be based on time available for evacuation. A timeline that indicates the allocated time for each phase of the evacuation process can be critical for coordination. The capacity of evacuation routes may vary depending on road conditions which can be standardized into categories such as, normal, disrupted, and blocked. Travel times can be computed for different scenarios of road

Fig. 8.7 Travel time estimation

	Road Condition		
	Normal	Disrupted	Blocked
Rural	High	High	Very High
Urban	Moderate	Moderately High	Very High
Highway	Moderate	High	Very High
Expressway	Low	Moderate	High

conditions and road classes (urban, rural and highway, expressway) as shown in Fig. 8.7.

Evacuation route should be determined based on the number of potential evacuees, closeness of the shelter to the route, predicted path of the disaster, and the travel time on the route. Traffic management strategy should be aimed at easing congestion, the ease of managing traffic flow, and the timely movement of evacuees. This may include "contra flow" on the expressways where *all* traffic lanes are directed away from the disaster site. Evacuation planning should include modeling and analysis of proposed traffic management strategies to predict evacuation feasibility. Evacuation centers should be predetermined and clearly categorized for suitability. Shelters should have a designated "safe haven", a section of the building that is reinforced to protect against specific hazards, such as fire, smoke or structural collapse.

Since hurricane Katrina, the importance of evacuation planning has increased several folds. Now sophisticated techniques are used to incorporate multi-modal transportation networks. Military airlift resources are used to facilitate evacuating people out of the affected area. More complex evacuation planning is being considered, such as using elementary schools as rally points for evacuation.

8.3.4 Analysis of Demand Shaping

As shown in Fig. 8.8, while the investment in demand shaping efforts, as discussed above, increases cost before a disaster, it substantially reduces the impact of catastrophe after the disaster (Bruno et al. 2011).

Assume, in the context of Fig. 8.8, that a *yearly* investment of $\$x$ in demand shaping (protection, preparation, and evacuation) results in post disaster saving of $\$y$, y is an increasing function in x, $y = f(x)$. If the probability of disaster is p, the expected saving will be $py - x$. Therefore, investment in demand shaping is justified if $y > x/p$.

Now consider investment every T years. Using Bernoulli trials, it can be verified that the probability of n disasters striking in T ($n \leq T$) years can be expressed as

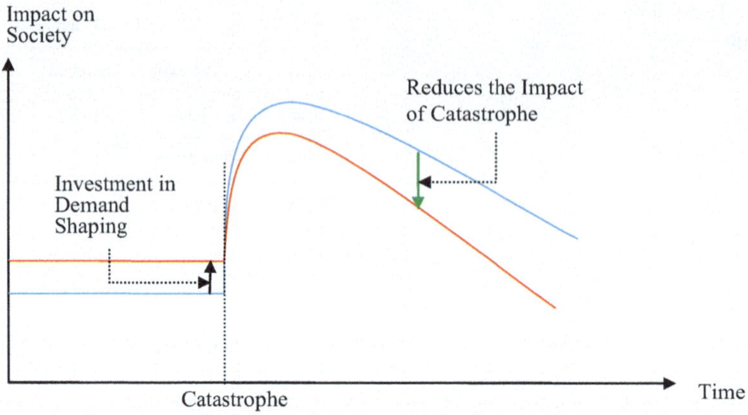

Fig. 8.8 Impact of preparation and protection

$\binom{T}{n}p^n(1-p)^{T-n}$. The expected payoff of this investment strategy can be written as $y\sum_{n=0}^{T}\left\{n\binom{T}{n}p^n(1-p)^{T-n}\right\}-x$.

In many scenarios the effectiveness of demand shaping, such as protection levees, decays in time. Therefore, the post disaster saving, from an investment in a particular year, would not be the same in all future years. Hence, we define the post-disaster saving for a disaster in year r as y_r, where $y_r = f(x,\theta_r)$. If θ_r is linear, a possible function would be $\theta_r = 1 - \frac{r}{T}$. Clearly, y_r is an increasing function in x and θ_r. We would also need to account for the disaster by the year r of its occurrence.

Consider an example with planning horizon, $T = 5$. To keep exposition simple, assume that no more than one disaster strikes in any period. It is clear that there can be from zero to five disasters by the end of five periods. As before, we assume the investment in demand shaping is x, and the probability of a disaster in any period is p.

Therefore, if there are no disasters in five periods, the post-disaster saving will be zero. If there is only one disaster, it can occur in any of the five periods, and the expected saving would be $p(1-p)^4(y_1+y_2+y_3+y_4+y_5)$. Next, for the case of exactly two disasters, there can be ten different ways two disasters can occur in five periods, as shown in Table 8.2.

Total expected post-disaster saving $= 4p^2(1-p)^3(y_1+y_2+y_3+y_4+y_5)$. Proceeding as in Table 8.2 it is established that for the scenarios of 3, 4, and 5 disasters the expected savings would be $p^3(1-p)^2(5y_1+6y_2+6y_3+5y_4+5y_5)$, $p^4(1-p)(3y_1+3y_2+4y_3+3y_4+3y_5)$, and $p^5(y_1+y_2+y_3+y_4+y_5)$, respectively, as established in Appendix 1. Therefore the total expected net saving is expressed as

Table 8.2 Occurrence of disasters

Year of disaster					
1	2	3	4	5	Expected post disaster saving
Disaster 1	Disaster 2				$p^2(1-p)^3(y_1+y_2)$
Disaster 1		Disaster 2			$p^2(1-p)^3(y_1+y_3)$
Disaster 1			Disaster 2		$p^2(1-p)^3(y_1+y_4)$
Disaster 1				Disaster 2	$p^2(1-p)^3(y_1+y_5)$
	Disaster 1	Disaster 2			$p^2(1-p)^3(y_2+y_3)$
	Disaster 1		Disaster 2		$p^2(1-p)^3(y_2+y_4)$
	Disaster 1			Disaster 2	$p^2(1-p)^3(y_2+y_5)$
		Disaster 1	Disaster 2		$p^2(1-p)^3(y_3+y_4)$
		Disaster 1		Disaster 2	$p^2(1-p)^3(y_3+y_5)$
			Disaster 1	Disaster 2	$p^2(1-p)^3(y_4+y_5)$

$$\pi(x,T) = \left| \begin{array}{l} \left\{p(1-p)^4 + 4p^2(1-p)^3 + p^5\right\}(y_1+y_2+y_3+y_4+y_5) \\ +p^3(1-p)^2(5y_1+6y_2+6y_3+5y_4+5y_5) + p^4(1-p) \\ (3y_1+3y_2+4y_3+3y_4+3y_5) - x. \end{array} \right.$$

Hence, protection is worthwhile if

$$\left. \begin{array}{l} \left\{p(1-p)^4 + 4p^2(1-p)^3 + p^5\right\}(y_1+y_2+y_3+y_4+y_5) \\ +p^3(1-p)^2(5y_1+6y_2+6y_3+5y_4+5y_5) \\ +p^4(1-p)(3y_1+3y_2+4y_3+3y_4+3y_5) \end{array} \right| > x$$

To maximize $\pi(x,T)$ for a given T, we compute $\pi(x,T)$ for different x, noting that y_r is a function of x, and obtain optimal $x(T)$. The whole procedure is repeated for different values of T to determine the optimal amount of investment, x, in demand shaping and its frequency (T).

8.3.5 Social Value

Social value can be defined, in the context of humanitarian relief, as the amount of resources a community is willing to part with to save a human life. It acts as a surrogate variable standing in for price in a traditional supply chain. In contrast to price, however, it shapes the *satisfied* demand and not demand per se. That is, the satisfied demand increases if the social value imputed by a community increases.

In any disaster, loss of life may occur instantly at the time of disaster (from drowning, fire, and suffocation in debris) or, later, from inadequate and/or delayed delivery of supplies. The value of saving a human life is not easy to determine. Landefeld and Seskin (1982) describe two approaches for valuing human life: human capital (HC), and willingness to pay (WTP). While HC estimates the future income potential of an individual, WTP assesses the amount a community would be willing to spend to save a human life. The authors also describe how statistical surveys can be designed for estimating WTP. In the context of disaster relief, WTP would appear to be more appropriate as a social value. Its value was estimated as $5.8 million (and rising) by FEMA (Burbank 2009). In structuring a strategy for humanitarian response, the social value must be weighed against the cost of relief supplies (food, shelter, clothing, and medicine), and the cost of infrastructure repair.

Another type of social value is defined in terms of the delay in receiving relief. If the victims receive supplies immediately after a disaster, it generates the maximum marginal social value of V (Vlaev et al. 2009). If not, they must wait for the emergency supplies to arrive. This generates the second type of marginal social value $v \leq V$. It follows that v would decrease in the duration of the wait time (for supplies), $\partial v/\partial t < 0$.

Note that while V influences the amount of supplies delivered, the decay in v in time together with the survival rate of the victims would influence how quickly supplies are sent to the disaster site. A plausible explanation for the relatively higher loss of lives in a poor community, for a given disaster intensity, is that the imputed marginal social values in rich and poor communities are not the same.

8.3.6 Customized Products

It is neither practical nor economical to supply disaster victims with items in the same shape and form that most households consume. For example, food is needed for immediate consumption after a disaster, but there is hardly any facility available for processing or preserving the food. Therefore, such items should be made available in pre-packed and ready-to-eat forms. They should also be long-lasting and contain the basic nourishment the victims require for survival. Such emergency kits can be designed for ease of delivery, including air drops. Many retailers now offer disaster relief supply in emergency kits, along with supplies that can be used to build a customized kit. Some of these kits, retailers claim, can last 25 years! Others have begun to supply kits for radios, flashlights, and ponchos.

For emergencies related to biological disasters, product modularization is essential for efficient distribution and dispensing of medical supplies such as antibiotics. To respond to a bioterrorism attack, resources delivered to local relief providers must be customized for both the suspected bioterrorism agent and the local population-characteristics. For example, planners should consider pre-packaging (Push Packs) the antibiotics in doses appropriate for both adults and children with labels written in several languages. As an alternative, the regional center could send all components of the medicine to the local site, where it could be customized for distribution with greater effectiveness. For example, push packs from the strategic national stockpile contain antibiotics, vaccines and other medical supplies. The local distribution centers, receiving the push packs, can dispense only the subsets of items that are necessary to combat specific emergency (PAHO 2001, NWCPHP 2013). As an alternative, the regional center could pre-pack most of the shipment but postpone labeling until the location and bioterrorism threat are known. In this case, the regional center would add the appropriate labels just prior to shipment (need to estimate how many of each type of label would be required) and have the locale-specific packing lists available. However, the postponement in labeling may delay the delivery of the push packs.

8.3.7 Dispensing Supplies

Supplies, available for humanitarian relief, are seldom sufficient. Therefore, special care must be taken in allocating supplies to victims at a disaster site. Developing a priority scheme for allocation in an effective way. Typical criteria for prioritization would be (i) supplies needed for *immediate* survival, (ii) supplies for the *neediest* (old, infirm, and minor), and (iii) supplies reaching as many victims as possible. However, these objectives can often be in conflict. In addition, equity in dispensing and safeguarding of supplies are essential for effectiveness. The dispensing system must be specific to the conditions of the population in need, as well as the capacity of the organization responsible for dispensing. The delivery process can become complex as it may not be possible to determine the needs with certainty in chaotic conditions, and supplies may not match the needs (surplus of some items while shortage of others).

Deliveries may be carried out by the relief providers or the intermediaries. While direct distribution endows greater control over the use of resources, it requires considerable logistical, administrative, and infrastructural capabilities. In addition, a structure must be in place for recording and identifying aid beneficiaries and aid amounts by date, including the use of coupons and vouchers. Recipients must be briefed on the dispensing system used. It may be advisable to use intermediaries to handle dispensing if working in unfamiliar places. For indirect dispensing to succeed there must be an unambiguous agreement with the intermediary about the distribution procedures and monitoring mechanisms to be used. There must be periodic reviews to ensure adherence of agreements.

Sharp (1998) suggests a three dimensional framework for determining distribution priorities (based on Drèze and Sen 1989) which combines institutional mechanism, level of targeting, and the criteria of defining the target group. The institutional channel determines whether the recipients is selected through the market mechanism, administrative procedures, or community decisions. The level of targeting determines whether selection is based on households, individuals, geography, or population. Finally, the criteria by which the target group is defined can be based on characteristics such as demographic and socio-economic, or nutritional such as production and food-availability measures.

Clearly cost of targeting is an important factor in the choice of a targeting method. While the market based mechanism is generally seen as an inexpensive option (because they avoid the need for direct screening of applicants for aid), the administrative systems – with their requirements of collection and analysis of data, assessment of potential recipients, and monitoring – can be expensive. However, the administrative system may be better suited for maintaining a sense of fairness in distribution.

8.4 Logistics

Logistics covers the gamut of activities from procurement and storage to the onsite delivery of goods, as discussed in Chap. 5. Therefore, logistics assumes a critical role in humanitarian relief, as supplies must be delivered to the needy quickly. Logistics must ensure that the availability of supplies is adequate, and the response to the need for relief is rapid. Obviously it must do so at a minimum cost. While procurement and warehousing ensure that supplies are available in right quantity, transportation enables quick delivery, and contracts and technology are needed as support services.

8.4.1 Procurement

The three imperatives of procurement for humanitarian relief are transparency, accountability, and cost effectiveness. Transparency in transactions goes a long way in developing trust with suppliers and building sound business practices. Accountability enables corrective actions, if needed. Cost effectiveness in terms of price, quantity, quality, timeliness, and delivery, ensures better utilization of resources. The procurement function must be able to leverage its knowledge of markets and suppliers in terms of products that are long-lasting, easy to handle, and have adequate nutritional value. It should be able to form collaborative relationships with suppliers and retailers for rapid transactions and deliveries. In many cases, it is possible to have the suppliers deliver the consignments directly to the disaster sites. The relief agency must continuously scan for new sources in local markets, and work with the ones that fit best in terms of humanitarian needs. As procurement is often dependent on the flow of funds from donors, the suppliers must be kept apprised of their payment status through information transparency.

The procurement process comprises assessment of needs from different disaster locations, establishing specifications, sourcing, placing orders, and managing suppliers for on-time delivery. Food, water, shelter, medicine, first-aid, and sanitation are some of the common procurement items for humanitarian relief. Real time procurement is triggered by requests from field-teams. The degree of specificity in a request depends on the situation on the ground, and the socioeconomic characteristics of the affected region or country. That notwithstanding, it should be possible to form a set of "core" needs, independent of the disaster type and region, that can be prepositioned or dispatched immediately to the sites. This can then be augmented based on the site-specific requests. This is similar in concept to core plus ad-on products discussed in Chaps. 3 and 6 for postponement decisions.

In real time, there will be little time available for traditional procurement functions such as gathering information, identifying and qualifying suppliers, and placing orders. Therefore, it may be necessary to purchase from local markets and from suppliers with good relationship. To speed up the process it may be necessary to relax the approval limits of procurers. Once the order is placed and the supplier has agreed to the contract terms, the supplier must be monitored to ensure on-time delivery. Expediting should be a continuous process in emergencies.

In proactive approaches the prepositioned inventory acts as a hedge against damages, enabling instantaneous relief for many. Relief organizations and national governments often make advance plans for life-sustenance (shelter, food, and medicine), support personnel, and equipment, all in anticipation. That notwithstanding, prepositioning of large inventories is fraught with high downside risk of asset underutilization. This suggests the need for a coordinated approach, integrating pre-positioned inventory with real time deliveries.

The relief provider must decide the quantities of supplies that must be procured proactively and allocated to potential disaster sites, quantities that must be rushed to the sites after a disaster has struck, and the speed of response in real-time. The decision framework is shown in Fig. 8.9.

Fig. 8.9 Procurement strategy

	Pre emergency Proactive purchase	During emergency Real time purchase
Prepositioning	Bulk purchase, Imports	As-needed purchase
Emergency supply		Local purchase, In- kind donations

Clearly, the prepositioned inventory can be bought in bulk, qualifying for price discounts. But they carry the risk of under-usage, deterioration, and pilferage, if disasters do not materialize. The supplies are stored in warehouses in strategic locations.

As emergency supplies are bought only after a disaster has occurred, they can be sent directly to the disaster site. The relative advantages of purchasing modes are shown in Table 8.3. This information is used in Fig. 8.9 for making decisions.

Table 8.3 Relative advantages of different purchasing modes

Purchase type	Advantages	Disadvantages
Local purchases	Fast delivery Lower transportation costs Supports local economy	Required quantity and quality not always available May cause a scarcity of provisions in the local market
Imports	Better quality and larger quantity Specialized items order	Increased delivery time Increased freight costs
Donations	Free or inexpensive Encourages collaboration Usually arrives quickly	Frequently arrive unsolicited May not be required Wasted time and resources Difficult to reject

Groups of buyers often collaborate for joint procurement to obtain economic advantage in purchasing. Buyers can pool their demand while purchasing same types of items or buying from same suppliers. For example, members of a purchasing organization called the Voluntary Hospitals of America aggregate their demand for medical supplies (Keskinocak and Savasaneril 2008). Joint procurement can be outsourced to a third party who, as an intermediary, pools demands from buyers to qualify for price discounts from suppliers (Clarke 2007). In certain scenarios, collaborative procurement may be limited to information sharing. For example, buyers may provide procurement information to others by sharing the preferred-supplier lists, or they may share their experience through blogs and discussion groups of best practices.

8.4.2 Procurement Contracts

Procurement contracts in a contemporary supply chain, as discussed in Chap. 4, can be of three types: cost-plus, fixed price, and incentive (Bolton and Dewatripont 2004).

As explained earlier, the relief supplies are procured for two "buckets": prepositioned inventory, and real time deliveries. Procurement can therefore be structured in the way discussed with respect to the Griffin Manufacturing case, discussed in Chap. 6. The supplies for prepositioned inventory can be procured in a proactive mode – a long-term contract with low-cost suppliers and ample delivery lead time. Such contracts are usually of the fixed-price type, where price is determined through a bidding process (reverse auction). The supplier bids a fixed price but must satisfy buyer stipulated specifications on the RFQ. The supplier bears all risk associated with cost escalation, but he/she prices his product appropriately to cover his production risk. Clearly, a fixed-price contract is easy to implement and inexpensive to monitor.

Supplies for real time delivery, on the other hand, must be procured from an agile supplier such as Griffin Manufacturing. The relief provider would prefer fixed price contract, as in an emergency with shortage of supplies prices tend to shoot up. As the procurement from the agile supplier will be uncertain, both in quantity and timing, the supplier would charge a premium price.

For emergency relief, contracts must reflect the uncertainties of disasters. This can be achieved in multiple ways. Instead of a fixed delivery time, the buyer can require supplies to be delivered at a time (to be specified by buyer) within a time window. The buyer may also insist that they should be able to change the order quantity, if needed; and vary the wholesale price within some bound. Buyer can buy insurance or financial options to hedge against demand volatility. Buy back contracts, where unsold items can be returned by the supplier at a discounted price, leads to sharing of demand risk by both buyer and supplier.

8.4.3 Transportation

Getting emergency supplies from origin to destination involves the combined use of different modes of transportation: air, road, rail, and water. There can be many more secondary modes representing various combinations of the primary modes such as air-drop combined with last-mile delivery by road. The modes of transportation have relative advantages over one another in terms of cost, capacity, and speed. Therefore the chosen mode of transportation must match the need at the disaster sites, and it should be feasible given the disaster conditions and cost. Route selection may depend upon the transportation mode, time available for delivery, and the delivery schedule. The challenge lies in making sure that the supplies arrive safely at their destination and on time.

The demand for logistics capacity at a disaster site is estimated using data such as the distance from warehouse to the disaster site, type of product (heavy, bulky, hazardous), and urgency. A transportation mode is feasible if it can deliver all needed goods on time, can travel on damaged roads, and can handle the type of products to be shipped. The cost of a transportation mode should be weighed against the social value of the lives saved. If the budget is limited, certain modes of transportation such as air-drop may become infeasible.

Land transportation segment is where most of the cost is incurred. Trucks are the common mode of land transportation. They are often required to handle rough roads, and travel long distances without frequent refueling. These road and fueling constraints reduce the effective vehicle capacity that can be used. For instance, trucks used for commercial purposes may not be appropriate in disaster zones. Few relief providers own their trucks because of the large capital expenses. They need to work with 3rd party logistics service providers. However, the assortment of available vehicles in an emergency may come from multiple 3PL sources, may need different fuels, may not be equally road worthy, and may not come with drivers.

Ocean carriers can move large quantities of supplies at low cost, but the shipping times are long. Therefore, they are ideal for transporting supplies for prepositioning inventory. However, the destination port must be equipped and able to handle the shipment. This is not always feasible in the environment of humanitarian relief. When Oxfam delivered a barge load of food to Cambodia the port had no forklifts, cranes, or conveyors (Long and Wood 1995). It took many days for the barge to be cleared but only a fraction of the supplies could be recovered because of spoilage. Air transportation, comprising cargo planes and helicopters, is the fastest mode for moving the supplies. However, high cost and lack of landing facility can be limiting factors. Helicopters are useful but they have limited capacity. Cargo planes such as the C-130 can be used for air drops – palletized supplies are pushed through the cargo door in the tail.

CDM International provides logistics and other services for emergency relief, infrastructure reconstruction, program management, procurement, and capacity-building (GHA 2011). It is a subsidiary of Camp Dresser & McKee Inc., and has built a considerable expertise in earthquakes, floods, hurricanes, landslides, droughts, and epidemics. CDM deployed specialists to perform assessments of water supply, sanitation, and health condition, for the floods in Mozambique in 2000; they compiled the data to identify the number of affected people. Similar projects were launched in response to food shortages after Hurricane Mitch in Nicaragua in 1998.

8.4.4 Warehousing

Warehouses endow flexibility in operations against uncertainty – storing unused material, in periods with no emergency, to be used when demand spikes due to a catastrophe. There are many other reasons for requiring a warehouse. For example, feasibility of a hub-and-spoke logistics network requires a warehouse at its hub. Similarly, warehouse facilities are needed anywhere the cargo changes from one mode of transportation to another, such as at a port or airport. For humanitarian relief, ideally, warehouses should be built close to the disaster sites, as transportation can be unreliable. This can however be risky as the warehouse can itself be damaged.

A distribution center is essentially a storage facility. The number and location of these centers depends on concentrations of population and the locations of the vulnerable members of the community. As discussed in Chap. 5, cross docking can cut delivery time to destination significantly. It may take less than 24 hour in a cross-dock compared to several days and even weeks in traditional warehousing. This is possible because the supplies are pushed through the distribution center rapidly. Supplies should be palletized based on the disaster locations before they are sent to a distribution center, where the pallets from different trucks are sorted and loaded on special vehicles for the last mile delivery. Similarly, drop shipping of real-time supplies, where the suppliers send the goods directly to the distribution centers, as directed by the relief providers (buyers), can eliminate warehousing time. However, this calls for an additional layer of coordination because the suppliers send the goods to the distribution center, and invoice to the buyer.

Long and Wood (1995) suggest four types of warehouses for humanitarian relief: General Delivery Warehouse is located at strategic points, where products may be stored for a long time. Slow Rotation Warehouse stores non-urgent or reserve stockpiles such as spare parts, equipment, and tools. Quick Rotation Warehouse stores emergency supplies that move quickly in and out, and require prompt distribution to the victims. They are located near the emergency zone. Temporary Collection Sites stockpile incoming supplies until a more appropriate storage space can be found. It is not possible to set up a formal system in such places. They can be used, for example, to sort and classify the in-kind donations.

Just like a commercial warehouse, warehouses for disasters must possess appropriate storage and retrieval systems, insurance, and theft protection. But they also need to have additional features to deal with an emergency such as storing items specific to a location in contiguous areas of the warehouse, rapid retrieval of needed items, and special storage for perishable items. Warehouses should be easily accessible to the incoming large trucks from suppliers and the specialized vehicles used for the last mile delivery.

Prepositioned supplies (in a general delivery warehouse) are used only when a disaster strikes. Therefore, the length of storage needed at such warehouses is uncertain. Balcik and Beamon (2008) discuss a facility location problem for humanitarian relief that integrates facility location with inventory decisions, and considers multiple items with different criticalities and response time requirements.

Some simple rules-of-thumb can be employed to estimate the warehouse storage space needed. Starr (2010) suggests providing 2 m^3 for every 1,000 kg of a bagged food item, and 1.4 m^3 for every 1,000 kg of vegetable oil in tins. In 2006, the Federal Emergency Management Agency (FEMA) contracted to have two large warehouses with 45,000 sq. feet and 58,000 sq. feet storage space, respectively, in Selma, Alabama and Cumberland, Maryland (Skinner 2009).

8.4.5 Scheduling

The main operational decisions related to distribution are the allocation of relief supplies and vehicle routing. Effective supply allocation among demand locations is vital due to the high cost of short-supplies or delayed supplies. Quantitative models can help in minimizing the cost of transporting the supplies to the disaster sites and/or minimizing shortages. Van Hentenryck et al. (2010), in their procedure, aim at shortening the "latest arrival times" of deliveries and the total amount of unmet demand. This integrates optimization in delivery quantity with that in delivery speed. Campbell et al. (2008) study vehicle routing that minimizes the latest arrival time of supplies. The authors find that their model also ensures faster delivery, though at a higher total transportation cost. The Federal Emergency Management Agency (FEMA) specifies the quantities to be distributed to the recipients. With these specifications and a definite availability of supplies, a cost minimization model for relief distribution may be appropriate.

Humanitarian relief supplies typically comprise multiple products such as food, water, blankets, medicine, and shelter. In their review, Torre et al. (2012) discuss models for the delivery of multiple commodities with differing transportation costs and demands. Balcik et al. (2008) explicitly model the difference between single-use perishable items and multi-use non-perishable items, with demand backlogging for non-perishable items and demand lost for perishable items.

Yi and Ozadamar (2007) discuss coordination of transportation of commodities from major supply centers to distribution centers in affected areas and evacuation of people from affected areas to shelters. Their objective is to minimize the delay in providing supplies and health care service. People and commodities are categorized into a priority hierarchy, where different types of vehicles are used to satisfy transportation needs. Salmeron and Apte (2010) analyze a problem similar to that in Yi and Ozadamar but they include multiple disaster scenarios.

Chakravarty (2013) discusses a 2-stage Bayesian optimization problem, where the real time delivery problem is solved first (stage 2) assuming disaster has occurred, and the results are embedded in the proactive model to determine the prepositioned inventory (stage 1) to cover all disaster scenarios. In this model, the relief provider decides the quantity of supplies that must be procured proactively and allocated to all potential disaster sites, quantities that must be rushed to the sites after a disaster has struck, and the speed of response in real-time. To model rapid-response, the response time (delivery time) is incorporated in defining the demand for supplies, as human survival is a function of time. The objective is to maximize social value which, as mentioned earlier, is a decreasing function of the response time. This enables the optimal delivery quantity to be expressed as a function of delivery time. The model establishes how the response option is impacted by the intensity of the disaster, and how the response time and quantity interact from a strategic perspective. The limited budget of the relief provider is also incorporated in a stochastic sense.

8.4.6 Technology

Technologies such as the Geographical Information Systems (GIS), Global Positioning Systems (GPS) and Remote Sensing Technologies play vital roles in disaster management. While these technologies are stand-alone type, they must be deployed in an integrated manner in humanitarian relief effort (discussed later).

Geographic Information Systems (GIS) is a tool that analyzes spatial information for interactive queries. It is capable of storing a vast amount of geographical data (including maps) and integrating them through editing, sharing, displaying, and analyzing. GIS is very useful in providing information on physical locations of shelters and other physical facilities, which can be useful in planning for evacuation routes, assembly points, and other related issues. Note that Google-Map is a generic form of GIS.

GIS maps can help determine whether the transportation and communication infrastructure is capable of handling disaster relief and, if not, the actions that must be taken to make them so. GIS technology can determine alternative routes to an affected region and therefore help provide quick response. It can facilitate "food drops" in affected areas, as the GIS generated maps can identify the specific areas of victim-clusters and the unique needs within these clusters. It can help coordinate critical relief agencies by establishing links between them.

It can create and manage appropriate databases for generating hazard-risk maps and vulnerability assessments of locations that can be used in disaster preparedness. This information can also be used to determine the desirable level of protection of a location or population. In disasters such as a chemical explosion, GIS can identify the devastated area and direct the relief providers to the available resources nearby.

The technology is complex and can be hard to train for. A large volume of information must be processed to get useful results from the system. This increases complexity.

Global Positioning System (GPS) makes it possible to pin-point the geographic location of people using satellites that orbit the earth, together with ground receivers. It is widely used and is sufficiently inexpensive. GPS is particularly useful during disasters because it operates in any weather, anywhere and at all times, and can achieve a high level of precision. In many instances GPS data is integrated with GIS to enable real-time operations during an emergency. It is used to track the location of emergency vehicles by identifying the GPS receiver attached to the vehicle, and displaying it on a GIS map. Volcanoes can also be monitored using GPS, and by measuring the deformation of the ground, inferences about volcanic activity can be made.

Remote sensing is the use of electromagnetic wave to acquire information about an object or phenomenon by a device (such as radar) that is not in a physical contact with the object. In the context of disaster management, these sensors are attached to aircrafts or satellites. Commercial and free software are available that permit viewing of data collected from multiple satellites. It is often difficult for relief providers to access areas that have been cut off by a disaster. Remote sensing can greatly assist this process as it enables users to "view" the affected area, without jeopardizing their safety. They can gather pertinent information and integrate it with GIS to facilitate timely response, rescue and relief efforts.

Hurricane forecasting is made easier with the use of remote sensing technology. Storm surges and coastal flooding, which often accompany hurricanes can be mapped to provide information on the level of flooding that can be expected. Earthquake is another emergency where it is hard to locate survivors quickly. In such hazardous operations, with falling debris from collapsed structures, remote sensing is very appropriate. It can significantly improve the timeliness and quality of aid, as search and rescue becomes less onerous. Note that remote sensing can be quite expensive to build and operate; and the data, especially involving small objects, can be difficult to interpret.

8.4.7 Command and Control System (CCS)

The command and control system detects and warns, establishes strategy and tactics, develops plans, allocates resources, assigns tasks, monitors processes, and reports results. It provides communication capabilities, enables collaboration, and enhances accessibility of sensor data. It can integrate location data for facilities, vehicles, and personnel. The set of humanitarian functions under the purview of a command and control system are shown in Fig. 8.10.

It can support different types of services including emergency medical services, response and recovery, and hazardous materials management. Specifically, it can be used for controlling evacuation and rescue operations, managing warehouses, testing the communication system, executing delivery of supplies, and informing

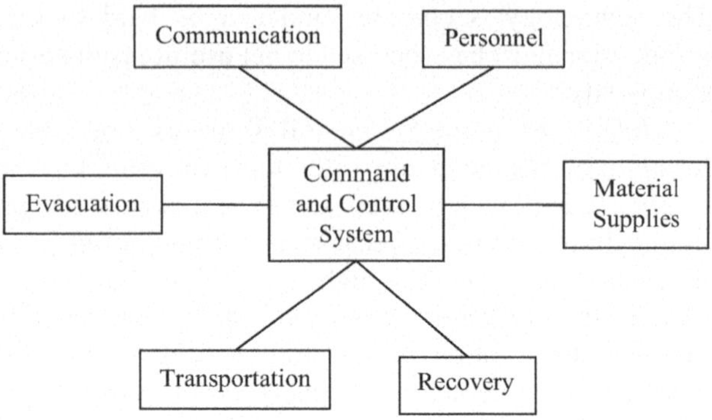

Fig. 8.10 Emergency management system

the public. For example, it can provide valuable information about road accidents to the emergency medical personnel using an automated collision detection and notification systems. Wireless sensors provide the exact location and characteristics of accidents as well as the relevant medical information regarding the vehicle occupants. Telemedicine systems provide links between responding ambulances and nearby emergency medical facilities, enabling doctors to advise paramedics on the treatment of patients in transit (discussed further in Chap. 10). Similarly, using roadside sensors, it can detect and track hazardous material shipments and notify if a shipment deviates from its intended route.

The command and control system sits on top of a technology backbone that manages and distributes data including real-time vehicle location, video, weather, infrastructure reports, and terrain information (Breimyer 2011). Incident data is displayed on a map, using a web-based platform. It is accessible to users through a web browser. It hosts a number of software and hardware tools including inventory control systems, scheduling and tracking, route planning, detection, and driver authentication. Databases exist to help monitor inventory levels of critical supplies and equipment on a continuous basis to ensure that adequate supplies are available to cover demand surges. Scheduling systems can monitor and coordinate response activities. CCS can provide assistance to commercial vehicle operators via electronic route planning services, and hazardous material compliance.

The technology can be used to provide early warning of disasters, and facilitate sharing of emergency information across multiple response agencies. The early warning system employs sensors in the infrastructure to detect disasters. It monitors the alert and surveillance systems to notify all responding agencies. Vehicle fleets can be tracked using GPS. CCS can be integrated with traffic management systems to enable sharing of emergency information.

8.5 Collaboration

In recent years, with increasing frequency and severity of disasters, the need for collaboration and risk sharing has increased tremendously.

8.5.1 Managing Diversity

For humanitarian relief, collaboration can be organized in many different ways. The Business Roundtable, an association of approximately 160 U.S. corporations, created the Partnership for Disaster Response (PDR) (Thomas and Fritz 2006), where member-companies donate resources (supplies, equipment, and expertise). The PDR, serving as a clearing house, matches its resources with the needs of relief providers that have relationships with PDR. Note that such large alliances are not unusual because the number of organizations involved in humanitarian relief is large. As Chia (2007) points out, over 40 countries and 700 non-governmental organizations (NGOs) provided humanitarian assistance in the aftermath of the 2004 Asian Tsunami.

Benefits of collaboration notwithstanding, a large alliance often proves difficult to manage. This is due to the diverse motivations, mandates, capacity, and expertise of the parties involved. While these organizations are vital in crafting an efficient response to a disaster, they may create confusion by competing for funds, resources, and infrastructure access (Thomas and Kopczak 2005). Therefore, it is imperative for the collaborating organizations to possess adequate expertise in managing diversity.

Diversity in a decentralized relief chain must to be managed through incentives and trust-building, or through imposition of behavioral restraints. Partners in an alliance must operate with a common purpose. While appropriate incentives can lead to economic equilibrium between parties through the market mechanism, behavioral restraints (through rule-making) require monitoring and corrective actions. In the specific case of a humanitarian relief chain, only a short period of time is available for the alliance to achieve its goals.

Even between different branches of the *same* organization collaboration can be hard to achieve. A prime example is the failure of adequate response by the federal, state, and local agencies during and after Hurricane Katrina in New Orleans. The crisis unveiled an alarming lack of synchronization between the command and control system, mentioned earlier, and the logistics functions. It led to the formulation of the national response framework (NRF) (http://www.fema.gov/national-response-framework) that provides the mechanisms for unity of effort between local, state, and federal officials. In most such cases, while the stakeholders are motivated by a common interests and goals that ensure unity of effort, they also respect each other's decision rights. The NRF suggests an approach for overcoming such problems.

8.5.2 Establishing Collaboration

In centralized system, a single agency has authority to coordinate the relief effort (Dolinskaya et al. 2011). In this setting, tasks are usually executed in a command mode, where the central agency controls the logistical resources, gathers information centrally, and makes decisions for all organizations involved in relief.

For instance, UN agencies often play the role of a decision maker and take charge of the logistical coordination decisions. Similarly, FEMA brings together government entities, using technology such as the Aidmatrix, to manage and share donations for disaster relief (FEMA 2013).

The United Nations has established the Office for the Coordination of Humanitarian Affairs (OCHA) that is especially focused on facilitating coordination of real-time response. OCHA conducts damage assessments in the disaster affected region, and this information is shared with other agencies. Specific coordination tools are developed by OCHA to address the unique relief needs based on the disaster type and location. It exercises coordination and control through the On-Site Operations Coordination Centre (OSOCC), set up in the region immediately following the disaster. It serves as a "local headquarters" for disaster management operations. The UN, NGOs and government organizations work with a central database of specialized personnel, technical specialists, relief supplies, equipment, and services available within the UN, called the Central Register (Kehler 2004).

In contrast, decisions in an alliance are decentralized and are conducted through consensus, where each relief organization makes its own decision about sharing information, expertise and responsibilities with others. Interactions involving two international relief providers are horizontal, in that they are concerned with resource sharing and joint decision-making (Balcik et al. 2010), whereas, interactions between a relief provider and a supplier is vertical. Collaboration can be achieved through an umbrella organization that creates the rules of collaboration. Kaatrud et al. (2003) describes an umbrella organization called UNJLC, which handles logistics issues. The services it provides include collecting information on infrastructure status, capacity availability, and disaster maps; providing information sharing tools; tracking and prioritizing movement of relief supplies; and pooling of assets.

Collaboration between organizations is not easy to establish in the short time period available after a disaster. Trust-building becomes a critical need. For example, two NGOs may be willing to collaborate in one region if they already have successful collaboration in another region. The primary reasons for caution are the possible cultural and operational policy differences between them. A medical doctor's experience in Haiti after the earthquake bears this out well (Dolinskaya et al. 2011). The physician posted a request for assistance with the transportation of medical equipment using the Aidmatrix web-based portal. A shipping company offered to transport the equipment, but only through an NGO registered in Haiti, which this doctor was not. The shipment had to be diverted through a *partner* NGO of Aidmatrix that was officially registered in Haiti.

8.5.3 A Framework for Collaboration

As mentioned earlier, there are two major causes of uncoordinated relief effort: independent decisions by organizations with different capabilities, and the sketchiness of information used for assessing relief-needs of regions. Organizations make decisions that are consistent with their individual resource-endowments and cost structures that can vary widely. For example, while some may possess large transportation fleets (air, land and water) or own efficient distribution systems (warehouses, and schedules), others may specialize in products (food, medicine, clothing, blankets, personal-hygiene, etc.) or have well-trained and motivated personnel.

Organizations find it difficult to obtain reliable information from affected regions because of the damage to the infrastructure – roads, rail lines, and communication lines. This is further complicated because of two other factors: regions are seldom endowed with equal resource to begin with (pre disaster), and disasters do not cause equal devastation everywhere.

As mentioned earlier, centralized supply chain decisions are effective in matching the aggregated needs of the regions with the aggregated resources of the service providers (relief organizations). Such a solution, while it avoids duplication, may not share the burden among the organizations in an equitable way. An organization, for example, may be asked to supply to a far-off region using logistics it does not possess. The centralized approach may also create a massive bureaucracy. In contrast, organizations making independent decisions, individually determine which regions to supply and how much. However, the individual approaches may fail to minimize the resulting mismatches between supply and demand. An effective way of doing so is to provide information transparency throughout the system, whereby any organization is able to "see" what the sister organizations are planning and executing – a framework for online collaboration.

The two dimensions of such a framework are the assessment of needs, and matching demand with supply. The assessment of needs is a continuing process of updating forecasts based on information obtained from the disaster sites. The initial forecast primarily uses historical information and scientific damage assessment, based on the location of the regions and the intensity of the disaster. The regional components of this forecast are updated by field workers of individual relief organizations. The demand forecasts for a region must be further updated by the total amount delivered (executed) to the region in the previous period, to obtain the yet unsatisfied demand. This must be done for each region and each product category. Note that while such an adjustment provides a better picture of demand, it does not contribute to avoiding mismatches in the current period's deliveries.

To match supply with demand, the organizations need to go beyond just information visibility; they need to compare and adjust their supply plans for the current period by resolving "conflicts", if any. For example, if two organizations have planned to supply antibiotics to the same region, they must ensure that it does not result in an antibiotic glut. Thus, the framework can be made operational through three pieces of information: demand forecast, delivery reports of the previous period, and delivery plans for the current period.

8.5.4 Implementation

The framework can be implemented through a private online exchange with access control, one for each country or region, shown in Fig. 8.11. In this exchange, the participants would be the relief organizations, the host government, suppliers of products and services, logistics service providers, and application service providers. The site will be secured for participants, but individual contributors (money and goods) will need to be provided access to complete their transactions related to making donations. Any participant may access another with a single click.

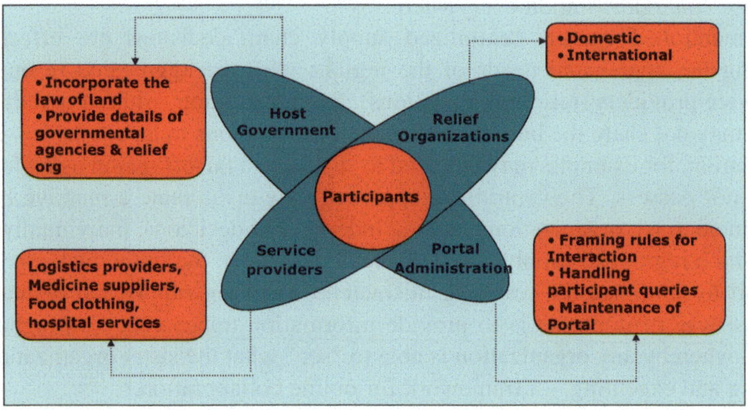

Fig. 8.11 Humanitarian relief portal

A relief organization posts two pieces of information: report of deliveries completed in the previous period, and plan of deliveries for the current period, all by region and by product category. The host government posts information on available and damaged infrastructure (road network, railroads, airports, communication, and power lines) and their capabilities, district level organization, availability of local warehouses etc. Vendors list the products that relief organizations may acquire from them; they also post price and shipping information. Logistics service providers list information on their delivery fleet, location, availability (updated daily), and pricing.

The exchange administration posts inventories available by disaster location, based on information from relief organizations. It facilitates the following processes: procurement and shipping of supplies from participating vendors, transactions between logistics providers and relief organizations, and transactions of donations including tax deduction.

The exchange creates consolidated demand information, offset by the amount delivered last period as posted by relief organizations, for each region and product category. The relief organizations use this information to determine their optimal delivery plans for the current period and post them on the exchange. The exchange identifies conflicts: two or more organizations supplying the same product to the same region, some regions not receiving adequate amount of supplies, insufficient capacities at the warehouses, and infeasible deliveries due to invalid infrastructural assumptions (damaged roads, airports not suitable for landing large aircraft, etc.).

It creates several alternative solutions by using algorithms, while ensuring that the organizations' current plans are not altered drastically; it lists the consequences of each alternative for the relief providers. The affected organizations are invited to a chat room, where the alternative solutions and consequences are displayed. The organizations may carry out multilateral negotiations, using the exchange-generated solutions as guidelines. The negotiated solution is then posted as the final plan of delivery. The exchange becomes the arbiter of unresolved conflicts. At the end of the period, the organizations obtain field-reports listing actual deliveries, and post them on the exchange (with comments) to be used for revising the unsatisfied demand for the next period.

The exchange creates value in multiple ways. Not only does it create better supply targets, it also minimizes the mismatch between supply and delivery. It is a facility for real-time interactions between relief organizations, vendors, government agencies, logistics service providers, and providers of other auxiliary services. It also provides a convenient way for individuals for making their monetary and in-kind contributions directly to the organizations of their choice. Additional functionalities can be built into the exchange for facilitating interactions with government agencies, large institutional donors, and for real-time tracking. A search engine specific to relief operations may also be added.

8.6 Fund Management

With a large number of funding sources, cost structures, mix of funds, intermediaries, and disaster locations, managing the funds is becoming quite complex.

8.6.1 Funding Sources

Funds for humanitarian relief may come from public or private sources, and their use can be domestic or international. Domestic funds (private or government sources) are raised and deployed in the same country. International funds, in contrast, could be raised from sources in multiple countries and used in disaster relief in any country. Examples include the UN funds, individual governments helping other governments, the Red Cross Society, non-governmental organizations (NGOs), and private donors. Note that a large chunk of UN funds comes from government sources involving many countries. This amounted to almost 76% of the UN's total fund in 2010. The remaining 24% came from other sources such as the European Commission, NGOs, and institutions such as the World Bank, and others. Donations may come in the form of money or in-kind such as food and clothing.

For example, Motorola, with annual sales of $27.1 billion, initiated a partnership with CARE International in 2004. CARE has an annual budget of approximately $800 million (Care International 2013). The partnership provides communications technology to remote areas. Motorola has also donated radio communications systems to Bangladesh, the Democratic Republic of Congo, and Peru. Another example is the World Economic Forum that established the Disaster Resource Partnership (DRP) in 2001 so as to link the relief providers with firms contributing to disaster relief (WEF 2013). DRP mobilizes medical emergency teams, relief supplies, communications, and logistics. Through the DRP, the World Economic Forum was involved in the 2004 tsunami, Hurricane Katrina 2005, and the 2006 earthquake in Indonesia.

Donors may provide funds that may be restricted or unrestricted. While restricted funds have conditions attached such as earmarking for specific programs, unrestricted funds can be awarded to any global or regional program on a competitive basis. Therefore unrestricted funds are more flexible and egalitarian. Donors may also provide funds in the form of direct grants to the affected countries.

According to the report of the UN secretary general (Secretary-General 2012), the largest 23 UN agencies (accounting for 87% of UN's expenses on humanitarian projects) had spent $14.118 billion on humanitarian relief in 2010. Of this amount, as shown in Fig. 8.12, the unrestricted fund (core resources) was $2.463 billion (17.5%), restricted fund (non-core resources) was $9.221 billion (65.3%), management fee was $2.174 billion (15.4%), and other expenses were $259 million (1.8%).

	Program Expenses	Management Fee	Other Expenses	Total Expenses
Core resources	$2.463 billion	$1.132 billion	$.243 billion	$3.840 billion
Share	64.1%	29.5%	6.3%	100.0%
Non-core resources	$9.221 billion	$1.042 billion	$.016 billion	$10.279 billion
Share	89.7%	10.1%	0.2%	100.0%
Total resources	$11.684 billion	$2.174 billion	$.259 billion	$14.118 billion
Share	82.8%	15.4%	1.8%	100.0%

Fig. 8.12 Fund allocation

Note that the unrestricted fund is rather small in relation to the restricted fund – just about a quarter. As the vast majority of the fund is restricted and earmarked for specific projects, there is not much *flexibility* in allocating funds to relief projects in real time. Second, the overall management fee at 15.4% is quite high and is not balanced; it accounts for a hefty 29.5% of the core resources but only 10.1% of the non-core resources.

Walker and Pepper (2007) describe a scenario where a fund of $1 million from a donor is channeled to a beneficiary through three intermediaries (a UN agency, an international NGO, and a local partner), each charging 10% fee. It can be verified that the final recipient only received $729,000. It is worth asking whether the $271,000 in accumulated transaction cost is justified. Because of the large number of intermediaries and its implied information asymmetry in the financial supply chain, management of funds is far from efficient. The intermediaries are often the source of delay and they charge hefty service fee. Management of risks related to raising funds and monitoring their flow are also important issues.

A recent innovation in providing relief is pooled financing that is designed to provide *quick response* and *flexibility* in fund allocation. It has expanded considerably over the last few years. A group of countries called the Development Assistance Committee (DAC) are the major donors to this fund, which was worth $733 million in 2009 (Randel 2011). Most members of DAC belong to the Organization for Economic Cooperation and Development (OECD) and they contributed 93.4% of the fund in 2009. The contribution from non OECD DAC countries was at 4.6%, while private sources donated 1.6%. Pooled funds constitute approximately 7% of the humanitarian aid from government sources (GHA 2012). Allocation of the fund to recipients is channeled through UN agencies, the World Bank, non-governmental organizations (NGOs), the International Red Cross, and other government agencies. In 2009, there were 20 major recipient countries.

8.6.2 Deployment of Funds

As mentioned earlier, donors may use multiple channels for funding: directly fund the relief providers, fund UN agencies, provide unrestricted funding to multilateral organizations, provide restricted funding to NGOs, and contribute to pooled humanitarian funds, as shown in Fig. 8.13. A large proportion (89%) of the global humanitarian assistance fund of $16.9 billion (in 2008) was *delivered* through intermediaries (GHA 2010). Local organizations that deliver funds to recipients include churches, local NGOs, and National Red Cross and Red Crescent Societies.

A delivery agency must manage a mix of funds comprising earmarked and non-earmarked contributions, and its own reserves. As funds may flow from a donor to a recipient via delivery agencies using multiple paths, fund flow decisions are complex. This decision can become convoluted; given that some agencies act as recipients as well as donors and implementers of humanitarian relief *simultaneously*. Note that the decisions for allocation of funds at different layers of the flow network can be different, depending on the sources of finance, disaster locations, earmarked funds, transaction fees, expertise and effectiveness of intermediaries, banking and government constraints, and the priorities of humanitarian relief.

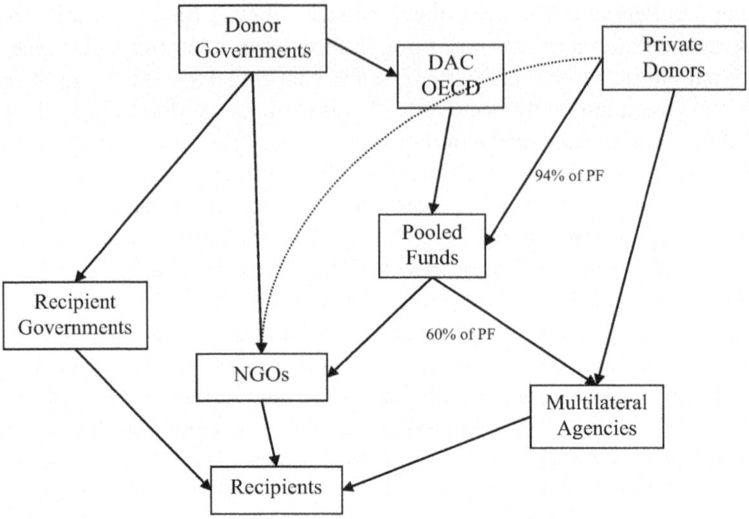

Fig. 8.13 International fund flow

8.6.2.1 Fund Deployment Model

Consider a donor with available funds R that it is considering allocating between two NGOs based on their relative performance. Assume that performance is measured by the quickness of an NGO's delivery of supplies to the disaster site. Assume that the *ith* NGO achieves a delivery time of t_i for a unit standard load, charges the donor a fee $\$\alpha_i$ per unit dollar of funds received, and incurs a logistics cost of $\$\beta_i$ per unit standard load, per unit time. Assume that donor's utility from the relief operations is inversely proportional to the delivery time.

It can be verified that funds allocated to NGO 1 would equal $Rt_2/(t_1 + t_2)$, and that to NGO 2 it would be $Rt_1/(t_1 + t_2)$. NGO 1 would deliver $\frac{LR(1-\alpha_1)t_2}{(t_1+t_2)}$ units of standard load, and incur a logistics cost of $L\{R(1 - \alpha_1)t_2/(t_1 + t_2)\}\{\beta_1/t_1\}$, where L is the number of standard loads of supplies that can be procured per dollar of the donor fund. Therefore, NGO 1's profit is expressed as

$$\pi_1 = \frac{R\alpha_1 t_2}{(t_1 + t_2)} - \frac{L\beta_1 R t_2(1-\alpha_1)}{(t_1 + t_2)t_1} = \frac{Rt_2}{(t_1 + t_2)}\left\{\alpha_1 - \frac{L\beta_1(1-\alpha_1)}{t_1}\right\}$$

Assume a simultaneous one-shot game involving NGO 1 (decision variable t_1), NGO 2 (decision variable t_2), and the donor (decision variable R).

NGO 1 will maximize profit in t_1 to obtain its profit maximization equation as

$$\alpha_1 {t_1}^2 - 2L\beta_1(1-\alpha_1)t_1 - L\beta_1(1-\alpha_1)t_2 = 0$$

Similarly, Donor 2's profit maximization equation in t_2 is expressed as

$$\alpha_2 t_2^{\,2} - 2L\beta_2(1-\alpha_2)t_2 - L\beta_2(1-\alpha_2)t_1 = 0$$

For equilibrium, the above two equations are solved simultaneously to obtain t_1 and t_2.

The donor's positive utility from relief operation increases if the number of standard load delivered per period increases. His negative utility increases in R. As the overall utility must be concave in R, we may express it as

$$U_D = \left\{(1-\alpha_1)\frac{t_2}{t_1} + (1-\alpha_2)\frac{t_1}{t_2}\right\}\frac{LRu}{(t_1+t_2)} - aR^2$$

u is the utility per unit load moved.

The value of R maximizing the donor's total utility is expressed as

$$R = \left\{(1-\alpha_1)\frac{t_2}{t_1} + (1-\alpha_2)\frac{t_1}{t_2}\right\}\frac{Lu}{2a(t_1+t_2)}$$

In Appendix 2, we discuss *sequential* decision making between the donor and the NGOs, and simultaneous decision making between the NGOs.

8.6.3 Government Sources of Funds

The United Nations Development Program administers the common humanitarian fund (CHF), and the Office for the Coordination of Humanitarian Affairs (OCHA) manages both the emergency response fund (ERF) and the Central Emergency Response Fund (CERF). CERF was launched in 2006 with a grant of $450 million (GHA 2011). The CERF provides the donors the opportunity to pool their funds on a global level to enhance the effectiveness and reliability of humanitarian assistance. Project proposals are reviewed by a humanitarian coordinator before they are sent to the CERF.

The aim of emergency response fund ERF is to provide rapid and flexible funding to in-country operations, mainly through NGOs. The pooled non earmarked contributions are allocated according to identified need of the country.

In many countries, governments often assume the role of funding disaster relief, as they can withstand the risk associated with low-frequency and high-severity disasters, better than NGOs. There are many examples of private-sector insurers working with governments to participate in risk exposures they would normally avoid. For example, FEMA in the US underwrites losses in flood insurance administered by private insurers.

The Consolidated Appeals Process (CAP), established in 1992, is a tool used by aid organizations to raise funds for humanitarian relief as well as to plan, implement, monitor, and coordinate their activities. CAP is a blueprint for action; it includes Common Humanitarian Action Plan (CHAP), a resource mobilization plan, program implementation and coordination, joint monitoring and evaluation (field visits, data collection), and reports on results. OCHA has the role of managing the CAP development and its launch every year. Donors may use the appeal as a catalog to select projects to fund, or they may contribute flexibly to an agency or to an appeal's pooled fund. The CAP projects are funded by the pooled Common Humanitarian Funds (CHFs). In contrast, CERF and ERF disburse funds to any new crisis that are not included in the CAP.

8.6.4 Non-Government Organizations

NGOs receive their funding from three sources: private donors ($3.7 billion in 2008), government agencies ($3.1 billion in 2008), and channeled UN funds ($900 million in 2008) (GHA 2010). Some of the pooled funds are also channeled to NGOs. International NGOs received $134 million in funding via CHFs and ERFs in 2010 (38.7% of the total); they cannot receive funds directly from CERF. Overall, NGOs were the second largest grouping that received 45.5% ($7.7 billion) of the total international humanitarian response in 2008 (GHA 2010). NGO funds raised from private donors including the public are considered to be one of the quickest ways of getting finance. In 2006, the major international NGOs were: Action contre la Faim, CARE, Caritas, Concern, Danish Refugee Council, Deutsche Welthungerhilfe, GOAL, International Medical Corps, International Rescue Committee, Medecins Sans Fron, Tieres, Mercy Corps, Merlin, Norwegian Church Aid, Norwegian People's Aid, Norwegian Refugee Council, Oxfam, Save the Children, Tearfund, and World Vision (GHA 2009).

As mentioned earlier, some of the funds have conditions attached to them in the form of earmarks for specific projects. Others may require transparency of transactions. This reduces the flexibility in how the funds are spent on humanitarian relief. Second, many NGOs lack the resources to sustain large reserves for operations, and they do not have direct access to flexible funds such as the CERF.

8.6.5 Donor Relationship and Campaigns

The donor community is concerned about enhancing the performance and accountability of donated funds. One option is to control the fund-deployment channels to influence the quality of humanitarian operations (Macrae et al. 2002). However, it is not easy to assess how increasing donor involvement would result in certain activities in certain regions being favored over others. The key question is whether the agencies would adapt their objectives and operations to conform to donor demands. For example, it is not certain how a donor may influence its relief chain partners' operations in low-risk areas and activities, where there are no established mechanisms for monitoring impact.

8.6.5.1 Accountability

Note that accountability includes the following: developing the terms of agreement on roles and responsibilities of all stakeholders, building provision for taking corrective action, reporting those actions, and responding to inputs from stakeholders (Raynard 2000). The author has highlighted the general lack of clearly defined responsibilities among humanitarian organizations, and the resulting problems of attribution. For example, if food aid did not get delivered to those in need, it is not easy to establish whether this failure is the responsibility of the NGO on the ground, the agency supplying the food, or the donor agency providing financial resources. If responsibilities have not been clearly defined at the outset, it is easy for one party to lay the blame for failure on another.

At present, there are no accepted performance metrics, and donors permit their partners to define and monitor performance on their own. While this allows intermediaries freedom in managing their operations, it makes it difficult to assess the overall effectiveness of donor-funds in humanitarian relief. To improve accountability, some donors have begun to develop a framework of agreement that require specific performances from the NGOs for certain commitments of funds from donors. Framework agreements have enhanced the predictability and timeliness of contributions to international organizations, and have also increased trust and mutual understanding. Note that while such agreements do not place specific restrictions on the use of funds (as it is with earmarked funds), they may require the NGOs to deviate from the strategies they would have adopted otherwise.

An online platform, described earlier, is a way of implementing collaborative agreements between donors and NGOs. Donors will be added to the exchange shown in Fig. 8.11. Note that the relief providers or NGOs post their report of deliveries completed by period, product, and destination, which will now be visible to the *specific* donors associated with a particular NGO. The NGOs would also be required to post the procurement invoices which the donor may access as needed. Using the above information donors can easily compute the NGO's performance, and assess its compliance with the framework agreement.

Appendix 1

Exactly three disasters in five periods

Year of disaster					
1	2	3	4	5	Expected post disaster saving
Disaster 1	Disaster 2	Disaster 3			$p^3(1-p)^2(y_1+y_2+y_3)$
Disaster 1	Disaster 2		Disaster 3		$p^3(1-p)^2(y_1+y_2+y_4)$
Disaster 1	Disaster 2			Disaster 3	$p^3(1-p)^2(y_1+y_2+y_5)$
Disaster 1		Disaster 2	Disaster 3		$p^3(1-p)^2(y_1+y_3+y_4)$
Disaster 1		Disaster 2		Disaster 3	$p^3(1-p)^2(y_1+y_3+y_5)$
	Disaster 1	Disaster 2	Disaster 3		$p^3(1-p)^2(y_2+y_3+y_4)$
	Disaster 1	Disaster 2		Disaster 3	$p^3(1-p)^2(y_2+y_3+y_5)$
	Disaster 1		Disaster 2	Disaster 3	$p^3(1-p)^2(y_2+y_4+y_5)$
		Disaster 1	Disaster 2	Disaster 3	$p^3(1-p)^2(y_3+y_4+y_5)$

Expected saving $= p^3(1-p)^2(5y_1 + 6y_2 + 6y_3 + 5y_4 + 5y_5)$

Year of Disaster					
1	2	3	4	5	Expected post disaster saving
Disaster 1	Disaster 2	Disaster 3	Disaster 4		$p^4(1-p)(y_1 + y_2 + y_3 + y_4)$
Disaster 1	Disaster 2	Disaster 3		Disaster 4	$p^4(1-p)(y_1 + y_2 + y_3 + y_5)$
Disaster 1		Disaster 2	Disaster 3	Disaster 4	$p^4(1-p)(y_1 + y_3 + y_4 + y_5)$
	Disaster 1	Disaster 2	Disaster 3	Disaster 4	$p^4(1-p)(y_2 + y_3 + y_4 + y_5)$

Exactly four disasters in five periods
Expected saving $= p^4(1-p)(3y_1 + 3y_2 + 4y_3 + 3y_4 + 3y_5)$

Appendix 2

Assume that the donor makes the first move and selects the optimal R for known t_1 and t_2. As in the text, we have $R = \left\{(1-\alpha_1)\frac{t_2}{t_1} + (1-\alpha_2)\frac{t_1}{t_2}\right\}\frac{Lu}{2a(t_1+t_2)}$. Next the two NGOs maximize profits in t_1 and t_2, subject to R set by the donor. NGO 1's profit, from the text, is $\pi_1 = \frac{Rt_2}{(t_1+t_2)}\left\{\alpha_1 - \frac{L\beta_1(1-\alpha_1)}{t_1}\right\}$. The first derivative of π_1 in t_1 is expressed as $\frac{\partial \pi_1}{\partial t_1} = \frac{Rt_2\left\{L\beta_1(1-\alpha_1)(2t_1+t_2)-\alpha_1 t_1^2\right\}}{(t_1+t_2)^2 t_1^2} + \frac{t_2}{(t_1+t_2)}\left\{\alpha_1 - \frac{L\beta_1(1-\alpha_1)}{t_1}\right\}\frac{\partial R}{\partial t_1}$, where $\frac{\partial R}{\partial t_1} = \left\{(1-\alpha_2)t_1^2 - (1-\alpha_1)t_2(2t_1+t_2)\right\}\frac{Lu}{2a(t_1+t_2)^2 t_1^2}$. It can be established that π_1 is unimodal in t_1, increasing then decreasing. Therefore, from the FOC, after simplification, we have $2L\beta_1(1-\alpha_1)^2(2t_1t_2^2 + t_2^3 + t_1^3) - (1-\alpha_1)\alpha_1(3t_1^3t_2 + t_2^3t_1) + (1-\alpha_2)\,\alpha_1(t_1^3t_2 - t_1^4) = 0$.

Similarly, from the FOC in t_2 we have $2L\beta_2(1-\alpha_2)^2(2t_2t_1^2 + t_1^3 + t_2^3) - (1-\alpha_2)\alpha_2(3t_2^3t_1 + t_1^3t_2) + (1-\alpha_1)\alpha_2(t_2^3t_1 - t_2^4) = 0$

The above two equations can be solved simultaneously for t_1 and t_2.

References

Balcik, B., & Beamon, B. (2008). Facility location in humanitarian relief. *International Journal of Logistics: Research and Applications, 11*, 101–121.

Balcik, B., Beamon, B., & Smilowitz, K. (2008). Last mile distribution in humanitarian relief. *Journal of Intelligent Transportation Systems, 12*, 1–63.

Balcik, B., Beamon, B., Krejci, C., Muramatsu, K., & Ramirez, M. (2010). Coordination in humanitarian relief chains: Practices, challenges and opportunities. *International Journal of Production Economics, 126*, 22–34.

Binder, A., & Witte, J. (2007). *Business engagement in humanitarian relief: Key trends and policy implications*. Study commissioned by the humanitarian policy group of the Overseas Development Institute (ODI) and Global Public Policy Institute (GPPI). http://www.odi.org.uk/resources/docs/375.pdf

Bolton, P., & Dewatripont, M. (2004). *Contract theory*. Cambridge: MIT Press. 2004.

Breimyer, P. (2011). *Next-generation incident command system*, Surveillance Systems Group, MIT Lincoln Laboratory. http://www.ll.mit.edu/publications/technotes/TechNote_NICS.pdf

Bruno, G., Watson, T., & Lee, A. (2011). *A vision for managing natural disaster risk.* World Economic Forum. http://www3.weforum.org/docs/WEF_VisionManagingNaturalDisaster_Proposal_2011.pdf

Burbank, K. (2009). An economist's role in disaster mitigation at FEMA. *Business Economics, 44*, 177–181.

Campbell, A., Vandenbussche, D., & Hermann, W. (2008). Routing for relief efforts. *Transportation Science, 42*, 127–145.

Care International. (2013). *Fighting poverty by empowering women and girls in the poorest communities around the world.* http://www.care-international.org/uploaddocument/news/publications/general%20information/english/ci%20annual%20report%202012_web%20version.pdf

Chakravarty, A. (2011). A contingent plan for disaster response. *International Journal of Production Economics, 134*, 3–15.

Chakravarty, A. (2013). *Humanitarian Relief Chain: Rapid Response under Uncertainty International Journal of Production Economics.* doi.org/10.1016/j.ijpe.2013.10.007

Chandler, D. (2010). *New analysis of the structure of silks explains paradox of super-strength.* MIT News, MIT Media Relations, March 14. http://web.mit.edu/press/2010/spider-silk.html

Chia, E. (2007). Engineering disaster relief. *IEEE Technology and Society Magazine, fall*, pp. 24–29.

Clarke, E. (2007). Let's work together. *Supply Management, 12*, 26–29.

Dolinskaya, I., Shi, Z., Smilowitz, K., Ross, M. (2011). Decentralized approaches to logistics coordination in humanitarian relief. In T. Doolen, & E. Van Aken (Eds.) Proceedings of the 2011 industrial engineering research conference, Reno, Nevada, May 21–25

Drèze, J., & Sen, A. (1989). *Hunger and public action.* Oxford: Clarendon Press. http://books.google.com/books/about/Hunger_And_Public_Action.html?id=OjCHQR-M0ocC

European Commission. (2003). *Disaster Preparedness and Prevention (DPP): State of play and strategic orientations for EC policy.* http://ec.europa.eu/echo/files/policies/dipecho/dpp_paper.pdf

FEMA. (2013). *Nationwide – working together – FEMA & the Aidmatrix foundation.* Video presentation. http://www.fema.gov/medialibrary/media_records/343

GHA. (2009). *Public support for humanitarian crises through NGOs.* http://www.globalhumanitarianassistance.org/wp-content/uploads/2010/07/2009-Focus-report-Public-support-for-humanitarian-crises-through-NGOs.pdf

GHA. (2010). *Global humanitarian assistance report.* http://www.globalhumanitarianassistance.org/wp-content/uploads/2010/07/GHA_Report8.pdf

GHA. (2011). *Pooled funding mechanisms and large-scale disasters.* http://www.globalhumanitarianassistance.org/wp-content/uploads/2011/04/Use-of-pooled-funds-following-a-natural-disaster.pdf

GHA. (2012). *Global humanitarian assistance report.* http://www.globalhumanitarianassistance.org/wp-content/uploads/2012/07/GHA_Report_2012-Websingle.pdf

Kaatrud, D., Sammi, R., & Wassenhove, L. (2003). UN joint logistics center: A coordinated response to common humanitarian logistics concerns. *Forced Migration Review, 18*, 11–14.

Kehler, N. (2004). Coordinating humanitarian assistance: A comparative analysis of three cases. Master thesis, Virginia Polytechnic Institute and State University. http://www.ipg.vt.edu/papers/kehler%20-%20majorpaper.pdf

Keskinocak, P., & Savasaneril, S. (2008). Collaborative procurement among competing buyers. *Naval Research Logistics, 55*, 516–540.

Landefeld, J., & Seskin, E. (1982). The economic value of life: Linking theory to practice. *American Journal of Public Health, 72*, 555–566.

Long, D., & Wood, D. (1995). The logistics of famine relief. *Journal of Business Logistics, 16*, 213–229.

Macrae, J., Collinson, S., Buchanan-Smith, M., Reindorp, N., Schmidt, A., Mowjee, T., Harmer, A. (2002). *Uncertain power: The changing role of official donors in humanitarian action.* Humanitarian Policy Group Report 12. London: Overseas Development Institute. http://www.odi.org.uk/publications/242-power-role-official-donors-humanitarian-action

NWCPHP. (2013). *Emergency distribution of pharmaceuticals.* Northwest Center for Public Health Practice. http://www.nwcphp.org/docs/mass_dispensing/pharmacy_print.pdf

PAHO (2001). *Humanitarian supply management and logistics in the health sector*. Washington, DC: Pan American Health Organization. http://www.paho.org/english/ped/HumanitarianSupply.pdf

Pederson, P., Dudenhoeffer, D., Hartley, S., Permann, M. (2006). *Interdependency modeling: A survey of U.S. and international research*. Idaho National Laboratory, U.S. Department of Energy. http://cipbook.infracritical.com/book3/chapter2/ch2ref2a.pdf

Queensland Government. (2010). *Queensland evacuation guidelines for disaster management groups*. http://www.disaster.qld.gov.au/Disaster%20Resources/Documents/Qld%20Evacuation%20Guidelines%20Consultation%20Draft.pdf

Randel, J. (2011). *Global humanitarian assistance report 2011*. Wells, Somerset: Development Initiatives. http://www.globalhumanitarianassistance.org/report/gha-report-2011

Raynard, P. (2000). *Mapping accountability in humanitarian assistance*. Report presented to the ALNAP Bi-annual Meeting. http://www.adpc.net/pdr-sea/eval/file22.pdf

Salmeron, J., & Apte, A. (2010). Stochastic optimization for natural disaster asset prepositioning. *Production and Operations Management, 19*, 561–574.

Secretary-General. (2012). *Analysis of funding of operational activities for development of the United Nations System for the year 2010*. Development Cooperation Policy Branch, Office for ECOSOC support and coordination. United Nations: Department of Economic and Social Affairs. http://www.un.org/esa/coordination/pdf/2012_funding_report-figures_and_tables.pdf

Sharp, K. (1998). *Development: Targeting food aid for disaster prevention in Ethiopia*. Relief and rehabilitation network paper #27. London: Overseas Development Institute. http://www.odihpn.org/download/networkpaper027pdf

Skinner (2009). *FEMA's acquisition of two warehouses to support hurricane katrina response operations*. Office of Inspector General. Washington, DC: U.S. Department of Homeland Security. http://www.oig.dhs.gov/assets/Mgmt/OIG_09-77_Jun09.pdf

Starr, L. (2010). *Distributing emergency food during a Pandemic*. Tango International. https://docs.google.com/viewer?a=v&q=cache:1BWBcu5ox_EJ:h1n1vax.aed.org/docs/food_security/session5_tool.doc+&hl=en&gl=us&pid=bl&srcid=ADGEESiKGD8-Qzn8gBnphQQTOBH_ttNpQmaLcdujZI61ZzISigAV5ud84DSYPmfXk65ZOHgo9v9O3ZIHykFhFJ3EXdSS7XECvf_t09dsNzPZ1KL2vJtFbF1W0qPvhWK2r-9WauU2oiuw&sig=AHIEtbQiakT0HtbpXhHCcbwl0z6ITGm5cw&pli=1

Thomas, A., & Fritz, L. (2006). Disaster relief, Inc. *Harvard Business Review, 84*, 114–122.

Thomas, A., & Kopczak, L. (2005). *From logistics to supply chain management: The path forward in the humanitarian sector* (pp. 1–15). San Francisco: Fritz Institute.

Torre, L., Dolinskaya, I., & Smilowitz, K. (2012). Disaster relief routing: Integrating research and practice. *Socio-Economic Planning Sciences, 46*, 88–97.

Van Hentenryck, P., Bent, R., & Coffrin, C. (2010). Strategic planning for disaster recovery with stochastic last mile distribution. In *Integration of AI and OR techniques in constraint programming for combinatorial optimization problems* (pp. 318–333). Berlin/Heidelberg: Springer.

Vlaev, I., Seymour, B., Dolan, R., & Chater, N. (2009). The price of pain and the value of suffering. *Psychology Science, 20*, 309–317.

Walker, P., & Pepper, K. (2007). *Follow the money: A review and analysis of the state of humanitarian funding*. Feinstein International Center. http://www.hapinternational.org/pool/files/ghd-iasc-final-paper.pdf

WBDG. (2012). *Natural hazards and security*. WBDG Secure/Safe Committee. National Institute of Building Science. http://www.wbdg.org/design/resist_hazards.php

WEF (2013). *Disaster management*. Video presentation. http://www.weforum.org/issues/disaster-management

Whoriskey, P., & Hsu, S. (2006). Levee repair costs triple. *The Washington Post*, March 31, pp. A01. http://www.washingtonpost.com/wp-dyn/content/article/2006/03/30/AR2006033001912.html

Wonders (2013). *Delta works, Netherlands*. http://www.7wonders.org/europe/netherlands/rotterdam/delta-works/

Yi, W., & Ozdamar, L. (2007). A dynamic logistics coordination model for evacuation and support in disaster response activities. *European Journal of Operational Research, 179*, 1177–1193.

Sustainable Supply Chains 9

9.1 Introduction

A pivotal moment for the sustainability of the global economic system was the year 2007, when the Intergovernmental Panel on Climate Change (IPCC) was awarded the Nobel Peace Prize. In global sustainability, the major concern has been the pattern of resource use that causes impairment to the environment. The negative impact of greenhouse gases and toxic chemicals on ecology can be horrendous. However, ecological preservation by itself cannot ensure that the global system will be sustained; it is only one of the pillars of sustainability. The other two are corporate social responsibility (CSR), and the global economy, as shown in Fig. 9.1. These three pillars together are known as the "triple bottom line".

While implementation of sustainability may increase cost in certain scenarios, it has in fact reduced cost in many others. Companies such as DuPont, Novartis, and 3 M have managed to cut cost through sustainability practices. The Economist (2008) reports that enhanced efficiencies in water and energy usage have led to a cost reduction of $2 billion at DuPont (since 1990), $50 million at Novartis in 10 years, and $82 million at 3M in 5 years. Moreover, at DuPont the greenhouse gas emissions went down by about 66%.

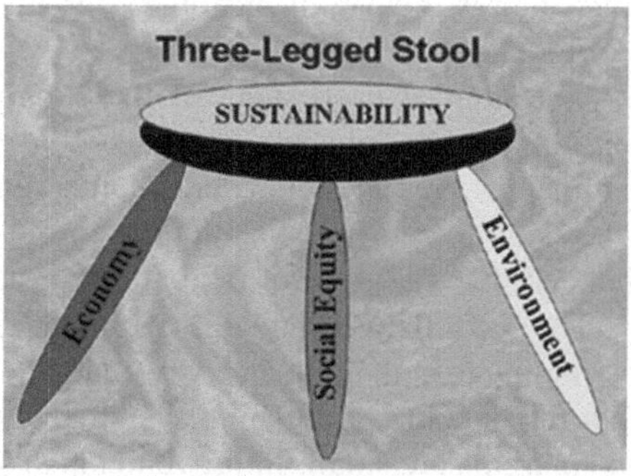

Fig. 9.1 Pillars of sustainability (Source: Web search)

A.K. Chakravarty, *Supply Chain Transformation*, Springer Texts in Business and Economics, DOI 10.1007/978-3-642-41911-9_9,

With increased globalization CSR has become a critical factor, especially where protection of brand value is a major concern. Companies are being assessed for their recruitment policies, workplace safety, and ethical practices. For example, the labor rights groups in Bangladesh have implicated major western retailers for not being willing to spend more to improve fire safety at a garment factory. These retailers were major customers of the factory that was burned down in a devastating fire in November 2012 and caused the death of 112 people. The retailers, including Wal-Mart, Gap, Target and Macy's, announced a plan that commits $42 million for worker safety that included regular inspections and an anonymous hot line for workers to report concerns about their factories. The plan also adds more than $100 million in loans to help local factory owners correct safety problems (Greenhouse and Clifford 2013).

In another example, the Anglo American company in South Africa has shown how to turn a problem into opportunity through CSR. In 2002, the company faced a major problem with rampant spread of HIV among its mine workers in South Africa. It found a novel way of reducing absenteeism in its mines by offering HIV testing and free anti-retroviral treatment. The company reports that the project is slowing the spread of the disease among workers, and is funding itself through reduced absenteeism.

Factors causing disruptions such as terrorism, infectious diseases, natural disasters, and human errors, can be catastrophic to economic sustainability. The likelihood and impact of such events are hard to predict and measure, as borne out by hurricane Katrina with an estimated economic loss of over $125 billion (Finkle 2005). Companies must be prepared with strategies for rapidly mitigating the impact of the disaster, to ensure business continuity. Another example is the recall of toys made in China by Mattel in 2007 that caused considerable disruption in Mattel's supply chain (Hoyt 2008). The recalls were for excessive lead content in toys, and exceeded 20 million toys. It sparked intense criticism of Mattel and its Chinese supply chain. The company's stock price fell 18% in 2007.

It can be seen that the supply chain of a company can have a major impact on each of the three pillars of sustainability. The supply chain provides the linkage for assessing impacts on the environment, on the society, and on economics; the decisions on resource use pattern, global reputation and brand value, and disruption risk entail the entire supply chain. In the context of Chap. 1, the supply chain must address managing customers, procurement, production, and delivery as shown in Fig. 9.2.

Note that at each stage of the supply chain in Fig. 9.2, there are social and environmental impacts. The stakeholders in the supply chain include logistics companies, researchers, consultants, governments, and risk managers. With a large number of stakeholders, the accountability of organizations to their stakeholders becomes critical at every stage.

The fact that sustainability practices can provide competitive advantage is beginning to sink in. In a survey conducted by the Economist (2008) it was revealed that the benefits of pursuing sustainable practices now outweigh the costs, even if the margins are small. Improvements in sustainable practices can help reduce costs

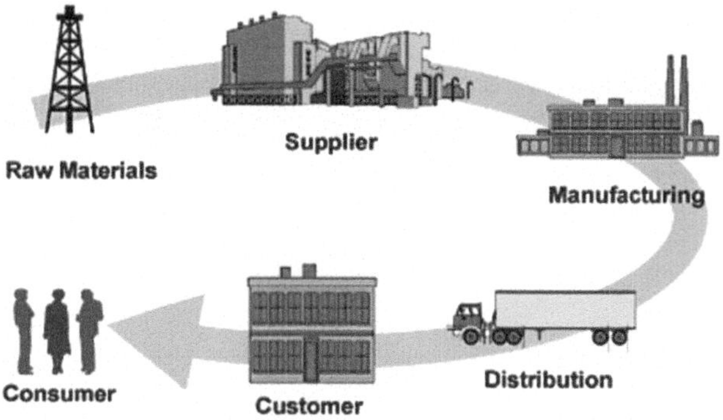

Fig. 9.2 The supply chain (Source: Web search)

(energy and water consumption, for example) and improve business reputation. It can also provide new business opportunities by developing novel sustainable products that match customers' preferences. One example is GE's low-energy light bulbs that added $20 billion to its profit (Economist 2008), while improving sustainability. However, in some cases the cost of implementation can be a barrier to adopting sustainability practices.

Perion based in Budapest, Hungary, makes batteries. It has created a new service for reprocessing car batteries. The service is based on its patented technology that has significant environmental benefits. This service sprang from its efforts to avoid heavy fines for hazardous discharges and to improve safety and environmental performance. It collected used batteries for reconditioning that brought additional revenue for the company.

It is clear that all stakeholders must act to produce common sustainability benefits. Individual companies need to figure out what such actions mean for them, how much to invest on the pillars of sustainability, and how to turn problems into opportunities through innovations.

9.2 Sustainability Drivers

The business case for the sustainability initiatives of a company depends on a variety of issues including industry, supply chain structure, stakeholder expectations, and organizational culture. Sustainability practices that respond to multiple drivers have the potential to maximize the value of the company.

As shown in Fig. 9.3, the major drivers of sustainability are ecological balance, resource availability, business reputation (brand value), climate change, and economic viability.

The benefits accrued from sustainability can be grouped as opportunity, efficiency, and reputation, as shown in Table 9.1.

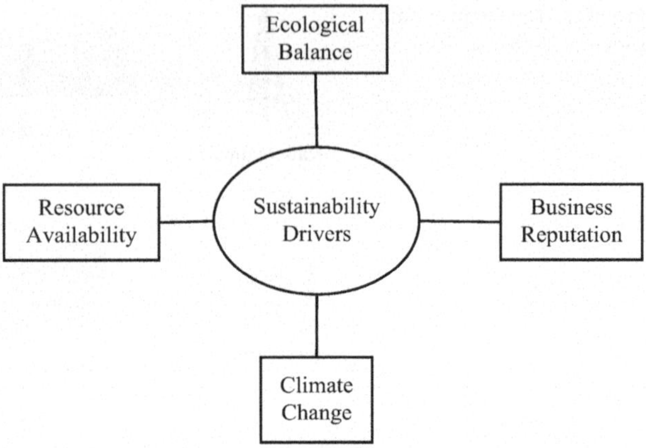

Fig. 9.3 Sustainability drivers

Table 9.1 Benefits of sustainability

Opportunity	Efficiency	Reputation
Increased competitive advantage	Energy efficiency	Brand reputation
Access to new markets	Materials or waste efficiencies	Positive perception of management
Innovation in products and services	Employee productivity	Regulatory compliance
Increased margins	Innovation in business models and processes	Ability to attract and retain top talent
Market share due to sustainability positioning		Enhanced stakeholder/ investor relations

9.2.1 Ecological Balance

Environmental changes such as deforestation, shrinking biodiversity, and greenhouse gas emission, indicate that developmental activity may be outpacing the regenerative capacity of the planet. Ecological footprint analysis is a way of comparing the demands on nature with the biosphere's ability to regenerate resources and provide services. It does so by assessing the productive land and marine area required to produce the resources a population consumes (food, housing, transportation, consumer goods and services) and to absorb the corresponding waste, using prevailing technology. This includes land used for gardens, agriculture, pasture, forest, and land of limited availability (untouched forests and non-productive areas such as deserts and icecaps). This analysis, in the context of historical trends, establishes a framework for identifying options, setting goals, and tracking progress toward stated goals.

In Fig. 9.4 the delicate ecological balance (or lack of) is shown. Improving recycling, adopting closed-loop manufacturing, and extending the lifespan of products, can help balance growth with its environmental impact.

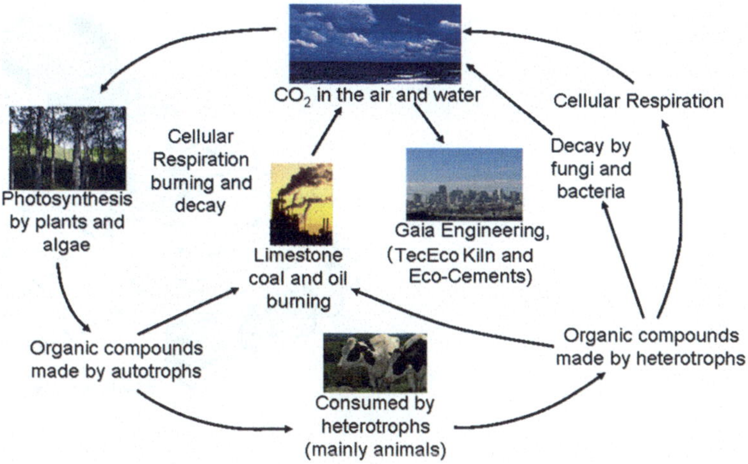

Fig. 9.4 Ecological balance (Source: Web search)

Environmental considerations are creating new business opportunities. Carbon footprint services are growing rapidly. Products in the wind, solar and bio-fuels markets have risen ten-fold over the 10 years (UNEP 2012). More efficient clean technologies are increasingly competitive and driving sales growth and duration.

9.2.2 Resource Availability

Brundtland (1987) defines sustainability as "development that meets the needs of the present without compromising the ability of future generations to meet their needs." It highlights business's dependence on human and natural resources, in addition to physical and financial capital. It emphasizes that economic activity must not irreparably degrade or destroy these natural and human resources.

Efficient activities have been widely adopted throughout the supply chain as they are generally easier to justify economically. However, there is still enormous scope for building on efficiency to doing "better" with less. Most companies are discovering that it is relatively easy to obtain resource efficiency at the process level.

Resources can be classified into three categories: abundant, renewable, and non-renewable. Abundant resources such as sunlight and wind do not diminish with use. Renewable resources such as timber and agricultural produce can be harvested and regenerated, using other resources such as land and water. While water can be considered an abundant resource, fresh (non-saline) and clean water can be very limited. Similarly, arable land is not plentiful, and forest-land is being denuded rapidly (Fig. 9.5). Fossil fuels such as the gasoline-crude and metal-ores are considered nonrenewable, as they are formed by geological processes over hundreds of years!

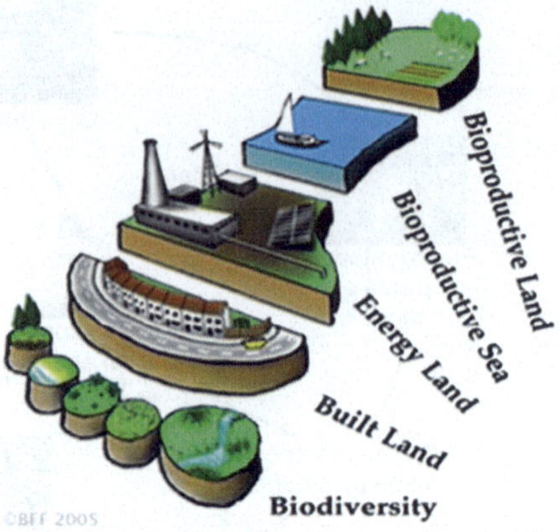

Fig. 9.5 Natural resources (Source: CBFF 2005)

Depletion and impairment of resources are two major concerns in sustainability. One way to slow down resource depletion is to "do more with less" and reuse. Reducing fuel consumption per mile in cars, reducing energy and water use per activity (by better design of products and processes), and minimizing the distance traveled in delivering goods, are examples of doing more with less; recycling is an example of reuse. Another way is to find substitutes such as bio-fuel for gasoline. It calls for innovation and R&D in identifying new products, processes, substitute resources, and creating transition to the new resource.

Use of certain types of resources causes impairment to the environment. For example, burning of fossil fuels accounts for almost 70% of all emission of greenhouse gases (GHG) such as CO_2. Use of timber (for fuel and furniture) accounts for 30% of deforestation. Consumption of food and fuel accounts for 25% of all GHG emission (Muilerman and Blonk 2001).

Decoupling implies that normal economic activities can be carried out without adversely impacting sustainability. However, a complete decoupling may not be possible, irrespective of innovations and improvements in efficiency – resource use efficiency cannot be increased indefinitely. Even if a short term decoupling is obtained, it may not be possible to preserve it in the long run.

9.2.3 Business Reputation

Sustainability is a key component of reputation for companies, customers and investors. A company's actions that improve its environmental and social performance go a long way towards creating a positive reputation for itself. Reputation, though not tangible, helps to increase sales, attract investment, and recruit workers. Measurement of reputation can be assessed through customer satisfaction surveys, and public opinion polls. Another proxy for brand value is the difference between the company's book value and its worth on the stock market. Reputation matters because it enhances the ability to mitigate risk, and provides the company the "license to operate". A good reputation also provides a cushion against minor errors. Reputation can improve when companies receive recognition and awards from organizations, governments, and rating agencies.

Jolyka Bolivia is the South American producer of tropical hardwood flooring products, which it sells to sophisticated consumers in US and Europe who care for sustainability. In 2000, Jolyka won the business plan competition for social and environmental benefits, which is a World Resources Institute program (Bassett 2003). This brought direct reputational benefits, including coverage in the media, as a viable and dependable green company. It also helped to raise new capital; Jolyka was visited by four investors and has renegotiated $2 million in long-term debt finance.

9.2.4 Climate Change

In the last two decades global insured losses from major extreme weather events have increased, averaging tens of billions of dollars annually. For example, the severe heat wave of 2010 in Russia caused massive wildfires and drought that destroyed crops (Gledhill et al. 2013). The resulting economic losses exceeded $15 billion. The export restrictions on wheat in Russia contributed to global price increases. According to the World Bank, the global mean temperature has increased by 0.8 °C in the last 50 years or so.

The two major factors that determine the extent of the adverse effects of climate change are the type of commodity and their concentration (Gledhill et al. 2013). Some commodities are more susceptible than others to changes in temperature and precipitation, rise in sea levels, and occurrence of storms and flooding. If a commodity is held by a small number of suppliers, disruption for any one major supplier can have global implications.

Climate change can cause adverse effects on businesses such as interruptions, and increased investment and insurance costs. The risk of adverse impact can be addressed by management of the three types of risk: risks to core operations such as physical plants, risks to the value chain, and risks from infrastructural disruption (Sussman and Freed 2008).

Core operations such as production facilities and buildings can be damaged. Physical structures and assets may require design and/or location change, affecting asset values. In the supply chain, the quality or quantity of inputs into production, or the demand for product can be affected. Supply of natural resources such as agricultural and forestry products may diminish. Water availability for production processes and irrigation may be restricted. Increased illnesses may increase health care costs and absenteeism. Demand for winter tourism and summer sports goods may fall. Disruption to utilities, especially electricity generation, water supply, and sewage can affect the supply chain. Flooding may damage transportation infrastructure or slow the delivery of supplies.

It should be apparent that the external and internal drivers are predicated upon a specific set of capabilities for sustainability. In Fig. 9.6, we show these capabilities using the framework developed in Chap. 1. Figure 9.6 embeds the three pillars of sustainability shown in Fig. 9.1, in its framework.

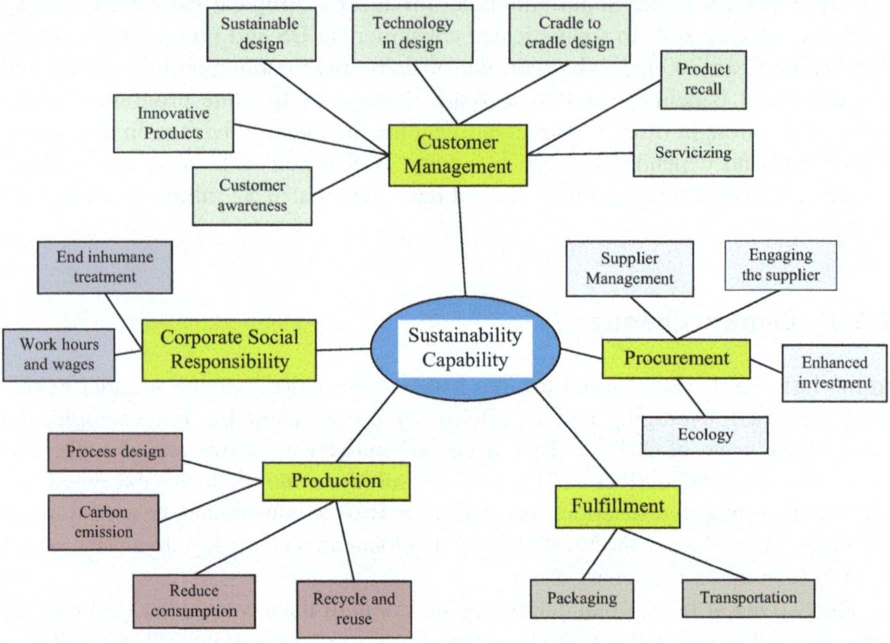

Fig. 9.6 Sustainability capabilities

We frame the sustainability capability clusters in the context of supply chain capabilities in Table 9.2.

Table 9.2 Supply chain sustainability capabilities

	Supply chain capability			
Sustainability concerns	Managing customers	Procurement	Production	Fulfillment
Ecology		XX	XX	X
Business Risk	XX	X	XX	XX
CSR	XX	X	XX	
Managing Suppliers		XXX		
Economics	X	XX	XX	X

Note that the sustainability concerns impact the procurement capability most and, to a lesser degree, the production and customer management capabilities.

9.3 Managing Customers

The role of customer management (discussed in Chap. 2) is to create value for the customer without compromising the value to the suppliers. In the context of sustainability, the customer management role becomes critical, as many customers are not aware what sustainability has to do with their bottom lines. Specifically, they need to understand whether the use of sustainable products and processes increases the their revenue and/or decreases their cost?

Innovative product redesign, cradle-to-cradle design, using new technology in design to minimize emission, product recalls, and servicizing, are some of the ways customer's value can be increased through sustainability.

9.3.1 Product Design

Design choices about durability, disassembly, and reusability affect the economics of product recovery. The challenge is to develop a framework for incorporating sustainability in design. Designers must appreciate the interactions between product attributes, sustainability, and consumer behavior (Edwards 2009).

9.3.1.1 Sustainable Product Design

Sustainable products must address the economic, environmental, and social effects. For example, dye for garments, pharmaceuticals, and cosmetics can adversely impact the biochemistry downstream of the supply chain. Material weight and packaging can have direct impacts on transportation costs and fuel use. Energy efficient products can impact energy consumption.

Fail-safe design ensures that the product either ceases to function or alerts the user when certain performance limits are violated. It prevents the user from making mistakes that cost time and effort. For example, in some car designs audio alert sounds if the headlights are left on. The same principle can be applied to sustainability. Some Singapore taxis are equipped with a bell that chimes when a certain speed is exceeded, alerting the drivers to slow down and save fuel.

Unsustainable products cause environmental impairment mostly when they are being used. For example, high consumption of energy and water, and creation of a huge amount of waste are some of the results of using unsustainable products. Sustainability drives have now led to the integration of renewable energy sources and innovative technology for energy efficiency into consumer electronics products. Clearly, the product design team needs to be aided with information on user profile, and the team should be trained to adequately assess the interactions between product features, user profile, and sustainability impacts.

The information needed by the design function, related to consumer choice and consumer action, as shown In Table 9.3 below.

The life cycle perspective that considers the pattern of material and energy use in procurement, production, and transportation, can identify the supply chain issues that can be tailored to environmental compatibility. This process can be framed in a

Table 9.3 Types of information communicated

Focus		Description
Consumer choice	Product attribute	Information about sustainable attributes of products
	Operational performance	Information about sustainable performance of corporate operations
Consumer action	Product use	Information on how to reduce negative impacts during product use
	End-of-life	Information about extending product life cycle and/or end-of-life actions consumers can take

quality function deployment (QFD) context, which can unearth significant opportunities to reduce the energy and material usage. In addition, products should be designed using material that can be easily reused or repaired when they have outlived their usefulness. Note that a bio-degradable product, per se, does not make it superior to a synthetic product in terms of sustainability; it may be more toxic or resource intensive to produce.

For example, consumers usually use cell phones for a life cycle of 12–18 months before it is replaced. Therefore, the phone should be designed for ease of recyclability to minimize disposal effort. The designer must select the amount and mixture of materials from which the phone is made to make disassembly easy. The design must also ensure that as few material types as possible are used during manufacturing (Donato 2008).

9.3.1.2 Innovative Products

Using an interdisciplinary approach that combines knowledge from microbiology, electrochemistry, materials, and environmental engineering, scientists have discovered a microbial fuel cell (MFC) that converts chemical energy present in organic waste biomass to electrical energy (Reddy et al. 2010). It provides new opportunities for the sustainable production of energy from biodegradable compounds.

Columbia Forest Products applied the principle that imparts adhesive ability to the blue mussel, to design a soy-based formaldehyde-free technology in the construction of hardwood and plywood products. The WhalePower Company designed wind turbine blades using the "Tubercle Technology" to help them be quieter, more reliable when winds fail, and perform better in turbulent winds (WhalePower 2013). The technology is based on the idea that humpback whales are agile in spite of their size because of bumps on their fins.

P&G discovered that by switching to cold water washing, US households would consume 80 billion fewer kilowatt hours and emit 34 million tons fewer CO_2. The company developed a cold water detergent that was received enthusiastically in Europe and the US. It's a win-win-win for P&G, consumers, and the environment (Martin and Rosenthal 2011).

Clorox teamed up with Sierra Club to develop a sustainable product called Green Works, which is a non-synthetic cleaning product. Clorox claims that these products are 99% natural plant-based ingredients mostly derived from coconut and lemon oils and are delivered in 100% recyclable packaging. The company spent $20 million in developing Green Works, and priced it with about 20% premium over conventional cleaners. Clorox captured 40% of market share in 2008 before the recession.

9.3.1.3 Technology in Design

Technology permits incorporation of ecological concerns in the product in an innovative way. Technology oriented changes such as the use of more energy efficient semi-conductors or better performing auto engines, can enhance sustainability. For example, "Green Lights" is a program that seeks more light from less electricity.

Evolving new technology is now enabling novel combinations of materials that produce products with specific quality and ecology implications. New guidelines are available that point the designers towards ecology friendly selection of material. Possible criteria for selection of material would be nontoxicity, naturally occurring such as cellulose rather than artificial such as chlorinated aromatics, material with existing recycling infrastructure, and recycled material (UNEP 2004). Solvent substitution, in which a toxic solvent is replaced with a more benign alternative such as a biodegradable solvent or a non-toxic solvent, is becoming more realistic. From a sustainability perspective, water based solvents would be preferable to organic based solvents.

Recycling of waste reduces dissipation of toxic material into the environment and minimizes unnecessary production. For example, rechargeable nickel-cadmium batteries can be recycled to recover both cadmium and nickel for other uses. Inmetco Corporation in Pennsylvania recycles such batteries using pyro-metallurgical distillation (Hendrickson et al. 1999).

Electronic identification can be attached to materials or products that reveal information about material handling and waste management. For example, plastic bottles used in consumer products usually have an identification symbol that can be used in recycling management.

9.3.1.4 Cradle to Cradle Design

Cradle to Cradle (C2C) product design eliminates disposal of used products in a landfill. The C2C products are either biodegradable or are returned to the manufacturer after use. In designing such products, it is necessary to understand how much of a product (and its components) would be recyclable. Recycling or a closed-loop implies reuse and, therefore, a reduction in the use of fresh components and raw material. In a closed loop, companies must rework their profit models. The closed loop cycle begins with collection of the used products, disassembly, restoration, testing, remanufacturing, inventory management, and ends with transportation.

Consider the sole of a shoe that can be made from a biodegradable material, while the upper parts made of recyclable material (Wikipedia 2013). When the shoe is worn out, it is returned to the manufacturer. The manufacturer separates the sole from the upper parts. The degradable material is disposed-off, while the recyclable material is reused to create the parts of another shoe. Another example of C2C design is a disposable cup (or bottle, or wrapper) made entirely out of biological materials. Interface, a carpet manufacturer, claims a 92% reduction in landfill waste in 2011, and a 75% reduction in greenhouse gas emissions, despite selling more products Rewrap (2013). They convert used fishing nets into carpet material, and recycle the nylon from their carpets. Rewrap makes the sleeves of laptop-covers using biodegradable wool and non-toxic dye with a minimum use of other materials.

9.3.2 Product Recall

Unless appropriate actions are taken by a company, defective and harmful products may reach customers. They may trigger incidents with unwanted outcomes, leading to product liability litigations, damage to the company's reputation, and even financial ruin. Often, it is relatively less expensive to recall the product and compensate the customers for damages, if any.

Note that recalls may cause disruptions in the supply chain of the company, which may also lead to losses. In the U.S. auto industry the annual cost of recalls for consumer products is estimated to exceed $700 billion (Wachs 2013). In recent years, some of the common causes of recall have been accelerator pedal problem (Toyota, Dodge Calibers, and Jeep Compass), brake fluid leaks (Jeep Wranglers), sliding-door hinge problem (Dodge Caravan), and windshield wiper fluid (GM) (Wachs 2013). Other examples of recall include toys by Mattel, blood-thinning drug Heparin by Baxter, and contaminated beef in California (Lefevre et al. 2010).

The impact of a product recall can be very detrimental to a company and causes its operations to be unsustainable. It requires reverse logistics, which is a major cost item that includes costs of customer returns, repairs, replacements, refunds, legal liabilities, disposals of returned products, and the cost of repairing brand image. According to Davidson (1984) the recall of the Ford's Pinto cost the company $125 million in punitive damages; recall of Pentium microprocessors in 1994 cost Intel $500 million.

Thus the company must choose whether it wishes to invest in designing the product to minimize the probability of recall or incur the cost of a possible recall. Johnson and Johnson, following a recall in 1982, redesigned Tylenol as caplets instead of capsules (Catalin Group 2012). The caplets were tamper proof, whereas, the capsules laced with cyanide triggered the recall.

Companies need to do rigorous testing for product performance and reliability under a range of operating conditions. They should create flexibility in the design process so as to be able to quickly redesign the product based on the reliability test. This may necessitate real time implementation of QFD with the design objective of sustainability. The possible marketing attributes would be: product attributes needing redesign, causes of failure, time to redesign, supplier implications, and new materials. Engineering attributes may include physical dimensions, input and output options, installation, custom component, and accessories – all geared to minimize another recall.

9.4 Servicizing

The objective of the product service system is not to just sell a product but to solve the customer's specific problem in using the product. These include switching the customer's current system over to the new product, and repair and maintenance of the product and its accessories. The customer's utility of using the product must increase. The customer pays for the functionality or result of the "product as a service" as in leasing, instead of buying the product itself. Typically the company that retains ownership of product is responsible for producing, maintaining and disposing off the product so that the whole value chain is made more sustainable.

PPG Industries sold paint to Chrysler auto plants. It charged Chrysler by the gallons of paint sold. Chrysler and PPG switched to a new contract where Chrysler paid by the number of cars painted (Rothenberg 2007). With this new contract, PPG was motivated to work closely with Chrysler to reduce the consumption of paint. PPG employees modified several functions at the plant, including material ordering, inventory tracking, and inventory maintenance. Through this increased service role, the company has helped Chrysler reduce material use. The simple analysis below helps explain it further.

Assume that the old and new ways of painting a car consumed x and y gallons, respectively ($x > y$). With a price and cost per gallon of p and c respectively, PPG's profit with the old system would be $(p - c)x$.

With the new system, assume that the price of paint for painting a car is fixed at P. It follows that for the new contract to be viable, we must have $(p - c)y \leq P \leq (p - c)x$. That is, PPG cannot do any worse than just selling y gallons in the open market, and Chrysler would not pay more than the cost of buying x gallons in the open market. Therefore, it would be a win-win proposition if $(p - c)x - P > 0$ and $P - (p - c)y > 0$. In addition, there is a sustainability benefit as fewer gallons of paint is produced and transported for painting a car, reducing GHG emission.

Another example is an energy saving company that optimizes customers' operations and gets paid according to the savings achieved. The customer pays according to the amount of service it uses. Similarly, the supplier of a chemical management service should accept the responsibility for managing chemicals of its customers and strive to reduce the associated costs and risks.

9.5 Procurement

Procurement of products, services, material, and components is an indispensable function in any business. Traditional procurement plans (discussed in Chap. 4) are based on unit price, quality, and usage rate of the item. This implies a tight control on items that are fast movers, critical, and/or expensive. However, procurement affects more than just the cost and quality of a product; it impacts the upstream carbon footprint. Therefore, for sustainable procurement additional factors must be considered. This includes toxicity of material (used in assembly or manufacturing), transportation, carbon footprint, packaging, and the social responsibility of suppliers.

9.5.1 Ecology

Driven by population growth and the purchasing power of people, global demand across all major primary resources is expected to grow: 33% (energy), 80% (steel), 27% (food) and 41% (water) (UNEP 2012). According to UNEP, the manufacturing sector is responsible for 35% of electricity use, 20% of CO_2 emission and 25% of primary resource extraction, directly impacting the environment and human health.

Advanced water treatment processes are now using innovative biological and physicochemical technologies that can improve the quality of waste water (Ngo et al. 2011). Pollutants such as nitrogen and phosphorous can be removed by biological processes. Physicochemical processes such as deep-bed filtration, floating media filtration, and membrane filtration, enable water reuse. Many large buildings in Japan have own treatment facilities to reuse waste water as flush water for toilettes.

Toxic substances can adversely affect human health and the environment. This includes heavy metals (lead, mercury), ozone-depleting substances, organic solvents, carcinogens, and phthalates (chemicals in plastics and solvents). To minimize the use of such substances the product would require a redesign with substitute material, increasing the cost of procurement.

The main function of packaging is to protect products during transportation, including transportation of raw materials from suppliers. When designing packaging, it is important to consider both the sustainability impact of the over-packaging and the impact of any product losses due to under-packaging. Sustainable packaging initiatives have focused on reducing the amount of packaging material used, reducing package weight, use of bio degradable material, increasing reusability, and using other sustainable materials.

The key ecological concerns in transportation are the use of non-renewable fuels, and the emission of greenhouse gases. Sustainability can be enhanced by measures such as using the right type of fuels in vehicles, using vehicles with high gas mileage, use of telematics that combines information about the driving-behavior and mechanical variables that affect fuel efficiency in the vehicles, and minimizing the number of trips per vehicle.

In China's Xinjiang Uygur region, Unilever is working with tomato producers to get them to switch to drip irrigation from flood irrigation (SPN 2011). It is a new farming technique that can achieve multiple objectives: create sustainable farming, ensure sustainable sourcing, save water and fertilizer, improve quality of tomatoes produced, and increase farmers' income.

9.5.2 Investment in Sustainability

Sustainability measures, in general, add to the cost of the material being procured. At the same time, as mentioned earlier, demands of products with sustainable components can in fact increase. However, not all items experience the same demand pattern. The interactions between the cost of sustainability and market response are shown in Fig. 9.7.

Fig. 9.7 Strategies for sustainable procurement

	Market responsive to sustainability	Market unresponsive to sustainability
High cost of sustainability	induce innovation to cut cost	high risk
Low cost of sustainability	Ideal	Low risk

It follows that the ideal materials are those with low cost of sustainability and strong market response. For these materials, investment in technology for increasing sustainability would be prudent. At the other extreme, with a tepid market response and high cost of sustainability, sustainability initiatives would be risky. Other reasons such as government regulation may force companies to bear the high risk for these substances. Finally, for substances with low costs of sustainability but unresponsive market, sustainability initiatives would be subjected to a low risk.

Consider the substances i in a product that has a demand of u. The supplier makes a sustainability investment in i. Assume the unit cost of i before and after the sustainability investment to be c_1 and c_2, respectively. Assume the demand enhancement of the product due to the improved substance i to be a random variable ξ (mean μ and variance σ^2) with probability mass function $f(\cdot)$. Clearly, the amount of i procured before the sustainability investment is u. Assume the purchased amount of improved substance i to be $u + q$.

The gain in the product's profit due to supplier's investment in sustainability can be expressed as

$$\pi(q) = p\left\{ \int\limits_{\xi \le q} (u+\xi)f(\xi)d\xi + \int\limits_{\xi \ge q} (u+q)f(\xi)d\xi \right\} - c_2(u+q) - (p - c_1)u.$$

As π is concave in q, optimal q is expressed as (follows from the first order condition), $F(q^*) = 1 - \frac{c_2}{p}$.

Hence, $\pi(q^*) = p \int\limits_{\xi \le q^*} \xi f(\xi)d\xi - u(c_2 - c_1).$

Clearly for the sustainability initiative to be feasible, it must be the case that $\pi(q^*) \geq 0$.

That is, $\int\limits_{\xi \leq q^*} \xi f(\xi) d\xi \geq \frac{c_2 - c_1}{p} u$.

Assuming $f(\cdot)$ to be a normal distribution, $F(q^*) = 1 - \frac{c_2}{p}$ is written as $\Phi\left(\frac{q^* - \mu}{\sigma}\right) = 1 - \frac{c_2}{p}$.

As the RHS in the above expression is constant, it must be the case that $\frac{\partial}{\partial \mu}\left(\frac{q^* - \mu}{\sigma}\right) = 0$ and $\frac{\partial}{\partial \sigma}\left(\frac{q^* - \mu}{\sigma}\right) = 0$. Hence, $\frac{\partial q^*}{\partial \mu} = 1$ and $\frac{\partial q^*}{\partial \sigma} = \frac{q^* - \mu}{\sigma} \geq 0$.

We can also express the feasibility condition $\int\limits_{\xi \leq q^*} \xi f(\xi) d\xi \geq \frac{c_2 - c_1}{p} u$ as

$$\sigma^2 \int\limits_{z \leq \frac{q^* - \mu}{\sigma}} z\phi(z)dz + \mu\sigma\Phi\left(\frac{q^* - \mu}{\sigma}\right) \geq \frac{c_2 - c_1}{p} u.$$

Thus, for items that are fast moving (high u), it is harder to justify sustainability initiatives.

Letting $LHS = \sigma^2 \int\limits_{z \leq \frac{q^* - \mu}{\sigma}} z\phi(z)dz + \mu\sigma\Phi\left(\frac{q^* - \mu}{\sigma}\right)$, we can establish that $\frac{\partial}{\partial \mu}(LHS) = \sigma\Phi\left(\frac{q^* - \mu}{\sigma}\right) > 0$. Therefore, the sustainability initiative becomes more feasible if the market becomes more responsive to sustainability (as LHS increases in μ).

We can rewrite the feasibility condition as $c_2 \leq \frac{p\sigma}{u}\left\{ \int\limits_{z \leq \frac{q^* - \mu}{\sigma}} \sigma z\phi(z)dz + \mu\Phi\left(\frac{q^* - \mu}{\sigma}\right) \right\} + c_1$.

Next, note that $\frac{\partial}{\partial \sigma}(LHS) = 2\sigma \int\limits_{z \leq \frac{q^* - \mu}{\sigma}} z\phi(z)dz + \mu\Phi\left(\frac{q^* - \mu}{\sigma}\right) > 0$. Therefore, it follows that sustainability becomes more feasible if the volatility in market response increases.

9.5.3 Supplier Selection

Supplier selection was discussed at length in Chap. 4. With the inclusion of sustainability as an objective, supplier selection becomes more complex. In a strategic sense, as shown in Fig. 9.8, the taxonomy can be described in two variables: supplier sustainability risk, and proportion purchased from a supplier.

The four categories of suppliers in Fig. 9.8 are based on high/low values of these variables. In terms of procurement, suppliers in categories D and A have high and low priority, respectively. Suppliers in B and C categories have medium priority. These categories are explained, next.

Fig. 9.8 Priority for supplier's sustainability improvement

Supplier sustainability level	Low	High
Low	C medium priority	D high priority
High	A low priority	B medium priority

Percentage of procurement from the supplier

9.5.3.1 Category A

Suppliers in this category are those who have already implemented sound sustainability practices and therefore pose a low sustainability risk to the buyer. However, the supplier's share of the firm's total purchases is low. Hence, the firm has a low priority for engaging this supplier in sustainability improvement. Because of the low purchase volume the firm would not be willing to pay a high price for sustainability of the supplier in this category.

9.5.3.2 Category B

Suppliers in this category differ from those in category A in terms of the volume of business with the firm. Because of the high purchase volume, the marginal value of improving the supplier's sustainability would be high to the buyer. Hence, improving supplier's sustainability would be of a higher priority to the buyer than those in category A. One way would be for the buyer to specify the sustainability criteria in its procurement specification, such as the RFQ. The buyer can drive his/her purchase cost down by qualifying for volume discounts.

9.5.3.3 Category C

In this category, the suppliers would not have implemented many sustainability initiatives and therefore the sustainability risk would be high. Note that with a low sustainability level (high sustainability risk), a large number of suppliers would belong to this category. Therefore, the firm would have many options, if it wanted to switch suppliers.

By and large, supplier's cost for improving sustainability increases more than linearly. Therefore, it is instructive to look at Fig. 9.9 to compare categories A and C.

As the supplier in category C is at a low sustainability level, the supplier's marginal cost (hence marginal price) of enhancing sustainability is lower than that the supplier in category A (as shown in Fig. 9.9). Therefore, the priority for improving the sustainability of a category C supplier would be higher than that for a category A supplier. Clearly, the starting point would be to invite a proposal from the supplier for improving sustainability. As the suppliers in category C (with a low endowed sustainability level) would need to climb a high cost curve for sustainability improvement (Fig. 9.9), supplier development projects would most likely fail. In such a case, with a large number of suppliers in this category, the buyer would have other supplier options to explore.

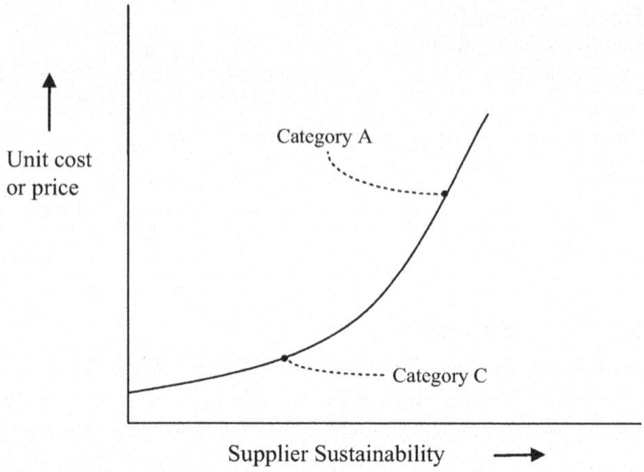

Fig. 9.9 Unit cost versus sustainability

9.5.3.4 Category D

Note that the sustainability level (and the marginal cost of sustainability) of a supplier in category D is the same as that of a supplier in category C, and lower than those in categories A and B (Fig. 9.10). However, with a larger purchase volume, the unit price of a category D supplier will be lower than that in category C. Therefore, the buyer would assign the highest priority to enhancing sustainability level of a category D supplier.

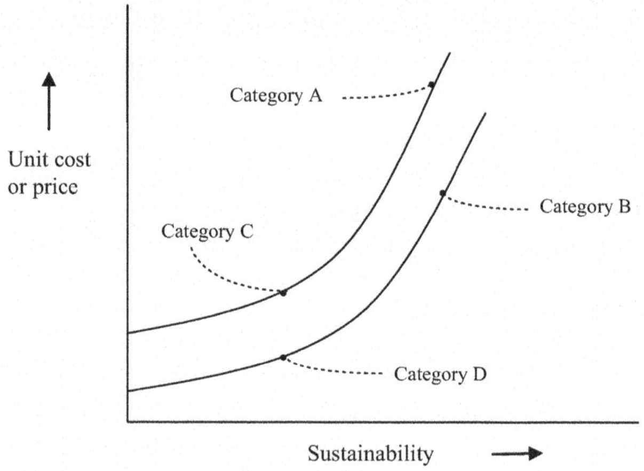

Fig. 9.10 Unit cost versus sustainability

The buyer needs to collaborate closely with the supplier to ensure that the supplier invests in both sustainability improvement as well as cost reduction. The supplier must develop a culture for cooperating with the buyer in improving sustainability. In addition, it may be necessary to monitor the supplier to ensure its improvements.

9.5.4 Supplier Engagement

The goal of engaging with suppliers is to develop a shared mindset about sustainability issues, and to craft a shared vision of sustainability. It involves setting expectations, monitoring, remediation, and partnership (methods of indirect development, as discussed in Chap. 4). To include sustainability, the buyer must communicate its expectation of the supplier in terms of material use, emission standards, and the work-environment. It does so through dialog where suppliers can raise their concerns about sustainability stipulations such as changes to orders and product specifications. The agreed upon standards, to be effective, need monitoring. It could be in the form of supplier self-assessment or as a compliance audit. Compliance audits conducted on-site can be invasive. Audits aim at gathering enough information to ensure understanding of factory conditions, and rely on procedures including: visual inspection of facilities, random on-site interviews, and daily work records. Remediation includes corrective action for non-compliance and suppliers' investment in capabilities. Cisco Systems coordinates supply and demand across its supply chain with an intelligent-software for early detection of conflicts and their resolutions (Lee and Whang 2001).

Capability building includes training for supplier personnel, improving working conditions, strengthening communication between managers and workers, and establishing resource networks (direct development approaches, discussed in Chap. 4). Suppliers can build capability through acquisition, learning, and through research. Acquisition can be in the form of a purchase of a new processing technology or a firm specializing in this technology. They could learn new expertise by doing it in-house (experimentation). Finally, they could invest in R&D to develop new processes and products.

Process capabilities can be built by creating routines. Routines may involve explicit knowledge as in verbal instructions and operations manuals, or implicit knowledge as in team-work.

9.6 Production

The manufacturing sector consumes a large amount of raw materials, and a large portion of the global energy supply. High resource consumption makes manufacturing a critical unit in terms of implementing sustainability measures. Manufacturing processes must be reconfigured not only to use fewer resources but also to understand how those resources are being used throughout the production process in terms of their social responsibility and environmental impacts.

9.6.1 Process Design

Manufacturing process redesign can be very effective in improving sustainability, as manufacturing processes strongly impact the social and ecological environment. Redesign implies devising new solutions and technologies that may improve process effectiveness. It requires knowledge of technology capabilities, and the effects of changes on sustainability.

Manufacturing processes for sustainability can be optimized in the context of life cycle analysis (Shonnard and Hiew 2000). It involves definition of the process boundaries and quantifiable sustainability impacts in the form of established metrics, incorporated into process design and optimization. It has been applied to determining waste treatment options, abatement of pollution, and designing the optimal recipe of solvents. Impact indices, such as ozone depletion potential to human toxicity and eco-toxicity, developed by the EPA, can be used. This method has been applied in a methyl ethyl ketone production plant to determine the effect of recycling on the environment (Shonnard and Hiew 2000).

9.6.1.1 Manufacturing Processes

Increased productivity, reduced plant operating costs, and reduction in work effort improves utilization of resources: energy, raw materials, human resources, information, and equipment. Taken together, the capability to meet the immediate needs of the plant has a positive impact for future generations as well and hence sustainability (Rockwell Automation 2009).

Manufacturing automation can reduce energy consumption, minimize scrap material, contributing to environmental stewardship. Automation can also be used to make cost reduction decisions, such as running plants during off-peak hours and selling surplus energy back to the grid, contributing to corporate responsibility.

Today's control systems were built for load curtailment and load shedding, generator control, thermal plant control systems, and energy monitoring. Although these were not developed for sustainability per se, they do contribute to energy conservation and control, and energy management solutions.

9.6.2 Reduce Consumption

One of the ways of achieving manufacturing sustainability is to reduce the generation of waste, reduce consumption of raw materials, reduce pollution, and reduce toxicity of material used in the manufacturing processes. In the metal removal processes, for example, the use of lubrication and refrigeration fluids can be reduced by redesigning the cutting process (Happala et al. 2013). Considerable progress has been made in dry metal cutting processes to avoid toxicity and pollution from lubricants. New fluids with reduced toxicity are being developed.

Wausau Tile reduced the use of natural raw materials by substituting them with used glass as a concrete aggregate (OECD 2011). Even though using broken glass increased cost, it was offset by developing new products, attracting new customers and reducing the negative environmental impact. The company also redesigned a number of their products to include large glass chips that were large enough to be architecturally and aesthetically valuable. It has recently introduced a new line of products that use post-industrial porcelain like sinks, bathtubs and toilet bowls as aggregates (Wausau 2013).

Other approaches include using optimal process parameters, different chemical compositions, appropriate lubricating strategies, and changing tool coatings. Food and glass manufacturing industries have successfully reduced the production of waste.

Energy, just as raw materials, must be managed as a resource. For example, in the processing plant of a large brewing company energy efficiency was low because of a number of factors: constant machine breakdowns, inadequate spare part tracking, and insufficient fault diagnostics. By implementing an integrated control system, the plant achieved a significant reduction in machine downtime, and gains in productivity. Energy consumption can also be reduced by integrating hybrid sources of energy. Consumer goods, such as concentrated detergents for laundry and dishwashing, offer benefits by reducing packaging and transportation requirements.

9.6.3 Recycle and Reuse

Recycling is one of the waste management techniques for decreasing landfill disposals. Leading companies strive for zero emissions, which means that all materials currently sent to landfills must be returned "upstream" to the materials and production cycle as feedstock for new products or services. They could also be replaced by materials that decompose naturally and are reintegrated into nature without negative impacts on the environment.

A supply chain with recycling includes reverse flows of products and components, from the end user to the OEM. Lead acid batteries are a good example of recycling of consumer durable products, with a high recycling rate. Because of the ever decreasing product life cycles of electronic gadgets, electronic components are a growing constituent of the waste stream. However, they currently attain only a 15% recycling rate (Pagell et al. 2007).

Note that recycling may occur with or without disassembly. If it is not feasible to disassemble the product, it is usually crushed and then the salvageable materials are reused. Consumer electronics are often refurbished by having their software upgraded. If the product can be disassembled before recycling, the components can be reused. Reusing durable products can offer substantial economic and environmental improvements. But products such as automotive parts and toner cartridges require extensive labor for disassembly, cleaning, and parts replacement. That notwithstanding, remanufactured cartridges can be produced at 20% of the cost of a new cartridge. Recovered components such as computer memory chips can also be sold to intermediaries.

Companies such as HP and Xerox have devised supply chain infrastructures to include recycling operations (Rose 2001). HP considers recycling to be a strategic opportunity and, therefore, they incorporate microchips into the cartridges to prevent unauthorized remanufacturing.

Intel operates a system that introduces a new manufacturing process technology in the first year of semiconductor development, allowing, for instance, reductions in semiconductor size and the subsequent manufacture of more semiconductors on a single wafer or placement of more transistors in an equivalent space. In the second year, this model introduces a new chip architecture or design with the same manufacturing technology. Each step provides the opportunity to establish objectives and strategies to reduce environmental impact, and Intel has worked with suppliers of semiconductor manufacturing equipment and materials to improve the environmental performance of various technologies using this approach.

9.6.4 Carbon Emission from Manufacturing

As mentioned earlier, carbon emission remains the greatest threat to sustainability. Carbon footprint is a way of measuring this emission. Governments and regulatory authorities have suggested trading in emission with a view to controlling it.

9.6.4.1 Carbon Footprint

Manufacturing is the single largest source of carbon dioxide emissions, as many processes are energy intensive. Industrial activities account for 30% of the emission, and power generation for 40%. Energy consumed directly during manufacturing is responsible for 90% of the total CO_2 emission. Manufacturers have several options available for reducing carbon emission: switch to non-fossil fuels; improve efficiency of manufacturing processes; switch to new technology, processes, and materials that emit less CO_2; and mitigate existing emission through initiatives such as reforestation, and use of cap and trade (OECD/IEA 2003).

The carbon emissions from in the major industrial sectors are shown in Table 9.4.

Table 9.4 CO_2 emission

Industry	Millions of tons per year
Petroleum	305
Chemical	311
Primary metal	213
Paper	102
Plastics	80
Food	95
Cement and glass	91

9.6.4.2 Emission Trading

Emission trading is a tool for mitigating greenhouse gas emissions, and it became part of the Kyoto Protocol agreement. It is a market-based tool that can be used by a national government to reduce pollution in the atmosphere. Under this scheme, the government sets a limit on the allowable emissions, which is gradually lowered over time towards an equilibrium target for the nation. Permits are issued to companies indicating their right to emit a specific amount of pollution. These permits can be traded in the financial market. The European Union Emission Trading System has implemented the most advanced scheme in which certain CO_2 emitting industries of the EU member states must comply with individual emissions allowances assigned via a national allocation plan (EU ETS 2013).

There are two sets of decision involved. The government needs to decide the upper limit of emission it would allow, with a provision to exceed the limit using permits that can be bought. Firms need to decide their individual emission levels, and the number of permits they would purchase or sell; the emission level may exceed the target set by the government.

9.6.4.3 The Trading Model

Consider the case where the government sets a cap, K, which is the upper limit of emission for all companies, and allocates a fixed number, A, of tradable permits to each firm. We assume that the initial allocation of A units is without charge. It follows that if the actual emission level of a firm, x, exceeds the cap, it must make up the difference of $x - K$ units from the allocated permits. If $x - K$ exceeds A, it must purchase additional permits from other companies. Firms with emission level below $K + A$, may offer to sell surplus emission rights to others. That is, the firm may sell permits if $x < K + A$. As a result, the total cost incurred by companies in achieving the environmental goal set by the government may be lower than that of a system without tradable permits. OECD (2002) reports application of cap and trade for controlling GHG emission. Individual firms must decide the optimal values of their emission levels x, and the amount of tradable permits, $y\ (= K + A - x)$, to sell. Note that y can be positive or negative, based on whether permits are bought or sold.

Consider two firms with initial and final emission levels of e_i and x_i units, respectively ($i = 1, 2$). Assume the firms trade y_i units of permits at a unit price p. Assume that the cost of reducing the emission level can be expressed as $C = c_i(e_i - x_i)^2$, the quadratic form captures the fact that it becomes progressively harder to reduce x_i, especially when x_i/e_i is low. Note that the firm will reduce emission level until the marginal cost $\partial C/\partial x_i$ equals p; at that point it will purchase tradable permits if $x_i > K + A$. Therefore, $-2c_i(e_i - x_i) = p$, implying $x_i = e_i + \frac{p}{2c_i}$ and $y_i = x_i - (K + A) = \left(e_i + \frac{p}{2c_i}\right) - (K + A)$. As $\sum_{i=1,2} y_i = 0$, we must have

$$p = \frac{2c_1c_2}{c_1 + c_2}\{2(K + A) - (e_1 + e_2)\}$$

Therefore, it can be established that for $i = 1, 2$; $x_i = \frac{2(K + A)c_{3-i} + c_ie_i - c_{3-i}e_{3-i}}{c_i + c_{3-i}}$ and $y_i = \frac{(K + A)(c_{3-i} - c_i) + c_ie_i - c_{3-i}e_{3-i}}{c_i + c_{3-i}}$.

Thus, under perfect knowledge, the trading price p will be set by the market, as established above. The emission levels and the number of permits bought/sold by each firm are also determined optimally.

9.7 Corporate Social Responsibility

More and more companies are realizing that embracing responsibilities toward customers, employees and the community, in addition to stockholders, enhances sustainability of companies. This is known as corporate social responsibility (CSR). Responsible companies are concerned with reconciling and aligning the needs and values of employees, customers, suppliers, communities, and shareholders. To ensure integrity the company must institute accountability and transparency in its supply chain activities.

In the apparel industry, H&M operates about 2,000 stores globally. The company uses close to 700 suppliers to manufacture H&M products such as its new eco-clothing line called Conscious Collection. Many of the suppliers are located in countries where issues such as child labor and workplace safety can be problematic. H&M works with suppliers to promote labor rights and reduce negative impacts on the environment, and has an audit team of 70 people that monitor suppliers' compliance with the H&M code of conduct. To improve working conditions, H&M also provides training to its suppliers (Swedish Institute 2011).

The treatment of employees is an important aspect of CSR. In general, the standards for working condition include the prohibition of human rights violations such as forced labor, child labor, very low wages, excessive working hours, lack of industrial hygiene, discrimination, safety violations, and high fatality rates. Manufacturers should recognize that high employee-standards offer an opportunity to gain strategic advantage. These standards are important to consumers, as evidenced by their willingness to pay for products that are created under humanitarian working conditions. Consumers have become increasingly critical of the working conditions under which products are created.

Manufacturers are also held responsible if any part of their supply chain comes under criticism for inhumane working conditions, even if they have little control over their suppliers. They do, however, have the option of leveraging the sourcing of components to influence supplier's behavior. To implement it, they need to understand how deep into the supply chain should the CSR conditions be upheld and verified. In general, supplier engagement (discussed earlier) ultimately leads to organizational and social learning. This knowledge forms the basis for long-term change and trust, but is not clearly quantifiable or predictable in the short term.

Levi Strauss was a pioneer in establishing guidelines for global sourcing. The guidelines outlined the company's commitment and includes assessment of conducting business in specific countries, and the terms and conditions for identifying business partners with CSR values.

Successful companies grasp the benefits that result from integrating CSR strategy into their business strategy. They appreciate their customers' CSR expectations and collaborate with consumers and business partners on their CSR initiatives by increasing the amount of information they provide about sourcing; and the composition and impact of their products, services, and operations. They also place critical importance on aligning philanthropy with business priorities.

9.8 Fulfillment

A major expenditure in any business is in delivering the orders. Order fulfillment consumes a large amount of fuel, as products must be transported from plants to distribution centers (DCs) and from there to customers. It therefore presents opportunities for strengthening sustainability. Companies can enhance sustainability by minimizing unnecessary handling, storage and re-handling of product, and the return of empty trucks after they make a delivery, as discussed in Chap. 5.

9.8.1 Transportation

The order delivery chain has undergone changes as a result of online ordering. What used to be bulk shipment of products from distribution centers to stores has slowly evolved to direct shipments of small packages from distribution centers to the customers. It implies more packaging. The major operations at a fulfillment center are shown in Fig. 9.11.

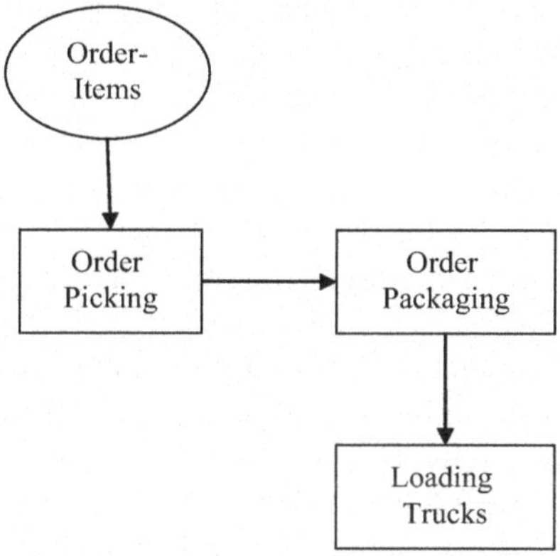

Fig. 9.11 Order fulfillment center

In addition to the fulfillment center, the order fulfillment chain comprises customer orders, delivery trucks and vans, local distribution hubs, and customers. Orders are received by the supplier electronically. These orders are processed, and sent to the fulfillment center where the ordered items are kept in inventory. At the fulfillment center appropriate items per order list are packed in parcels and prepared for shipment. The parcels are then sent to the distribution hub closest to the customer. From the hub, the parcels are delivered to the customer, the so called "last mile" delivery (Fig. 9.12).

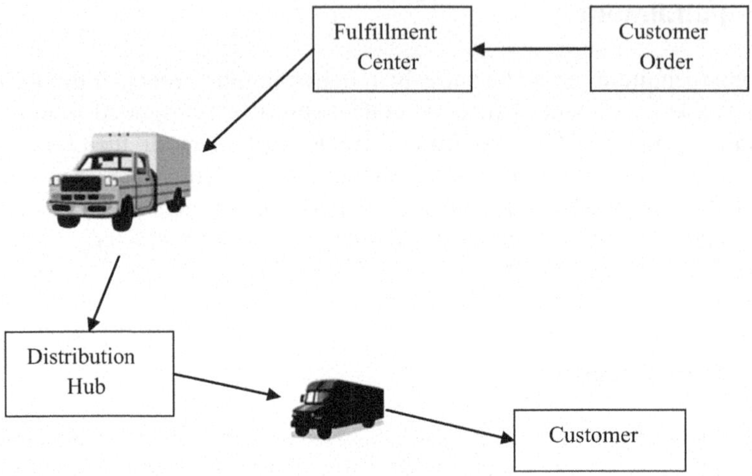

Fig. 9.12 Customer order fulfillment

9.8.1.1 Distribution Hub to Customer

The deliveries from local distribution hubs typically consist of mixed loads, which require careful scheduling for drop-shipping. There is a great deal of concern about the environmental repercussions of the expanding last-mile delivery. As vans emit a large amount of GHG per mile on urban roads, interest in the use of electric vehicles is increasing, especially in the online grocery sector. Several leading online retailers now use third-party networks for deliveries. These deliveries are often made by small vehicles and are of much shorter duration than typical van-based delivery rounds (Edwards et al. 2010).

Deliveries are usually planned based on the number of drops per trip, and the factors that have significant influence on the level of GHG emissions include: the likelihood of completing a delivery in the first attempt and the returns of unwanted or damaged goods. In case of home deliveries, delivery may fail if the parcel recipient is not at home. It is estimated that the incidence of failed deliveries can be as high as 30% of all deliveries.

Customers return items for different reasons: wrong product, mismatch with needs, and damaged products. Almost 30% of all consumer products bought online are returned. The delivery company's strategy for collecting the returns would clearly impact sustainability. It is estimated that a standard van emits 21.7 kg of CO2 per 50-mile trip (Edwards et al. 2010). Numerous factors influence emissions from home deliveries. They include: drop densities; the distance and nature of the delivery round; the type of vehicle used; and the treatment of failed deliveries and returns.

Delivery companies aim to maximize drop densities, avoid dedicated trips for collection of returns, and use low emission vehicles, such as electric vehicles. The use of parcel deposit boxes at homes can eliminate failed deliveries; consolidation of orders to a particular address in a single delivery can cut vehicle-miles; and adoption of flexible delivery-pricing can promote fuel-efficient speeds by activating off-peak delivery hours. In addition, the companies may offer incentives to customers to pick up their parcels at their distribution hubs.

In 2011, UPS restructured their pickup and delivery operations to eliminate about 85 million driving miles, 8.4 million gallons of fuel, and 83,000 metric tons of CO_2 emission (UPS 2011). The restructuring involved assignment of pickups and deliveries to the most efficient vehicles at each facility; routing vehicles on shortest paths to destinations; and dynamically re-routing drivers in response to changing customer needs for pickup and/or delivery location. For their air fleet, UPS has developed a new approach where pilots take a continuous glide path while descending rather than reducing altitude in discrete steps. It reduced fuel consumption and noise levels.

9.8.1.2 Fulfillment Center to Distribution Hub

After making a delivery to a hub, very often the truck returns empty to the fulfillment center, burning millions of gallons of fuel with tons of emissions. Companies need to minimize the empty miles on deadhead routes. Ideally, vehicles should be scheduled so that whenever they drop off loads, they should also pick up complementary loads in the vicinity of the drop-off points. In addition, collaboration between two companies with complementary deliveries – one delivering loads from A to B while the other from B to A – can minimize empty travel. Creating a web portal with appropriate algorithms that match suppliers' shipment-needs with available empty backhaul truck capacity in the country is the next step in this vein.

9.8.1.3 Palletization

In traditional deliveries, manufacturers send products in large lot sizes to distribution centers (DC), each product individually packed as pallet-loads. At the DC the pallet-loads are broken up and re-palletized to match individual customer's orders, as discussed in Chap. 5. Re-palletization at the DC, clearly, reduces the total distance travel for fulfillment, reducing emission and enhancing sustainability.

Consider the example where a manufacture packs two products individually in lots of 100 units. Assume the DC needs to satisfy three customers with the following order pattern:

Customer	Product A	Product B
1	35	50
2	40	0
3	25	50

Clearly pallet A must be split three ways and pallet B two ways. The resulting pallet for customer 1 would have 35 units of A and 50 units of B; pallet for customer 2 will have only 40 units of A; pallet for customer 3 will have 25 units of A and 50 units of B. Note that without palletization the vehicle carrying product A will visit all three customers and the one carrying product B will visit customers 1 and 3. With palletization three vehicles will be dispatched, one to each customer. Thus, there will be considerable saving in distance travelled and in GHG emission.

If the manufacturer had created customer-specific pallets to start with, much of the handlings and re-work at the DC would be avoided, delivering additional sustainability and cost savings from the total supply chain perspective. Note, however, that in that case the manufacturer would need to develop expertise in customizing pallets.

9.8.1.4 Technology

Teaming up with Kinko, FedEx developed a new way of transporting documents. Instead of using physical infrastructure, the company perfected digital transmission between two cities, printing the document at the destination and delivering them using its vans at the destination city. The cost to FedEx shrank, and it could project itself as eco-friendly.

9.8.2 Packaging

Packaging is a part of order fulfillment (Chap. 5), as it provides protection to the product during transportation. While consumers are interested in sustainability, they value convenience as well. For example, the ready-to-eat and single-serving packages increase customer's convenience. But they also increase the volume of packaging, and therefore impair sustainability. Companies have responded with innovative sustainable designs for packaging, materials, and formats. The objective is to reduce the volume of packaging material, while shipping more products per truckload. It also decreases fuel consumption, and GHG emission. In addition, manufacturers are attempting to cut packaging weight by cutting back on the use of materials (plastic bottles, for example) to reduce transportation costs.

Companies are trying out new packaging materials such as paperboard, which has a coating made of recycled PET bottles (Deloitte Development 2008) that provides a protection barrier. Note that switching to recycled paperboard also reduces the amount of water needed in the production process.

Other initiatives being tested are substitution of petroleum-based resins by starch-based adhesives, using polylactides (derived from corn) as a renewable input for the production of packaging material, and replacing oil with eco-friendly and less expensive sugar cane. Finally, note that the use of foaming adhesive can cut the amount of adhesive needed, as it has low density.

9.9 Industrial Symbiosis

Symbiotic relationship between two or more firms is based on using the outputs of one firm as inputs to another firm. This has consequences for both the ecology and the economic efficiency. If the "waste" produced by a company can be used by another, the net production of waste in the system is minimized. For example, the filter-sludge produced as waste from a sugar refinery and the white-sludge produced from alkali production can both be used as inputs in a cement mill.

In general, economic efficiency is obtained when a company uses the byproduct of another firm, saving transportation cost, which permits selling of the product below market prices. For example, molasses, a co-product from sugar refinery can be used in producing alcohol (Chertow 2007). Bagasse, another co product from sugar refinery, can be used as an input in a pulp plant to produce paper. These are examples of symbiosis. Chertow (2000) defines industrial symbiosis as engaging "traditionally separate industries in a collective approach to competitive advantage involving physical exchange of materials, energy, water, and by-products. The keys to industrial symbiosis are collaboration and the synergistic possibilities offered by geographic proximity".

Piluso and Ricciuti (2008) classify the symbiotic flows as intermodal (between two firms), internal recycle (a firm reusing its recovered components), raw material (net exogenous input), product stream (net output of usable products), and the net output of waste. They show how an input–output model can be set up, based on Leontief's work (Leontief 1936) on large-scale systems. They use the input–output model to compute the total net waste production, percentage of input raw material, and the percentage of intermediate products that end up as waste.

The decision-analysis framework of Piluso and Ricciuti (2008) provides a methodology for firms to assess their current status of sustainable operations, and to determine the best options for future sustainability. The analysis can help managers determine how modifications in their decisions would impact performance.

Chertow (2007) describes how industrial symbiosis evolved at the industrial park in Kalundborg (Denmark), shown in Fig. 9.13.

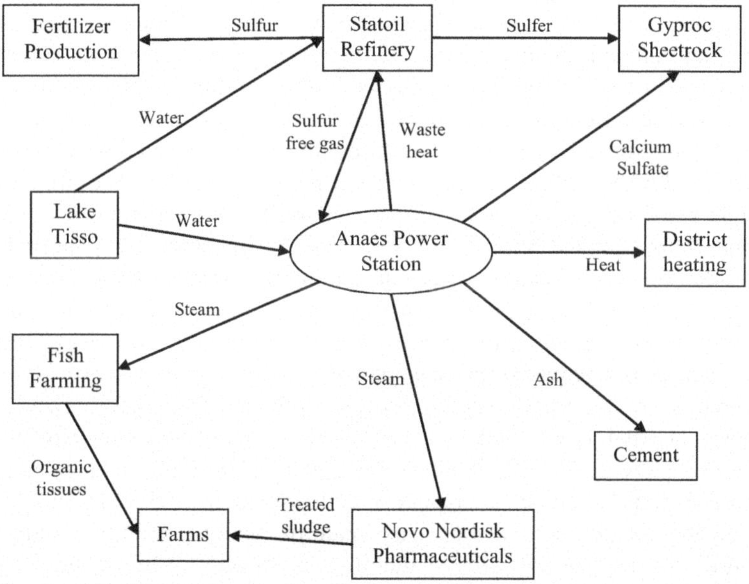

Fig. 9.13 Kalundborg symbiosis (Source: Chertow 2007)

Several companies cooperate in exploiting "wastes" from other factories, with four major players: Statoil Refinery, Asnaes Power Station, Novo Nordisk Pharmaceuticals, and Gyproc Sheetrock manufacturer. The coal-fired electric power plant supplies the oil refinery with waste heat from its steam turbines. The oil company removes polluting sulfur from the gas released by its refining processes, and the released gas can then be burned by the power plant, saving 30,000 tons of coal per year. The removed sulfur is sold to a nearby fertilizer plant. The power plant also precipitates pollutants from its coal smoke in the form of calcium sulfate, which is consumed as a substitute for gypsum by the sheetrock company. Ash removed from the same smoke goes to the cement factory. Other surplus steam from the power plant warms the biotech pharmaceutical plant and 3,500 homes, as well as a seawater fish farm. High-nutrient sludge from both the fish farm and the pharmaceutical factory's fermentation vats are used to fertilize local farms.

9.10 A Systematic Approach

Sustainable processes enhance resource availability on which a company depends for its long-term viability. The resources can be of different types: financial, material, societal, and human. Sustainability is the aligning of these processes, so that goals for different resource types are not in conflict. Clearly, there is a need to control resource availability so as to satisfy the needs of the future. In addition, companies must protect their brand name so as to stay viable in the market place, and ensure sustainability. As Carter and Rogers (2008) point out, the pattern of resource use (PRU) can have a significant effect on sustainability. In turn, PRU is influenced by corporate social responsibility (CSR), actions of firm and its suppliers, and disruptions in business. Economics, of course, is a major consideration in determining how the three pillars of sustainability are put together.

In essence, sustainability is a 2-period resource-use optimization, where "now" is period 1 and "future" is period 2. To address this problem we need to know (a) the future demand for resources, (b) discovery of new resources, and (c) the impact of the current resource-use pattern on future availability. While there is uncertainty in assessing future demand and new discoveries, the resource-use pattern is more controllable. The role of market forces in shaping resource use through pricing and incentives, and the role of technology in enabling changes in resource use must also be factored in. For example, the internet has revolutionized how many economic activities can be performed simultaneously.

The analysis requires understanding of a set of interactions, summarized in three questions: (1) what is entailed in a transition from resource-use pattern (A) to resource-use pattern (B); (2) what is the level of satisfaction derived from a resource-use pattern, given the resource endowment in period 1; and (3) what would be the impact of a resource-use pattern, using period 1 resources and potential discoveries, on resource availability in period 2. While science can help in assessing the impact of a use pattern on future resource availability, it cannot be known with certainty. Deriving satisfaction from a specific use pattern is primarily a function of customer preferences and life styles. Finally to understand how a transition from pattern A to B can be achieved, one must comprehend the role of science and technology in addition to the market forces and customer behavior.

The underlying structure of a supply chain becomes relevant, as it is a major factor in assessing the market forces that determine resource use. For example, activities such as minimizing disruptions, improving working conditions, minimizing transportation, reducing packaging, and requiring suppliers to undertake environmental and social programs, make more resources available in a future period.

Letting A_{it} as the availability of resource type i in period t, we can express the resource consumption pattern in period t, $P_{i,t}$, as $A_{i,t+1} = P_{i,t}(A_{i,t})$. It follows that $A_{i,t+2} = P_{i,t+1}(P_{i,t}(A_{i,t}))$. Therefore, the cost of the 2-period transition is expressed as $C_{t+2} = g_{t+2}(P_{i,t+2}, P_{i,t})$. One can create an n -period simulation with different consumption patterns $P_{i,t}$ and an initial resource endowment A_{it} to obtain the resource availability, and the cumulative cost of transitions, after n periods. Letting $S_{i,t}(A_{i,t})$ to denote community satisfaction level with a resource endowment of A_{it}, in period t, we can express the sustainability problem as,

$$\text{Maximize} \sum_{i,t} S_{i,t}(A_{i,t}), \text{ subject to } C_t \leq b, \quad \forall\, t.$$

References

Bassett, R. (2003). Developing value: The business case for sustainability in emerging markets. SustainAbility and Ethos Institute. greenbiz.com/sites/default/files/document/O16F30165.pdf

Brundtland, G. (1987). *Our common future*. Report of the World Commission on Environment and Development. http://conspect.nl/pdf/Our_Common_Future-Brundtland_Report_1987.pdf

Carter, C., & Rogers, D. (2008). A framework of sustainable supply chain management: Moving toward new theory. *International Journal of Physical Distribution and Logistics Management, 38*(5), 360–387. Emerald.

Catalin Group. (2012). Product recall: The developing story. A Catlin Group Limited Report. http://www.catlin.com/flipbook/product-recall/files/inc/382055422.pdf

CBFF. (2005). CBFF launches project to support sustainable forest resources management in Gabon. http://www.cbf-fund.org/en/node/299

Chertow, M. (2000). Industrial symbiosis: Literature and taxonomy. *Annual Review of Energy and Environment, 25*, 313–337.

Chertow, M. (2007). "Uncovering" industrial symbiosis. *Journal of Industrial Ecology, 11*, 11–30.

Davidson, D. (1984). Managing product safety: The ford pinto. In K. Goodpaster (Ed.), *Ethics of management*. Boston: Harvard Business School.

Deloitte Development. (2008). Sustainability from the boardroom to the breakroom. maonline.org/downloads/research-and-reports/SustainabilityBoardroom.pdf

Donato, J. (2008). Design products for sustainability. British Plastics Federation. bpf.co.uk/Sustainable_Manufacturing/Design/Designing_Sustainability.aspx.

Economist. (2008). Doing good: Business and the sustainability challenge. The Economist Intelligence Unit. http://graphics.eiu.com/upload/Sustainability_allsponsors.pdf.

Edwards, S. (2009). A new way of thinking: The Lowell center framework for sustainable products. Lowell Center for Sustainable Production, University of Massachusetts Lowell. http://www.sustainableproduction.org/downloads/LowellCenterFrameworkforSustainableProducts 11-09.09.pdf

Edwards, J., McKinnon, A., & Cullinane, S. (2010). Comparative analysis of the carbon footprints of conventional and online retailing: A "last mile" perspective. *International Journal of Physical Distribution and Logistics Management, 40*, 103–123.

EU ETS. (2013). A "cap and trade" system. European Commission Climate Action Policy. ec.europa.eu/clima/policies/ets/index_en.htm

Finkle, J. (2005). Economic development after Katrina: Lessons from national post-disaster response and recovery. http://www.iedconline.org/Downloads/GulfCoast/Katrina_Lessons.pdf

Gledhill, R., Hamza-Goodacre, D., & Low, L. (2013). Business-not-as-usual: Tackling the impact of climate change on supply chain risk. PWC. http://www.pwc.com/gx/en/governance-risk-compliance-consulting-services/resilience/publications/business-not-as-usual.jhtml

Greenhouse, S., & Clifford, S. (2013, July 10). U.S. retailers offer plan for safety at factories. *New York Times*. http://www.nytimes.com/2013/07/11/business/global/us-retailers-offer-safety-plan-for-bangladeshi-factories.html

Happala, K., Zhao, F., Camelio, J., Sutherland, J., Skerlos, S., Dornfeld, D., Jawahir, I., & Clarens, A. (2013). A review of engineering research in sustainable manufacturing. http://pasi.eng.wayne.edu/wp-content/uploads/2013/06/Haapala3.pdf

Hendrickson, C., Conway-Schempf, N., Lave, L., & McMichael, F. (1999). Introduction to green design. Green Design Initiative. Pittsburgh: Carnegie Mellon University. http://gdi.ce.cmu.edu/gd/education/gdedintro.pdf

Hoyt, D. (2008). *Unsafe for children: Mattel's toy recalls and supply chain management* (Case # GS 63). Stanford: Stanford Graduate School of Business, Stanford University.

Lee, H., & Whang, S. (2001). E-Business and supply chain integration. SBSCMB-W2-2001. Stanford Global Supply Chain Management Forum. Palo Alto: Stanford University.

Lefevre, C., Pellé, D., Abedi, S., Martinez, R., & Thaler, P. (2010). Value of sustainable procurement practices. PWC. http://www.pwc.com/en_GX/gx/operations-consulting-services/pdf/value-sustainable-procurement-practices.pdf

Leontief, W. (1936). Quantitative input–output relations in the economic system of the United States. *Review of Economic Statistics, 18*, 105–125.

Martin, A., & Rosenthal, E. (2011, September 16). Cold-water detergents get a cold shoulder. *New York Times*. http://www.nytimes.com/2011/09/17/business/cold-water-detergents-get-a-chilly-reception.html?pagewanted=all&_r=0

Muilerman, H., & Blonk, H. (2001). Towards a sustainable use of natural resources. http://ec.europa.eu/environment/enveco/waste/pdf/muilerman.pdf

Ngo, H., Vigneswaran, S., & Sundaravadivel, M. (2011). Advanced treatment technologies for recycle/reuse of domestic wastewater. In *Wastewater recycle, reuse, and reclamation* (Vol. I). Oxford: Eolss Publishers. http://www.desware.net/sample-chapters/d02/E2-14-01-01.pdf.

OECD. (2002). *Implementing domestic tradable permits: Recent developments and future challenges*. Paris: OECD.

OECD. (2011). Sustainable manufacturing good practices. oecd.org/innovation/green/toolkit/sustainablemanufacturinggoodpractices.htm#isothane

OECD/IEA. (2003). Policies to reduce greenhouse gas emissions in industry – Successful approaches and lessons learned. Workshop report, OECD and IEA Information Paper. Paris. http://www.oecd.org/env/cc/2956442.pdf

Pagell, M., Wu, Z., & Murthy, N. (2007). The supply chain implications of recycling. *Business Horizons, 50*, 133–143.

Piluso, G., & Ricciuti, R. (2008). Fiscal policy and the banking system in Italy have Taxes, public spending and banks been procyclical in the long-run?. CESIFO Working Paper No: 2442, October. http://www.econstor.eu/bitstream/10419/26487/1/587095326.PDF

Reddy, L., Kumar, S., & Wee, Y. (2010). Microbial fuel cells (MFCs) – a novel source of energy for new millennium. In A. Méndez-Vilas (Ed.), *Current research, technology and education topics in applied microbiology and microbial biotechnology* (pp. 956–964). Badajoz: Formatex Research Center.

Rewrap. (2013). Cradle to cradle principle. http://www.rewrap.eu/webshop/sustainable-13-inch-laptop-sleeves/

Rockwell Automation. (2009). Manufacturing execution systems for sustainability: Extending the scope of MES to achieve energy efficiency and sustainability goals. Milwaukee. emea.rockwellautomation.com/enduser/docs/paper11.pdf

Rose, C. (2001). Design for environment: A method for formulating product end-of-life strategies. Dissertation Abstracts International, *62*(01), 504B (UMI No. 3000092)

Rothenberg, S. (2007). Sustainability through servicizing. *MIT Sloan Management Review, 48*, 83–91.

Shonnard, D., & Hiew, D. (2000). Comparative environmental assessments of VOC recovery and recycle design alternatives for a gaseous waste stream. *Environmental Science and Technology, 34*, 5222–5228.

SPN. (2011). Sustainability purchasing trends and drivers. unilever.com/sustainable-living/sustainablesourcing/fruit-vegetables/index.aspx

Sussman, F., & Freed, J. (2008). *Adapting to climate change: A business approach*. Arlington: Pew Center on Global Climate Change and Center for Climate and Energy Solutions (C2ES). http://www.c2es.org/docUploads/Business-Adaptation.pdf.

Swedish Institute. (2011). Sweden leads by example in corporate responsibility. sweden.se/eng/Home/Business/Ethics_and_Business/Facts/CSR/

UNEP. (2004). *Voluntary environmental initiatives for sustainable industrial development: concepts and applications*. United Nations Publication, ISBN Number: 92-807-2480-0. http://www.unep.org/pdf/rowa/ROWA92-807-2480-0.pdf

UNEP. (2012). The business case for the green economy. http://www.unglobalcompact.org/docs/news_events/upcoming/RioCSF/partner_deliverables/The_Business_Case_%20for_The_Green_Economy.pdf

UPS. (2011). Logistics at the core. Corporate Sustainability Report 2011. responsibility.ups.com/community/Static%20Files/sustainability/2011_UPS_CSR_Report.pdf

Wachs, S. (2013). How to reduce warranty costs using tools that predict product and process performance. http://www.winspc.com/what-is-spc/ask-the-expert/420-how-to-reduce-warranty-costs-using-tools-that-predict-product-and-process-performance

Wausau. (2013). Eco premier. http://www.wausautile.com/pages/ecoPremier.cfm

WhalePower. (2013). *Promising test results from first humpback whale inspired air foils*. http://www.whalepower.com/drupal/?q=node/9

Wikipedia. (2013). Cradle-to-cradle design. http://en.wikipedia.org/wiki/Cradle-to-cradle_design

Healthcare Delivery Network 10

10.1 Introduction

The healthcare systems in both mature and emerging economies are struggling to satisfy an increasing demand for healthcare, as health expectations rise with living-standards. People are spending more for their healthcare – healthcare expenditure, as a percentage of GDP, is rising. Therefore, cost containment is important, but so are enhanced quality and productivity. For sustainability, it must also address accessibility, timely care, patient-preferences, lifestyles, expensive-to-treat diseases, and proliferation in treatments and technologies.

Patients have *traditionally* paid directly to clinics and hospitals for services received and medicines bought. However, because of the high out-of-pocket costs, they sought medical care only sparingly. Preventive healthcare (such as vaccination) was administered through federal programs. Hospitals were run as non-profit organizations, and received a majority of their funding from the federal government in the form of grants, and some from private donations.

This simple business model became untenable, as the expensive-to-treat diseases drove many households bankrupt. Fortunately, the fact that such diseases afflicted only a small portion of the population enabled incorporation of risk sharing in the business model through 3rd party insurance companies. Today, the major stakeholders in the healthcare system include providers (hospitals, labs, and clinics), insurance companies with a diverse set of health plans, drug suppliers (retailers and manufacturers), medical device manufacturers, payers (employers and the government), research institutions, and patients as shown in Fig. 10.1.

The business environment, in which the healthcare network operates, can be described in terms of the knowledge-intensiveness that makes healthcare supply driven, payer's inability to control outcomes, risky investment in new drugs and treatments, defensive practice of medicine, paucity of investment in preventive healthcare, industry-fragmentation, and market imperfections.

In a system with decentralized decisions, fragmentation is caused by a lack of coordination among its constituents. Patients with serious conditions require care

A.K. Chakravarty, *Supply Chain Transformation*, Springer Texts in Business and Economics, DOI 10.1007/978-3-642-41911-9_10,

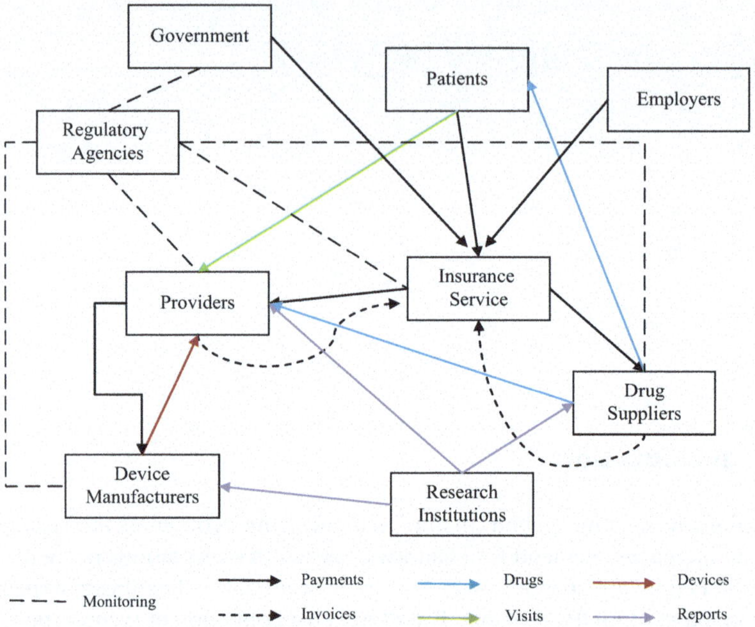

Fig. 10.1 Healthcare network

from multiple clinics and health professionals, who do not often work as a team. For example, a surgeon might use a high-sugar intravenous therapy after an operation on a diabetic patient, without consulting the diabetic specialist (Elhauge 2010). Multiple health plans with different eligibility requirements from different insurance agencies cause the payment system to be very fragmented and chaotic, resulting in unnecessary delays in reimbursements to the providers and/or patients.

The healthcare system is very knowledge intensive, as it relies on a sophisticated knowledge-base involving diagnostics, tests, imaging, treatments, drugs, and medical devices. Options provided to patients are mostly determined by this knowledge base. This causes the system to be supply-driven, suppliers being the providers. This "push" mode of service can create mismatches – creation of services that are not needed by patients. To be effective, the medical knowledge base must be integrated with the institutional knowledge, and patient-knowledge. The healthcare professionals should be able to interpret their medical knowledge for the specific medical history of a patient, and determine the optimal diagnostic or treatment procedures that would be feasible at specific institutions (Adams et al. 2008). The knowledge base is central to creating centers of excellences – the Cleveland Clinic with expertise in cardiac treatments draws in a large number of patients with heart disease.

The government and/or the employers pay a lion's share of the total healthcare costs, but they have little direct control over outcomes: quality (errors and satisfaction level), efficiency (value per unit cost of services), resource utilization (patient volume per unit capacity), patient's access to medical services (affordability, travel distance, up time), and sustainability. Most such measures are controlled by providers. As payers do not pay directly to the providers (initiated by patients), they have only limited leverage. At the same time, while the patients choose the care providers, they are not motivated to control cost (moral hazard). Though the individuals are the ultimate payers, their payment is routed through government taxes and employer benefits, and administered by the insurance companies.

The drug development process is novel. It is knowledge-intensive, prolonged, costly, and suffers from a high failure-risk. It takes many years to develop a successful new drug, and the average cost of development is very high – estimated as $1 billion (Pharmaceutical Research 2007). It follows that this cost will increase if the proportion of drugs failing in clinical tests increases. Therefore, given the high cost and uncertainty, the drug development process must be controlled tightly.

Defensive medicine is the practice of ordering a large number of procedures and treatments that may not be necessary. Providers contend that they engage in this practice to safeguard themselves against malpractice suites. It follows that defensive medicine can be a significant driver of inflation in healthcare cost. For cost containment, medical malpractice reforms may be necessary to ensure providers' acceptance of changes in the practice-pattern of payment and delivery.

Disease prevention practices aimed at wellness centers, healthy diets, and workouts can cut health-care costs. Community-based programs often address environmental and social causes of health problems. According to Edelson (2011), spending $10 per person annually on prevention can save more than $16 billion in the US, a return exceeding 500%. Prevention programs can reduce the rates of diabetes, high blood pressure, heart disease, kidney disease, stroke, some forms of cancer, arthritis, and lung disease. While most people are supportive of it, they are concerned about its affordability. If it can be funded through healthcare plans without increasing the cost to the individual, most would be for it. Payers would benefit from fewer cases of acute and chronic diseases, wellness centers would benefit from increased revenue, patients would benefit from longevity, but traditional providers such as hospitals and primary care would stand to lose financially.

The healthcare market is not perfect for many reasons including asymmetric information, adverse selection, moral hazard, and externalities. The patient (consumer) often does not know what medical services he/she may require, and does not know how to assess the quality of services provided. The supplier (providers) determines the services the patient needs (diagnostic and treatment). Therefore, unless it is a standard service, the information asymmetry inhibits the patient from shopping around for price and quality. The second reason is the insurance market – healthy people prefer not to buy health insurance (adverse selection), and insured patients may agree to unnecessary procedures because purchasing insurance decreases their price sensitivity to healthcare services (moral hazard). Finally, there is an externality effect in health care especially where infectious diseases are concerned – vaccinations of others benefit "my health". This may lead to vaccination services being under-consumed.

Healthcare operations, in many ways, are similar to humanitarian relief operations discussed in Chap. 8, in that in both systems supplies are required to satisfy critical needs. However, in contrast to humanitarian relief where generic supplies (food, water etc.) are adequate, healthcare requires the supplies (drugs, treatments etc.) to be specific to the disease, and customized diagnosis is performed by qualified physicians in special facilities. Although the time to deliver supplies is an important consideration in both systems, it is more critical for humanitarian relief. For healthcare the critical capabilities are diagnosis (customized) and deployment of appropriate treatment.

We have noted that while the impact of disasters can be reduced through protection, preparation, and evacuation, the likelihood of the disaster incidence cannot be controlled. That is not the case in healthcare where the incidence of diseases (as well as the impact) can be controlled through preventive care which decreases the demand for diagnosis and treatment. In addition, the advent of the internet and mobile technology provides novel options for the diagnosis and other aspects of healthcare.

10.2 Healthcare Stakeholders

The healthcare stake holders are shown in Fig. 10.1. Major stakeholders are providers, payers, patients, and suppliers of pharmaceuticals and medical-devices. Other stakeholders include regulatory bodies, universities, and research institutions.

Healthcare providers include hospitals, family clinics, retail clinics, wellness centers, concierge medicine, medical tourism, mobile care, ambulatory care, and home care. *Hospitals* account for about 50% of the healthcare costs. They are often structured by specialties such as surgery, emergency care, pediatrics, obstetrics, gynecology, cardiology, nephrology, gastroenterology, neurology, and ophthalmology. They are staffed with physicians, surgeons, nurses, technicians, and other service providers representing a wide range of expertise. Managerial issues range from organizing the flow of patients, utilization of beds, scheduling (nurses, doctors, and surgery), procurement of equipment and supplies, storage and dispensing of drugs, and claims management.

Concierge medicine offers patients a variety of amenities in multiple tiers, above and beyond what regular providers cover. Services range from same-day appointments and 24/7 access to physicians (base tier); to house calls, home delivery of medicine, and care at vacation homes (top tier). Concierge practices charge membership fees for payment of services and amenities. The fee varies by location and amenities offered (Spooner 2007).

Ambulatory care is the provision of healthcare diagnostics and therapeutic services in an outpatient setting. Ambulatory care facilities include clinics, medical offices, and mobile vans. Services that are usually provided include diagnosis, treatment, testing, minor surgery, preventive care, prescription, counseling, and management of chronic conditions.

Home Health Care is the provision of services, such as intravenous treatment, feeding, diagnostic testing, and other medical therapies at home, by licensed health care workers. Over ten million individuals in the US, who suffer from acute illness, chronic health conditions, permanent disability, and terminal diseases, receive care from home health providers. Home health care was judged to rank highest in terms of outcomes and cost effectiveness for patients who have undergone hip or knee replacements.

Wellness centers provide services in prevention, intervention, and innovation. Prevention includes screening, vaccination, diet, exercise, and regular physician visits. Intervention includes early identification, early treatment, timely care, and disease management. Innovation includes new ideas on techniques, creative exercises, and new diets.

Retail clinics are found in retail centers such as large discount stores, grocery stores, and retail pharmacies. Examples are Wal-Mart, Target, CVS, and Walgreen. They are staffed by nurse practitioners who provide healthcare service for common medical ailments. These clinics create value in several ways: convenience, access, and low cost (Keckley 2008). They are convenient because no appointments are needed; operating hours extend to evenings and weekends; and wait times do not usually exceed 15 min. They are easily accessible because they are located in communities where customers can visit several times a week. They are inexpensive as service costs are much lower than that at a primary care center.

Insurance companies serve an important role by absorbing the demand shocks in the market. They act as the agents of the payers and attempt to implement cost control measures on payer's behalf. However, because of information asymmetry – providers know a lot more about matching patients with treatment regimens than the insurers – cost control measures lead to dissensions and loss of trust between providers and insurers. Insurers attempt to fix payment rates by standardizing treatments and drugs for specific illnesses. The problem is that not all patients with the same illness respond the same way to a specific treatment. In addition, the proliferation in health plans, with distinct eligibility and deductibility criteria, causes confusion to the providers.

Drug manufacturers supply medicine to providers and retailers, sometimes through 3rd parties such as distributors and general purchasing organizations (GPO). Pharmaceutical companies must decide their drug portfolio based on multiple factors: product type (mature, emerging, and generic), length of patent protection, drug approval process, failure risk, and cost. Emerging drugs are subject to uncertain demands, and use "push" production strategies. Demand for the drug is established working with providers who are not the final consumers. Emerging and innovative drugs, while risky, help expand company's market share. Mature drugs, on the other hand, are at the end of their patent protection cycle. They need to be managed for the transition from a "high-margin and low volume" market of branded products to a "low-margin and high-volume" market of generics. Drugs are distributed to providers such as hospitals, and made available to patients through retailers. Because of the rigid expiration dates, regulated by the FDA, retailers may insist on buy-back contracts. All these factors put tremendous pressure on the pharmaceutical industry.

Payers can be the government, employers, or the individuals. While in the US payers are the employers and the government (Medicare and Medicaid), in Europe it is mostly the national governments. The US employers make a fixed monthly payment per employee to private insurance companies, and the government directly finances Medicare (for the elderly) and Medicaid (for the poor). In many European countries, government pays to insurance companies who are regulated by the government.

Medical devices range from simple tongue depressors and bandages to complex programmable pacemakers and sophisticated imaging systems. Primary products are cardiovascular, orthopedic, vision care, defibrillators, physiologic monitors, electrocardiographs, vital signs monitors and diagnostic devices (Lantronix 2010; Larew 2011). Typically, individual physicians or surgeons decide which devices to procure. The cost of the device, borne by the provider, is passed on to the payers through fee-for-service. Medicare lumps the cost of the device with facility fees. The introduction of a new device often requires the providers to create entirely new treatment procedures, cost codes, coverage policies, and payment methodologies. Providers depend on reimbursement to cover the cost of process reengineering and, therefore, reimbursement policies can influence adoption of new medical devices.

Medical device companies need to design differentiated products based on the specific needs of providers. One approach is customer segmentation based on such factors as the willingness-to-pay, switching behavior, growth strategies, and investments. Different groups involved in purchase decisions may be motivated differently. While physicians primarily care about device-effectiveness in caring for patients, most administrators like to ensure low cost and effectiveness. Skoskiewicz et al. (2012) define four provider segments as traditional (physicians dominated), price driven (administrator dominated), partners (collaborate with suppliers to make improvements), and centers of excellence (build best in class capabilities). The attributes of the device offered to a segment must fit with the needs of the segment. Obviously, the margins would differ by segments.

Regulatory agencies implement a comprehensive regulatory framework that determines where and how competition takes place in health care markets. Regulations are developed and enforced by federal, state, and local governments. In addition, a large number of private regulators have oversight responsibilities in accrediting providers, licensing examinations, and certification of specialists. Similarly, the pharmaceutical industry is governed by its own set of rules. For example, a manufacturer must obtain a patent from the Patent and Trademark Office (PTO), and receive permission for clinical testing from the Food and Drug Administration (FDA). Next, the manufacturer must register on the formularies of private pharmacy benefit management companies (PBMs), which administer reimbursement plans. The drug can only be dispensed by approved pharmacists (Field 2008).

10.3 Value Creation and Business Models

As the healthcare system is extensive and complex, it presents opportunities for different business models to emerge. The possibilities include, emergence of niche providers, patient activation, innovations enabling process automation, shift from resource utilization to final outcomes, reconfiguration for accountability, new technologies such as electronic medical records, bundled payments, new medical devices that necessitate process reengineering, and government vs. private ownership of services and/or payments.

10.3.1 Shift from Utilization to Outcomes

The practice of health care service at hospitals and clinics is mostly geared to process compliance and resource utilization. A shift from routine compliance to *outcomes* that emphasizes the final result would induce changes throughout the care-delivery system. For example, the provider-structure may have to be based on patient-centric concepts, requiring a major reengineering of operations. An outcome-oriented organization will focus on patient pathways and configuration of clinical processes that lead to desirable outcomes. Processes will be evaluated based on its positive contribution to the outcome and not the productivity (e.g., number of lab tests per week) of individual processes (Bohmer and Lee 2009).

If the pathway for generating a desirable outcome is *well structured*, it can be implemented as is. In cases where there are no such structured pathways, the relevant knowledge will have to be created (innovation). New pathways can be recorded as *process flow* diagrams that can be used for execution by physicians. Thus, in addition to their clinical roles, physicians must also be knowledgeable about how to design, monitor, and improve the care system. This knowledge, resident in physicians and absorbed by the institution, can become its *competitive advantage*.

10.3.1.1 Healthcare Value

Healthcare value created for the *ith* patient over a period can be defined as $\sum_{j=1}^{n} v_{ij} / \sum_{j=1}^{n} c_j$, where v_{ij} is the value of the *jth* health outcome (out of n possible outcomes) to the patient, and c_j is the cost incurred in delivering that outcome. The health outcomes may be captured in terms of whether the diagnosis was right and/or rapid, whether the treatment was right and/or non-invasive, whether the providers made errors, whether recovery was quick, whether a long term care was mandated, and whether the problem recurred (Porter 2012). The average value created by the *jth* outcome will be $\sum_{i=1}^{m} v_{ij}/c_j$, where m is the number of patients using the system in the period. For example if, with $j =$ rapid-recovery, the value of $\sum_{i=1}^{m} v_{ij}/c_j$ is not high enough, processes leading to rapid recovery would be reexamined.

10.3.2 New Provider Models

As mentioned earlier healthcare facilities such as hospitals, clinics, specialist centers, and test and imaging centers, are known as providers. Lately, wellness centers have emerged as legitimate providers in the value chain. An increasing number of these providers are responding to a changing healthcare "market place". The important drivers in this marketplace are (i) the changing expectations of healthcare customers, and (ii) the ability to treat complex medical conditions with advanced diagnostics and treatments. The demand driven process-adaptation is, in fact, occurring throughout the care delivery value chain.

Over the years, there have been radical changes in surgery (implants, transtympanic (ear), robotics), imaging (X-ray, MRI) and laboratory medicine (bacteriology, hematology), among others (Rechel et al. 2009). Hospitals need to deliver high-quality care, while ensuring access and close collaboration with primary care and other services located outside the hospital (Black and Gruen 2005). In addition, services typically provided in hospitals are now often moved to ambulatory care. Thus, a key challenge will be *flexibility* for quickly adapting to the changing needs.

New provider-business-models are emerging that address cost, quality, and access gaps across different forms of care delivery. Many of these models are focused on acute care, which will remain important but, with the advent of ambulatory care, they may not produce as much value in the future. There will be a need to coordinate and integrate many of these independent care-systems, in addition to extending care into preventive and chronic services. One scenario has the providers evolving into community-based outpatient clinics which operate as a multi-agency one-stop facility. These clinics would reduce the number of beds, and waiting areas. There would be increased use of remote IT monitoring, disease prevention, and psycho-social support. The patient will be an engaged co-producer with rights and responsibilities.

The healthcare value chain could evolve to include providers of alternative medicine. People are increasingly revealing their preference for the integration of eastern and western medicine, and other alternative therapies such as the practice of acupuncture and yoga. Spending in complementary and alternative medicine in US exceeded $27 billion in 1997 and US $20.3 billion in Japan in 2004 (Adams et al. 2008). The Japanese market is expected to exceed US $43.1 billion in 2013.

10.3.3 Consumer Driven Model

Increasingly, patients are making better lifestyle choices that have pronounced influence on the health status of individuals. It is known that approximately 80% of coronary heart disease, 90% of diabetes, and up to 70% of cancers could be prevented or significantly delayed through lifestyle changes – proper diet, regular exercise and abstinence from smoking (Adams et al. 2008). In a consumer driven system, the patients and providers jointly determine the appropriate care. For example, providers can point out costs, benefits, and risks of viable options (medication side effects, invasive procedures, success rates), and patients can choose the options they prefer.

To minimize information asymmetry between providers and patients, there is a need for new services to help patients make informed choices on clinical options, lifestyle choices, and financial viability. Individuals may need relevant information for decisions such as choosing appropriate providers, treatments, tests (predictive and diagnostic), and medications (Adams et al. 2008). In a system where private benefit plans exist, families may need advice on optimizing their choice of benefits. Individuals also need to understand the risks and predicted outcomes of their lifestyle choices, and what constitutes a healthy lifestyle. Finally, people may need advice on how to pay for health services. The new services can proactively devise holistic healthcare, financial, and retirement plans, and advise people of their financing and insurance options.

10.3.4 Globalized Healthcare

The health care industry is not immune to global competition. While *wealthy* patients from emerging economies have long traveled to mature countries for advanced medical care, there is now a flow in the reverse direction. An increasing number of foreign patients are now receiving healthcare in emerging countries. These patients are seeking quality care at affordable prices.

An estimated two million Americans traveled abroad for treatment in 2005 (Herrick 2007). A majority went to Mexico and other Latin American countries. Approximately 250,000 foreign patients sought care in Singapore, 500,000 in India, and as many as one million in Thailand. The cost savings for patients seeking medical care abroad can be significant. For example, the Apollo Hospital in New Delhi, India, charges $4,000 for cardiac surgery, compared to about $100,000 in the United States (in 2012). Hospitals in Argentina, Singapore or Thailand charge from $8,000 to $12,000 for a partial hip replacement, which is about half the price charged in Europe or the US. Hospitals in Singapore and India charge $18,000 and $12,000 respectively for knee replacement that costs $30,000 in the US (Herrick 2007).

Patients seeking medical service abroad can be categorized as value shoppers, experience seekers, leisure or business travelers, and payer-directed. Value shoppers are the price-sensitive patients who look for the best value for money. Experience seekers are less price sensitive consumers. Leisure and business travelers seek healthcare that range from cosmetic to business services. Payer-directed patients are those that seek healthcare in other countries as advised by their employers, private payers, and governments.

10.3.5 Electronic Health Records

Electronic health record (EHR) is a computerized patient-centric history of an individual's health care that includes treatment records from multiple providers, clinical information, and administrative tasks for patient management. It is built from data in electronic medical record (EMR), which is specific to a provider. EHR enables linking of clinical information with payment systems, and can retrieve data from multiple sources of care that the patient might have used. Because they are interoperable, they can be accessed at any point of care. It makes it possible for patients to obtain accurate and reliable information about the cost and quality of healthcare. Note that interoperability and data transparency require national data standards.

EHR creates value in the healthcare business model in multiple ways. From the provider's perspective, it becomes easier to coordinate referrals. Pharmacist can access prescription information in EHR to fill online prescriptions. Physician can retrieve all relevant healthcare information on a patient from a single source (IBM 2003). Using web portals the system can schedule patients, activate health assessment tools, and provide consumer health information. It can help establish treatment standards for diseases based on best practices, and these standards can be accessed in real time by practitioners. It can generate summary statistics of patient-care. Examples are the percentage of diabetic patients receiving regular eye and foot examinations, and percentage of children receiving immunization on schedule.

Patients can demand information on provider quality. They may challenge recommended treatments as they can better assess provider's performance. EHR enables matching of E-prescriptions against the patient's record for harmful interactions or allergies, and patients can make lifestyle choices to alleviate health problems.

10.3.6 Bundled Payment

A bundled payment is a single-fee paid to a provider or group of providers for a defined episode of care. For example, a single payment for the episode "knee replacement surgery" would cover pre-admission test, hospitalization, surgery, and rehabilitation (DHCFP 2011). Bundled payment would replace fee-for-service payment that entails paying multiple providers (or multiple procedures performed by the same provider) individually for their services. Because multiple providers share a bundled payment, there is a built-in incentive to coordinate their services. In addition, it includes a bonus payment to the providers for minimizing avoidable complications from their treatment, and for quality improvement.

Such a model may bundle services with three components: a base-payment, a patient-specific severity adjustment, and an allowance for potentially avoidable complications. The base payment is determined by calculating the "average" cost of treating a given condition from beginning to end using commonly accepted (based on evidence) clinical guidelines. The severity adjustment is arrived at through a multiple regression model that takes several factors into account, including patient characteristics, morbidity demographics, geographic location, and provider specialty. The allowance for potentially avoidable complication (PAC) is calculated as a $r(x - y)$, where r is a fraction, x is the cost of treating patients with reengineered processes to avoid complications, and y is the average cost of base payments. The ratio r is established through negotiations between the stakeholders.

McKesson (2010) provides a hypothetical example of a bypass surgery for a patient with uncontrolled diabetes. In this scenario, fee-for-service payments would entail the following charges:

	Hospital	Surgeon or physician	Total
Surgery	$47,500	$15,000	$62,500
Diabetes treatment	$12,000	$2,000	$14,000
Post-surgery infection			$25,000
			$101,500

Under Bundle payment, the possible payments may look as follows: severity-adjusted payment of $61,000 for the hospital and $13,000 for the surgeon/physician. In addition, there would be $15,300 allowance for PAC. The total payment would be $89,300. Compared to fee-for-service, the cost to the payer is reduced by $12,200. If we assume a base payment of $50,000 for surgery and $10,000 for diabetes, we would have $x = 101,500$ and $y = 60,000$, so that $r = \frac{15,300}{101,500-60,000} = 0.369$. Note that negotiating for PAC is the same as negotiating for r.

By preventing readmission the hospital and surgeon/physician are paid $76,500 under fee-for-service, whereas, the bundle payment stays at $89,300. Thus, the providers are paid $12,800 more under bundle payment and it is the bonus for preventing readmission for infection treatment.

10.3.7 Government and Private Ownership of Services

The great healthcare debate is on whether providers should be private or run as government owned clinics, and whether the healthcare insurance industry should be regulated. The crux of the debate is whether healthcare is a private consumption good or a social good. That is, should a sick person be treated as a consumer shopping around for the best service at low cost or someone that the community must take care of? The answer is not easy, as is evident from its country-specific nuances.

Reinhardt (1993) presents a framework using the dimensions of healthcare ownership, and healthcare finance, to differentiate the healthcare systems of nations. As shown in Table 10.1, healthcare financing is available from government financed insurance or private insurance. Private insurance comes in two forms: regulated and unregulated. Regulated insurance companies must conform to the statutory framework laid down by a national government. Similarly, the delivery of healthcare can occur at government owned or privately owned providers.

Table 10.1 Healthcare framework

	Healthcare finance		
Healthcare ownership	Government financed insurance	Regulated private insurance	Unregulated private insurance
Government ownership of providers	Canada, UK (hospitals) US Veterans	Germany	US Private Healthcare
Private ownership of providers	Canada, UK (clinics) US Medicare/ Medicaid	Germany	US Private Healthcare

Note that while the healthcare systems in Canada and UK use socialized financing, they do not necessarily practice socialized medicine as patients have the option of obtaining healthcare at private clinics. The Medicare and Medicaid plans of the US healthcare can access private hospitals as well as clinics. In contrast, healthcare delivery to US military veterans is the responsibility of government owned providers that are also financed by the government (tax payers).

In Germany, health insurance is provided by a privately managed insurance plan that is regulated by federal statutes. This insurance plan covers almost 90% of the population. The health care delivery system is a mixture of private and public providers. Thus, the German health care system is not socialized medicine, and it is financed by private insurance regulated by the government.

Unlike Germany and Canada, the American healthcare fits into multiple columns in Table 10.1. Almost 40% of the US healthcare expenditure is borne by government financed insurance (for Medicare, Medicaid, and VA), 45% by private funds (private health insurance), and the remainder by out-of-pocket expenses. Through a series of reforms such as “managed care” and, the more recent, “affordable care act”, the American private healthcare has been converging towards the German healthcare.

Reinhardt (1993) points out that Canada and the European countries prioritize patients for resource allocation based on medical judgment and length of the “queue” for service. Price is negotiated between the associations of payers and providers, and they are strictly adhered to. In the American system, in contrast, resource allocation is based on price charged and the patients’ ability to pay – the so-called market driven health care. A market based system creates surplus capacity because of the uncertainty in demand, which drives up prices. As pointed out earlier, with information asymmetry, patients (as consumers) are unable to determine what services they need and the quality thereof, and that creates imperfections in the consumer driven healthcare market.

10.3.8 Pharmaceuticals

Pharmaceuticals can be categorized into several market segments: acute care, chronic care, patient centric, drug-test combos, prevention, and emerging markets (Keeling and Schrader 2012). The paradigm is evolving from blockbuster drugs to specialized drugs with short life cycles, highly complex, small batches, pharma-economic performance, and price transparency.

Customized medicine needs complete information such as patient characteristics, disease state, and responses to treatments, to help prevent the onset of a disease and aid in early detection. Patient's unique physiology can then help determine the best preventive or therapeutic approaches. Drug developers need to know what medicine worked, for whom, and why, and apply that knowledge consistently across the patient population. Personalized healthcare is still in its infancy and would likely pose major challenges for scientists, payers and providers as it makes progress. Providers will require new knowledge and skills for (a) predicting or assessing therapeutic capabilities of molecular tests and (b) major investments in IT-related capabilities such as access to both phenotypic and genotypic information combined with robust analytics capabilities (Adams et al. 2008).

Chronic disease management (75% of U.S. healthcare expenses) will continue to grow as more conditions, including heart disease and certain cancers, go from an acute to a chronic category. The boundaries between different forms of healthcare are blurring as clinical advances are making it possible to treat previously fatal diseases, as chronic cases at worst. Treating chronic conditions with remote-monitoring, apps on smart phones, and location-based services, changes the nature of care to continuous response. Makers of generics, on the other hand, might position themselves as high-volume, low-cost providers, with the consumer products industry as the model.

PWC (2011) suggests two business models: collaborative and integrated. The collaborative model comprises a network of separate organizations linked by a shared purpose and infrastructure. The integrated model, on the other hand, comprises a network of entities owned by a single parent company. It is appropriate for generic drugs, driven by efficiency and low cost of manufacturing and distribution. PWC also identifies two variants of the collaborative model. In the virtual version, a company outsources most or all of its activities that are coordinated by a "hub"; contractors are assigned work on a project basis. In the venture version, the alliance manages a portfolio of investments, driven by innovation and capital growth; members share intellectual property and risk is spread across the portfolio.

10.3.9 Medical Devices

Medical devices such as fetal heart rate monitors, patient monitoring, life support devices, and pulse oximeters, are in everyday use around the globe. In future, the healthcare industry can expect to see many advanced medical devices in use: nanoparticles, magnetic drug homing devices, micro-needles, and polymer capsules (Friend et al. 2011).

Other new technologies offer great promise. A "smart cap" for insulin injection pens can record the time since its last use. Similarly, "Smart bottle" can remind the reader the amount used up, and it can communicate with patients, pharmacies, and physicians; "Chip-on-a-pill" technology can also record the actual usage. Finally, ingestible chips can send a signal to an external device. Once in the digestive tract, the chip transmits vital signs and confirms the dose taken. Genomic and combinatorial measurements are transforming the ability of physicians to adjust medication to individual patients.

10.3.10 Accountable Care Organization (ACO)

ACO is a group comprising one or more hospitals and physicians. Physicians use the hospital facilities but are *not necessarily* employed there. They are brought together, under one "umbrella". Public and private payers can hold the ACO accountable for the quality of care it provides, without it sacrificing high resource utilization. Unlike an HMO where the payers make most decisions, the ACO is responsible for setting performance metrics, healthcare processes and pathways, alignment of incentives, and collaboration media and formats. In addition, providers can choose to accept from among a variety of payment structures that offer flexibility in risk exposure.

Kaiser Permanente runs an ACO that brings together providers such as the Mayo Clinic, Cleveland Clinic, primary care physicians, specialists, and hospitals; they work together to manage the overall care of patients. The KP ACO has chosen to *own* the hospitals and *employ* salaried physicians. Providers within the systems collaborate to improve quality and efficiency by exchanging patient information, and developing and adhering to practice guidelines (Health Affairs 2010).

According to Deloitte (2010), ACO should demonstrate several capabilities that include *processes* for improving quality of care, structure for reporting costs, and a legal framework for distributing bonuses to participating providers. To break even, they need at least 5,000 Medicare beneficiaries, and should have contracts in place with a core group of specialist physicians.

Different revenue models are possible for the ACO: shared saving, comprehensive care payment, episode payment, and hybrid payment (Miller 2009). Shared saving is an incremental payment. While the ACO is paid on a fee-for-service basis, it receives a bonus in the form of a share of savings generated from reducing the total healthcare spending for its patients (e.g. reducing hospitalization-days through a better care program). The payer spends less than it would have, and the provider earns more than it would under a pure fee-for service system. The provider's income does not decrease for keeping people healthier!

Comprehensive care payment is a single payment (like bundled payment) for all of the healthcare services needed by an individual for a fixed period of time. The amount of the payment is adjusted based on the type and severity of the conditions the patient has, and on the quality of care delivered. If the cost of care goes up, the excess cost is divided into two parts: risk factors such as aging population, and increase in procedures (higher proportion of bypass surgery for mild coronary blockage). The ACO would be accountable for the latter, but not the former. While simplifications in billing and payments for ACOs and payers are permitted, the ACO would need to provide additional information on patient conditions to support severity and risk adjustments.

In a hybrid model, the ACO receives a single payment to cover all costs associated with ambulatory care services for the patients in its care. It also receives a bonus payment based on shared savings (or penalty) in costs of inpatient care services. The ACO is not fully at risk for inpatient care, as it would be under a comprehensive care payment. The ACO is far more accountable for inpatient care than that under a fee-for-service system. It also gives the ACO the flexibility to try different treatments than that under a shared savings approach.

10.4 Capabilities

The healthcare delivery network comprises two distinct supply chains: service supply chain and material supply chain. Service supply chain includes healthcare services such as admissions, diagnosis, lab test, surgery, treatment, and recuperation. Material supply chain involves procurement, storage, and deployment of healthcare products – pharmaceuticals, sensors, wearable devices, test equipment, and communication devices. As shown in Fig. 10.2, the materials supply chain includes all the entities, except the employers; and the service supply chain excludes drug manufacturers, device manufacturers, and retailers.

Next, we discuss the healthcare drivers shown in Figs. 10.3 and 10.4, both external and internal (as outlined in Chap. 1). In what follows, we include both the service and material supply chains in the discussion of drivers, and delineate them when we discuss delivery network capabilities.

The major drivers of healthcare are the healthcare-need, patient preferences, technology, and alternative medication. *Healthcare Need* can be defined by two factors: age of the population and the frequency of disease-incidence. For example in a population with a large percentage of older people, one would expect the healthcare to be primarily treating chronic cases (cancer, heart disease, diabetes etc.) rather than acute cases (pneumonia, cholera). About 84% of people who are

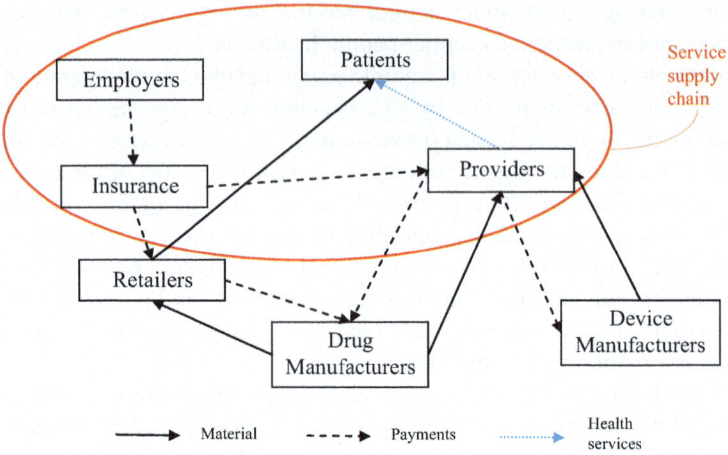

Fig. 10.2 Service and material supply chains

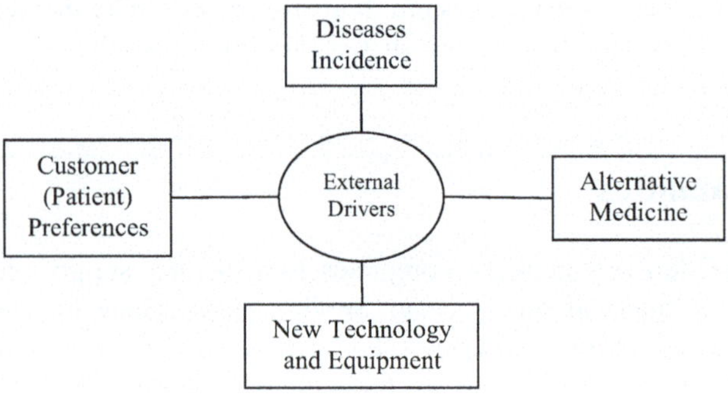

Fig. 10.3 Health care external drivers

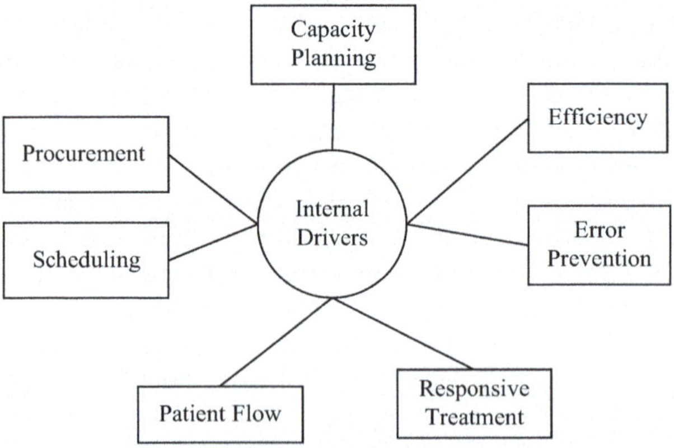

Fig. 10.4 Internal drivers

over 65 suffer from chronic conditions, compared to 38% of those ages 20–44 (CHWS 2006).

The rise in new *disease incidence*, such as prevalence of obesity, is causing changes in diagnosis and clinical guidelines, and new capabilities in treatment. As shown in the Table 10.2 below, Thorpe and Ogden (2008) illustrate how cost and diseases incidence have risen, by examining the increases in cost per case and disease incidence for the period 2000–2005.

Table 10.2 Disease incidence

Condition	Increase in cost per case (%)	Increase in disease incidence and prevalence (%)
Circulation disorder	185.9	−143.3
Skin disorder	90.6	6.8
Childbirth	88.5	8.8
Trauma	76.5	17.0
Kidney	51.1	23.2
Cancer	47.2	45.7

Complementary and Alternative Medicine (CAM) are becoming increasingly popular worldwide. These therapies include acupuncture, ayurveda, chiropractic, herbal medicine, homeopathy, diets, and massage. The percentage of CAM patients has increased to almost 42% of the US population. This trend suggests an increasing demand for CAM that may alter the health care delivery model (Frenkel and Borkan 2003). For example, the Memorial Sloan-Kettering Cancer Center offers integrated medicines that include music, massage, reflexology, acupuncture, meditation, self-hypnosis, herbal medicine, yoga, and tai-chi (CAM 2005).

Technological advances and medical breakthroughs can produce new therapies that enable treatment of conditions, not possible before. For example, coronary angiography enabled physicians to observe blood flow and assess blockages. Similarly, new technologies are making it possible to perform bypass surgery on patients, considered frail. Technological advances in new drugs that prevent rejection of donor kidneys, have substantially improved the success rate of kidney transplantations. New metals and plastics have led to the development of strong and long-lasting materials that can be used for making better artificial joints. As a result, surgery for hip and knee replacements has increased significantly (Orszag 2008).

Individual patients express *preferences* in terms of structural, procedural, and outcome aspects of health care. Choice of HMO, desire for certain types of information, and interactive decision-making styles are some of the examples of structural preferences. Procedural preferences, on the other hand, determine the treatment and test options – surgery, physiotherapy, ECG, and MRI. The outcomes of health actions such as the speed of diagnosis, germane treatment, rapid recovery, and relapse-prevention, are the third category of preferences. Such preferences must be incorporated in designing clinical protocols and/or providing convenience. For example, patient preference data such as electronic health record (EHR) can reveal the percentage of patients who like to avoid invasive procedures. However, as preferences vary among patients in a population, it is not cost effective to provide services for all preference combos. Providers must decide which of the combos they would cover, and promote their services to the relevant population segments, accordingly.

Internal drivers, shown in Fig. 10.4, act as motivators for reducing patient wait-time, enhancing patient safety, improving procurement efficiency, increasing availability of beds, improving scheduling of operating rooms, enhancing emergency room effectiveness, optimizing nurse schedules, reducing lab turnaround time and errors, improving imaging responsiveness, and managing inventories of drugs and supplies.

The *capacity plan* establishes the size of needed resources – staff-size, number of operating rooms and diagnostic units, number of beds, and pieces of equipment. Efficient *procurement* requires specialized knowledge of medicines, devices, consumables, and the suppliers of products. It involves careful selection of suppliers, sourcing for new suppliers, price negotiation, inbound quality monitoring, and capability for testing (equipment, protocols, and procedures). It also requires the ability to maintain a transparent process, and skill in contract management.

Medical errors include diagnostic errors, equipment failure, botched-up surgery, avoidable infections, and transfusion of incompatible blood. In addition, transcription errors by medical personnel can lead to incorrect treatments, erroneous procedures, and wrong medication. Errors can be reduced by instituting appropriate protocols, monitoring procedures, and training routines.

Internal drivers evolve as part of an organization culture put together by a far sighted leadership team. For example, rapid response system (RRS) which is a multidisciplinary team strategy, has led to better anticipation and prevention of in-hospital cardiopulmonary arrests (Sakai and Devita 2009). Early detection and response to physiological deterioration is another program that can improve outcomes for infants, children, and adults.

It should be apparent that the external and internal drivers are predicated upon a specific set of capabilities in the healthcare delivery network. In Fig. 10.5, we show these capabilities for both the service and material supply chains, using the framework developed in Chap. 1. The cluster of capabilities, labeled "demand management" refers to external drivers, whereas the other four clusters labeled "fund management", "provider management", "operations and supply chain", and "E-healthcare", are relate to internal drivers.

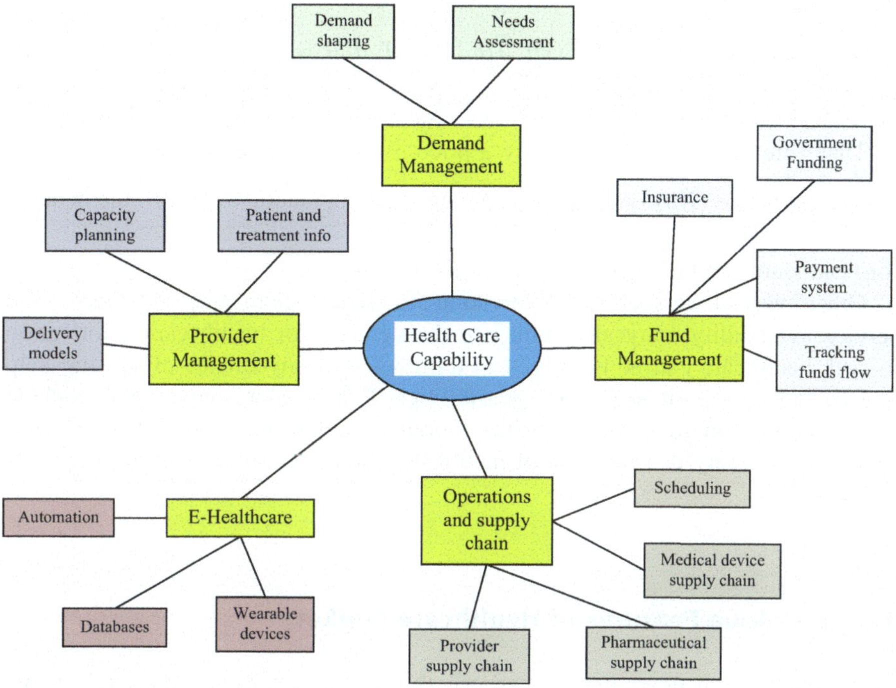

Fig. 10.5 Healthcare capabilities

10.5 Demand Management

In this section, we elaborate how demand for healthcare can be assessed, and how demand can be shaped by managerial measures.

10.5.1 Healthcare Market and Demand

As mentioned earlier, the demand for healthcare depends on such factors as price, income, age, and preferences. In most cases, the need for healthcare arises only when the person is sick. The two major determinants of a sick person's use of healthcare services are the price of the service and household income. We define $f(p|I, \theta)$ as the probability mass function that a random individual will pay the amount p for the use of healthcare, where I and θ are the income and health status of the individual, respectively. Note that $I = 1$ ($I \in [0,1]$) is the highest income level, and $\theta = 1$ ($\theta \in [0,1]$) is the health level of the sickest person in the population. Therefore, the probability mass function of payment x by the population as a whole is expressed as $g(x) = \int_{I=0}^{1}\int_{\theta=0}^{1} f(x|I,\theta)h_I(I)h_\theta(\theta)dId\theta$, where $h_I(I)$ and $h_\theta(\theta)$ are the income and health-status probability mass functions of the population, respectively.

Therefore, the willingness to pay a price p for healthcare can be expressed as $1 - G(p)$, where $G(p) = \int_{x \le p} g(x)dx$. Assuming a population of size R, the demand for healthcare can be expressed as $D = R\{1 - G(p)\}$.

Other variables, such as motivation to seek healthcare, resources, and service availability, may also influence the demand for health care. Motivation to seek healthcare can be described in terms of the convenience of seeking help and the awareness of healthcare possibilities. Individual resources may include assets in addition to income, health insurance, and welfare eligibility. Finally, service availability is a function of facility location, and investment in provider's expertise.

10.5.2 Unique Features of Healthcare Market

In relation to consumer goods markets, the healthcare market possesses some unique features. First, as Nahata et al. (2005) point out, the demand for health care is inelastic at high prices, is elastic for middle price range, and becomes inelastic again at low prices. The third-party-payment system such as health insurance is the primary reason for this unusual behavior in the demand function, according to these authors. The authors propose an inverse demand function of the type, $p = 2D^{-2.5} - 10D + 50$, that meets this requirement.

Second, as mentioned earlier, the fact that providers have discretionary power over patients and decide what treatments a patient should undergo, causes provider-induced demand. Fuchs (1978) found that when the number of physicians in the population increased by 1%, they took care to *increase* demand for their service by 0.3%, by putting patients through additional treatment regimes. As a result, the physicians' income did not decrease even though the number of physicians in the population increased. This contradicts the basic tenet of market competition in economics.

Third, there exist many conflict of interest scenarios in the healthcare market. To a consumer, prevention is much more important than treatment; but to the providers, driven by profit-maximization, healthcare treatment is much more lucrative than prevention. Prevention reduces the volume of patients available to a traditional provider thereby decreasing its income. As a result prevention and wellness centers have received only lukewarm support from traditional providers. Preventive medicine that reduces the number of acute and chronic patients can fragment hospital services into many ambulatory care facilities, another cause of concern for providers.

Fourth, competition between providers can drive up prices. This can happen when providers invest in the latest and very expensive technologies to *recruit* patients, and charge the cost of investment to the patients. Another example of conflict of interest is a provider investing in a piece of expensive equipment such as a CT-Scan. To recover the cost of investment, the provider may order unnecessary treatments for patients.

10.5.3 Demand Shaping Through Prevention

Traditionally, the focus of healthcare providers has been the diagnosis and treatment of the sick. This is a pull (reactive) system that attempts to satisfy the consumer only when he/she falls ill. While a reactive response to a consumer product can lead to efficient resource utilization, it may result in high costs in healthcare. An extreme example of a pull system is the emergency room (ER) – patients delay seeing a physician for various reasons and, when it becomes a life threatening case, they are rushed to ER. Demand at an ER can be uneven and the treatment very expensive.

In a preventive system, on the other hand, the emphasis is on reducing the demand for acute care (Adams et al. 2008). This approach, based on lifestyle, diet, and exercise, helps prevent or delay the onset of serious illness such as coronary conditions, cancer, and diabetes. Therefore with a preventive system working alongside traditional providers, the potential for reducing healthcare costs will be greatly enhanced. It is worth noting that an increasing number of chronic conditions are strongly influenced by the behavioral choices people make. For example, obesity that leads to many chronic health conditions is largely preventable.

Adams et al. (2008) discuss three common approaches for prevention by building capabilities: educating consumers, motivating them to make proactive decisions, and inducing life-style behavior change. Consumer education may include advice on wellness, nutrition, prevention, and disease management, as well as information on health care services available.

The healthcare system can make relevant information available to help consumers make proactive choices. For example, consumers may be informed about provider performance; physician qualifications and experience; hospital infection and readmission rates; the risks, benefits, and costs of procedures; and mortality rates by physician and procedures.

Incentives for behavior change include nicotine-replacement therapy, insurance premium discounts for healthy behaviors, and creating awareness for vaccinations. Incentives need to be designed to counteract the moral hazard effect by discouraging unnecessary or inappropriate use of health services.

10.6 Provider Management

In addition to ambulatory care, other healthcare providers include primary care, hospitals, labs, and home care facilities, as mentioned earlier. Although ambulatory healthcare facilities make up 88% of all facilities, hospitals account for 46% of revenues and employ more than 42% of the industry (Carlson and Russ 2007).

In relation to the traditional hospital, emerging providers such as mobile care, medical tourism or global healthcare, and telemedicine have induced many innovations. They have spawned new capabilities with novel technologies. These providers occupy important niches in the healthcare delivery network, impacting patient-flow with creative business models. These providers use advanced information technology that enables rapid processing of digital images to monitor and manage patient care in real time. In addition, miniaturization has enabled user-friendly diagnostic kits, which would have required a whole laboratory before.

10.6.1 Hospital Management

Hospitals manage a diverse set of tasks – patient care, diagnostic facilities, surgery, and research. The primary concern in the design and operations of a traditional hospital is resource utilization; patient care often takes a secondary role, unfortunately. Admitted patients are moved to different departments for diagnostics or treatments only when the appropriate resources become available. Thus, patients could be spending most of their time waiting for procedures and diagnostics. In effect, beds and wards become the means of creating "holding areas" for patients.

Flexible approaches are now emerging that reduce the length of hospital stays, and combine short inpatient stays with outpatient visits. Case managers, a new healthcare role, help patients navigate the system. In some scenarios, healthcare services may be owned and operated by contractors. For example, laboratory services in a typical hospital may be owned and run by a 3rd party, dialysis run by a private service, housekeeping run by a contract cleaning service, and food services operated by independent vendors. This requires new capabilities: creation of flexible processes, creation of innovative service roles, and coordination of multiple independent service providers.

The structural rigidities in hospitals are giving way to new initiatives. The conventional wisdom of using scale-economy to justify large institutions with multiple departments, operating theaters, X-ray machines, laboratories, and medical specialists, is being challenged. New capabilities that enable unique configurations are emerging. For example, services are being grouped by case severity and patient needs rather than specialty. This may imply grouping by critical, acute, and ambulatory care (Edwards et al. 2004). Second, services are being decentralized to improve access and to provide customized care, through a network of dispersed hospitals. Third, new health delivery models are enabling replacement of laboratory functions by test-kits that can be used at the bedside, use of ultrasound scanners at primary care, and augmented skills at primary care.

Fourth, virtual service centers that use the internet and remote sensing devices are enhancing the utilization of medical personnel.

10.6.2 Ambulatory Care

Primary care, the mainstay of ambulatory care, is the entry point to the health care system for many patients. The ambulatory care network operates as a set of decentralized physician practices, where a wide range of processes, operations, IT systems, and personnel are in use. The structural complexity of many care networks has increased significantly in recent years in that they now need to manage a diverse collection of physicians with differing loyalties, referral patterns, and behaviors. They are also subject to volatility in volumes, and care complexity.

Effective coordination of ambulatory and inpatient services, such as the shared protocols for referral and management, is needed in reducing overuse of hospitals for nonemergency care. As payers begin to provide incentives to ambulatory care to prevent hospitalization, hospitals would need to rationalize their service offering. For example, they would need to prepare for patient-transitions from inpatient to outpatient care.

Care for chronic illnesses requires an integrated approach for primary, secondary, and community providers. A particular challenge is how to build capability for managing long-term medical conditions with care at the community level. Comprehensive approaches involving program restructuring are often necessary, and cost-effectiveness depends on the characteristics of the health and social care systems.

10.6.3 Capacity Planning

Hospitals maximize financial contribution of their resources through efficient capacity allocation. While hospital revenues are increasingly based on services, bed utilization (occupancy) remains the primary metric in hospital capacity planning. However, with a varying case mix, complexity of treatment, limited operating rooms, random patient arrivals, and growing importance of the consumer-driven perspective, bed utilization has ceased to be an adequate measure of predicting revenue and performance. A major problem with bed utilization as the objective is that it "encourages" administrators to increase the length of stay of patients in order to increase revenues. It is also unproductive as it increases operational costs. New care models are emerging that are based on performance, and patient pathways.

In many scenarios performance can be measured by the metrics of healthcare outputs, such as rapid recovery and error-free diagnosis. The output measures by themselves may not be appropriate, if the resources used to produce the outputs (such as the number of hospital-beds, and number of surgical facilities) are not used efficiently. Therefore, the ratio of output to input, also known as productivity, would be a better metric of performance. However, the value of the ratio (output/input) becomes arbitrary at best, as there can be multiple attributes of inputs (and outputs) with different units of measurement. A weighted average would be more appropriate, provided the weights attached to the attributes are logically coherent.

Data Envelopment Analysis (DEA) fills this void (Cooper et al. 1999). The strength of DEA is that it lets each hospital choose the set of weights that would maximize its performance ratio. However, it is ranked against a virtual hospital which has its attribute values as the convex combination of the attribute values of all real hospitals. As each of the real hospital chooses the best set of weights for itself, the virtual hospital becomes the benchmark that cannot be outperformed.

To evaluate a hospital, say k, we express its performance as $\theta_k = \sum_{i=1,n} u_i y_{ik} / \sum_{j=1,m} v_j x_{jk}$, where x_{jk} is the value of its jth input, v_j the corresponding weight, and m is the number of input attributes. Similarly, y_{ik} and u_i are the value and weight, respectively, of its ith output, and n is the number of output attributes. Clearly, hospital k would choose the vector of weights u and v to maximize its performance, θ_k. Note that the performance of any hospital cannot exceed 1, and this is ensured by limiting the performance of the virtual hospital (the best) not to exceed 1. Hence hospital k's capacity planning problem can be expressed as $\underset{u,v}{Max}(\theta_k) = \sum_{i=1,n} u_i y_{ik} / \sum_{j=1,m} v_j x_{jk}$, subject to $\sum_{i=1,n} u_i y_{ir} / \sum_{j=1,m} v_j x_{jr} \leq 1,\ \forall r, r = 1,R,$ where R is the number of hospitals in the sample.

Note that $\underset{u,v}{Max}(\theta_k)$ is evaluated for one hospital at a time, whereas, all R constraints must be satisfied for each k.

10.6.3.1 Patient Pathways

We may define capacity in terms of the treatment pathways of patients (Rechel et al. 2010), and the resources that constrain them (bottlenecks). The constraining elements could be the number of beds, operating theatres, diagnostic equipment, and the specialist staff. To identify a bottleneck resource, the pathways must be tracked to examine whether one or more pathways "intersect" at a particular resource causing possible bottlenecks. The objective would be to find the shortest path for each patient within the network, while minimizing delays at bottlenecks. This approach has major implications for patient satisfaction and hospital capabilities, as it focuses on the patient.

10.6.4 Provider and Patient Knowledge

In delivering healthcare, two types of knowledge are important: provider's clinical knowledge, and the patient-specific knowledge (case history). Providers (physicians) need to know the pattern of what treatments, medication, and surgery worked for which patients and for which disease. They can then apply the knowledge appropriately and consistently. In many cases providers may not have easy access to the relevant clinical knowledge, and are forced to rely almost exclusively on their individual experiences. Similarly, patient-specific information may not be readily available, if the patient received treatments from multiple providers. The interactions between the two knowledge types can be summarized, as in Fig. 10.6.

Patient Specific Knowledge	High			Personalized Care
	Medium		Experience Based Care	
	Low	Trial and Error		
		Low	Medium	High
			Clinical Knowledge-base	

Fig. 10.6 Clinical and patient-specific knowledge (Source: Adams et al. 2010)

Personalized care becomes feasible if the providers can readily access both the clinical knowledge-base, and the patient-specific database including the genomic data. At the other extreme, when providers must rely on individual clinical experience and patient's case-history comprises what the patient tells the physician, healthcare delivery becomes an exercise in "trial and error". In the middle, we have experience-based care, when population information (if not individual patient's) is available and the physicians can exchange clinical experience with colleagues. E-healthcare, discussed later, is enabling development and maintenance of both clinical and patient-specific knowledge. Note that information technology creates new capabilities, such as quick access to a vast clinical knowledge-base and information on patient profiles. It enables greatly improved diagnosis and treatment of complex conditions. Healthcare practice is moving along the "diagonal" from the bottom-left cell to the top-right cell, in Fig. 10.6.

10.7 Fund Management

As a large chunk of national budget is absorbed in providing healthcare, it is becoming imperative to minimize waste in the system. At the same time, too many controls can stymie innovations in the system.

10.7.1 Funding Sources and Flows

In the healthcare business model of today, providers are paid primarily by the employers through insurance companies or by the government through Medicare and Medicaid (as in the US). Private insurance supports inpatient and outpatient hospital care and physician services. Medicare is a program for the elderly, financed through payroll taxes, premiums, and general revenues. Medicaid, administered by the states with matching funds from the federal government, covers the poor. Payers have attempted to control cost escalation through provider contracting, discount price negotiations, risk-sharing payment methods, and managed care. Hospitals still capture the largest share of the payments.

In is a transaction oriented system payers are motivated to minimize their cost by tightly controlling claims-processing against insurance plans. To discourage monopolization and price fixing, different measures have been put in place by individual states in the US. While this promotes fairness in competition, it makes it difficult to introduce standardization. In addition, if there is no synchronization of medical records between different service providers service fragmentation can occur, leading to redundant tests and imaging.

10.7.2 Expense Mapping

The flow of funds in the US healthcare in 2005 is shown in Fig. 10.7 (LeCuyer and Singhal 2007). Providers need to collect reimbursements and grants from multiple sources, each with its unique accounting system. Insurance companies may not reimburse all cost incurred by patients, as they may disagree with procedures performed. If so, providers attempt to have the patient pay the difference. Finally, billing codes used by private insurers and the government (Medicare and Medicaid) can differ significantly, causing confusion. They need to be rationalized. Recently, there have been reports in the media that some providers are using the electronic medical records to up-code and increase the claim amount (Abelson and Creswell 2012).

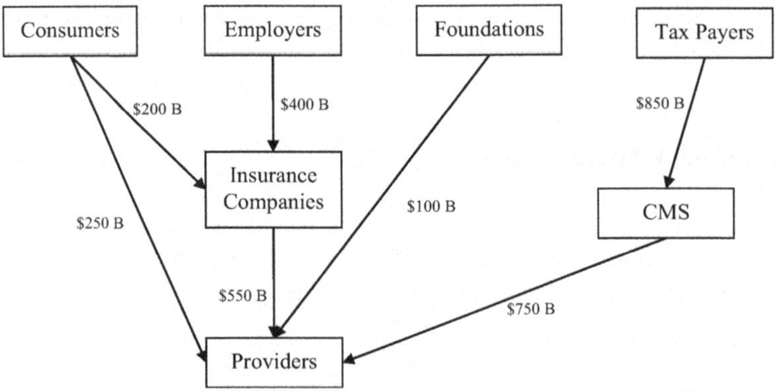

Fig. 10.7 Healthcare expense mapping 2005 (Source: Office of the Actuary, US Center for CMS)

High transaction costs, fragmented processing, and inefficient payment practices, increase the provider's cost. Batch processing of transactions, such as billing and wholesale pricing, is a major cause of such inefficiency. Providers submit claims to payers for settlement. The amount not covered by the payers is then billed to the consumers. Only about 50% of insured patients and 15% of self-pay patients (worth $250 billion in 2005) care to settle their bills on time. The growth in self-pay patients (estimated to grow to $420 billion by 2015), and higher-deductible insurance plans, have made the situation worse. Unless a better transaction processing system, similar to that used in the retail industry, is implemented the providers will be under pressure to raise payment rates.

10.7.3 Provider Payment Schemes

A payment scheme is a mechanism used to transfer funds from the payers to the providers. It can have a significant effect on how health care resources are allocated and services delivered. A well designed scheme facilitates access to health services and ensures high quality care, while promoting efficient use of resources at low cost. Therefore, it accomplishes far more than just funds transfer. As outlined in Langenbrunner et al. (2009), the methods of payment include fee-for-service (FFS), capitation, per diem, case based as in bundled payment, and pay-for-performance (P4P).

10.7.3.1 Fee-for-Service

The provider is reimbursed for each individual service performed (diagnostics, surgery, tests etc.) for every patient. This is often called "retrospective cost-based" payment, as providers submit bills to payers for incurred costs, only after completing the relevant services. Note that the payer bears full risk as it must pay for all service, whether or not necessary. The temptation for charging for unnecessary services (not verifiable) has been discussed earlier; it increases the costs of healthcare. The patient usually does not object to additional and/or expensive services as he/she is not asked to pay for it *directly*. The built-in moral hazard and patients' acquiescence make this method of payment unsustainable. It has some advantages, however. It can be easily developed and implemented, and it more accurately reflects the work actually done and the efforts expended.

10.7.3.2 Capitation Payment

The provider is paid a predetermined amount for a defined set of services for a fixed period. The salient point is that the payment to a provider is not linked to his cost of inputs. If the provider incurs costs that are greater than the capitation amount, the provider takes the loss. Thus, some risk is transferred from the payer to the provider. On the other hand, if the provider's costs are lower than the capitation amount, it can usually retain and reinvest this surplus. Note that once the capitation amount is agreed upon, the provider may be tempted to reduce the cost and quality of service by "cutting corners".

10.7.3.3 Per Diem Payment

Per Diem rate is usually based on bed occupancy rate. The base average rate is calculated as the ratio of the total hospital costs and the total number of bed-days, over a period, and is charged as $ per bed-day. It is then adjusted for factors such as case-mix, patient characteristics, and the medical-specialty in question (discussed earlier). However, the measure may trigger negative incentives. For example, providers may "transfer" outpatient and rehabilitation services to inpatient setting and boost the average length of stay, to increase bed occupancy.

10.7.3.4 Case-Based Payment

Case-based payment reimburses to hospitals at a predetermined fixed rate for each treated hospital case. Cases are grouped based on similarity in clinical characteristics and resource requirements, with a fixed rate for a group. For example, grouping criteria can be department (clinical specialty), diagnosis, and procedures. The choice often depends on the nature of clinical data available for computing the cost per case. As the rate for a case group is pre-determined, it is beneficial for hospitals to admit inexpensive cases within a payment category so that they get reimbursed higher than the actual cost for the case. In addition, they may split expensive cases into multiple stays, or decline to admit them. To counteract these adverse incentives, procedures are being developed for (i) minimizing the variance of case expenses within a group, and (ii) differentiating cases of different resource intensities. The ideal system would be transparent so that the payers know which services are included in a case group and how actual quality and prices compare across hospitals.

10.7.3.5 Bundled Payment

As discussed earlier, a bundled payment provides a single payment for an entire episode of care during a set period (treatment for a specific medical condition). Implementation of bundled payment includes, bundle construction and pricing, billing and payment distribution, care redesign, and quality monitoring (Mitre Corp 2012).

Bundles should be designed and priced according to the evidence-based historical price and cost data. Bundle definition should include information on episode triggers and end points, patients and providers that would be included in the bundle, and factors that cause a patient or service to be removed from a bundle. The price must be lower than the total FFS payments for the episode, but sufficiently high to ensure adequate compensation to the providers.

Once a bundle has been created, billing and payment systems need to be modified using the bundle as the unit of billing. Patient registration process would need modification, to identify patients by bundle. Patients must be made aware of bundle-billing, claims, and distribution processes.

Care redesign is a primary objective of bundled payment. IT capabilities may prove useful for process support and data access, communications, and analytics. The use of electronic health records may also prove beneficial to support robust care redesign. Quality monitoring activities benefit from IT-supported analysis of data from many sources.

10.7.3.6 Pay for Performance (P4P)

The inability to control costs has pushed large employers to look for pay for performance (P4P). Most P4P systems include bonus payments for better outcomes. The DEA model, discussed earlier, can be used to assess the relative weighted average performance ratings of providers, which can then be used to determine the rates of payments. In general, performance is defined as the ratio of the value of the output and the cost of delivering that output. The health outcomes may be captured in terms of metrics such as the frequency of error-free diagnosis and treatments, speed of diagnosis, availability of non-invasive procedures, and the quickness of recovery (Porter 2012).

10.7.4 Comparison of Payment Methods

In general, the payment methods can be categorized by the type of payment scheme (cost of inputs and performance of outputs) and whether the rates are determined or charged ex ante/ex post, as shown in Table 10.3.

Table 10.3 Payment-model comparison

	Rate determination	Payment	
	Ex ante	Ex post	Ex ante
Input cost based	Fee for service	Fee for service	
Output performance based	Case based	Case based	
	Bundled payment	Bundled payment	
	P4P	P4P	
	Capitation		Capitation
	Per diem	Per diem	

The choice of a payment system is clearly influenced by the objectives such as quality of care, cost, and patient-throughput. However, with multiple stakeholders, the objectives may often be in conflict. In general, the ex post elements of the payment system perform better in terms of access, provider risk, revenue, and quality, while the ex ante elements do better in establishing the optimal levels of services, efficiency, and cost.

Input based methods tempt the providers in increasing the number of inputs so as to earn higher revenues. An output based method, on the other hand, creates incentives to increase the number of services. Input-based payments, unlike case-based payments, do not foster competition because the money does not follow the patient. Resource allocation is not responsive to changes in demand and utilization, and is often determined by historical patterns and political priorities.

10.8 Operations and material Supply Chain

Managing operations, and coordination of the material supply chain are vital in supporting the service delivery chain, discussed earlier. First, we examine the operations effectiveness in hospitals. This is followed by an analysis of the supply chains of pharmaceuticals and medical devices.

10.8.1 Operations Scheduling

At the operational level hospitals must improve patient flow, provide timely treatment, and maximize resource utilization. Healthcare services may be executed on a patient in multiple sequential steps and require coordination of physicians, medications, and diagnostic equipment, so as to minimize bottlenecks. Delays can be reduced through scheduling, process improvement, and information management.

To appreciate the level of complexity, consider scheduling of treatments for patients, subject to a pre-defined sequence of procedures, completion times of specific procedures, and availability of medical equipment and physicians. Note that each treatment comprises a number of procedures that must be performed in a specific order. Each procedure must be completed in specific time duration and may require a set of medical equipment. Each piece of equipment has an implied capacity and it must be operated by a physician or a trained personnel. The procedures, a physician is qualified to perform, are determined by his/her field of specialization.

The dynamics governing a hospital implies that the models used for decision making must incorporate the complexity, uncertainty, variability, and resource limitations. Complexity arises from rules requiring some beds to be kept free for emergency patients, and intra-service conflicts in scheduling patients. Uncertainty is caused by emergency patients who arrive randomly, and who must be admitted with minimum delay. Another source of uncertainty is the length of stay (LOS). LOS for geriatric care has very different characteristics from that of pediatric care. Finally, providers must manage limited resources such as beds, operating theatres, the workforce, and critical care.

Harper (2002) describes a hospital (unnamed) scheduling process with respect to its existing capacity, provision of beds, forecast of patient case-mix, and the surgical demand (inpatient and outpatient). The surgical patients were categorized into three classes based on case-severity that determines the needed medical staff and patient's length of stay. Two objectives were established in consultation with the hospital staff: maximization of operating theatre utilization, and maximization of bed utilization. The author created a model for simulating hospital services and considered several scheduling policies including, first come first served (FCFS), longest operation times first (LOT), and shortest operation times first (SOT). The best policy was found to be LOT: schedule large operations first followed by smaller operations later in the time cycle. An LOT policy showed that the hospital could improve efficiency by maintaining current throughput, while saving patient-time. The model was also used to examine the day-to-day scheduling of patients. This helped to even out busy and slack times in bed utilization, while minimizing the number of patients declined admission.

10.8.2 Material Supply Chain

The healthcare material supply chain is vast, complex, and very inefficient. The supply chain attempts to provide the right medical products to the right customers (providers, retailers, and consumers) at the right time. To operate successfully, the healthcare companies in the supply chain – pharmaceuticals, medical equipment, biotechnology, hospitals and clinical services, GPOs and distributors – are adapting to the challenges stemming from customer demand volatility, low visibility, and diverse processes.

Material supply chain management, as discussed in Chaps. 1, 4, and 5 is focused at processes such as order placement, logistics, sourcing, and storage. With an outcome-based patient centric approach, the material supply chain needs to reinvent itself. For example, instead of making decisions based on demand forecasts, it needs to target supplies of pharmaceuticals and equipment that contribute positively to outcomes. In addition, it needs to track how advances in pharmaceuticals and medical devices impact the treatment regime and, consequently, the service supply chain. With medicine being personalized, drugs would need to be ordered in complete kits and *tracked* for individual patients. With new expensive medical devices substituting some of the tasks performed by nurses and physicians, monitoring and scheduling of devices become critical tasks. Thus, during the transitions, hospitals need to manage two types of supply chain: traditional, and emerging.

As discussed in Chap. 1, capability of a traditional supply chain is described in terms of executing functions such as planning, procurement, contracting, storage, and inventory. In a hospital with an inventory of thousands of supplies – drugs, surgical instruments, medical devices, bandages and syringes – the supply chain functions become a critical undertaking. They consume as much as 30% of total operating budget (Schwarting et al. 2011). The supply chain must also account for the inventory held by vendors and wholesalers, as well as the value of inbound supplies, the shipping cost, and shrinkage during transportation. Note that the performance metrics in traditional supply chain include product availability and cost. As a consequence, it often emphasizes inventory optimization and information visibility.

The traditional material supply chain is experiencing disruption from two distinct effects in healthcare: personalized medicine, and consumer-driven health services. Personalized medicine makes it possible to have the right treatment for the right individual at the right time (Hu et al. 2005). Genomics is the technology that identifies the basic difference between two individuals at the molecular level. Molecular diagnosis can serve as the basis for matching medicines to individual patients. As a consequence the positive outcome of the drug will increase, but the total consumption and sales of individual drugs will decrease (by minimizing prescriptions of unmatched drugs). This would most likely lead to increase in the price of the drug. Therefore, while the total revenue from a drug may or may not increase, the total cost of manufacturing will certainly decrease (manufacturing a smaller total quantity of drugs). It has implications for the supply chain, as the order quantity of the drug will change, and they will be targeted to specific patient segments.

In a consumer driven healthcare, individuals must become more active in their own health decisions to ensure high value care. In this transformed system, patient and providers jointly decide the healthcare needs. For example, physicians may provide information on the benefits and risks of alternative treatments, and patients make choices based on this information and their personal preferences. The transformed system may alter the material supply chain significantly, as it would cease to be forecast driven. The needs of the patients, determined in real time, must be satisfied. This would require the supply chain, including procurement and logistics, to be responsive – purchase in small batches, just in time, from sources close to the hospital. Another effective supply chain strategy would be to use vendor managed inventory (VMI) with consignment orders, where the hospital pays for inventory only when it uses them (not when it orders them). Real time response may also require suppliers to work closely with physicians who determine items to be ordered. Finally, the supply chain processes must be designed to support the physicians in understanding the trade-offs involved between patient outcomes and costs, in selecting products such as diagnostic equipment.

Pharmaceutical companies have traditionally relied on mass production of profitable drugs with long production lead times, use of special purpose equipment, sluggish product changeovers, and are subject to numerous regulatory requirements. Plants are often underutilized as they are designed for *specific* drugs with uneven demand. These policies have led to high inventories and late deliveries that hurt profitability of manufacturers.

These companies are now experiencing rapid advances in therapeutic drugs and gene profiling, threatened by shorter patent-protection, and face increased competition from generics. They are also beginning to deliver products "direct to customer", as in the computer industry. As a consequence, manufacturers are reconfiguring their facilities for niche drugs with small batch production (Koester and Rash 2005). They are also reengineering their fulfillment capability, so as to transition from "selling to wholesalers in large quantities" to "selling to a large number of diverse customers". Such a transformation may require changes to plant layout and material handling as well as small batch ordering, warehousing, and fulfillment. In addition, new regulatory compliance must be integrated into manufacturing processes. It involves understanding the regulatory changes, designing operating procedures, and implementing process changes. Building a fulfillment capability to satisfy the evolving demand would require investments in information technology, rationalized packaging, network optimization, 3PL assessment, and customer service.

Note that ordering and servicing needs of providers, retail pharmacists, and consumers can be different from those of the wholesalers and GPOs. Providers require frequent deliveries, packaging in smaller lots, configured treatment-kits, and deliveries to a large number of locations.

Evaluation and procurement of *medical devices* are now garnering more attention from healthcare providers. Smart implants, nanotechnology and RFID in diagnostics and monitoring, neuro-stimulation devices, device-based drug treatment, imaging, and other similar technologies are altering treatment plans and the needed resource of human expertise (Rehan et al. 2009). While the primary concern in procuring medical devices remains patient safety, there is now a growing emphasis on process reengineering and cost effectiveness. Rising costs are compounded by increase in medical procedures that use advanced medical devices. Between 1991 and 2004 the cost of orthopedic implant devices rose by 132%, and knee replacements increased by 40% in 4 years from 2000 to 2004.

Health safety and quality standards affect medical device supply chain: development, production, and distribution. Regulations define which devices can be marketed in which regions, and ethical considerations can influence whether certain devices would be adopted at all. Medical device companies are developing partnerships in their supply chains to meet challenges in personalized medicine and convergence of devices, drugs, and diagnostics.

10.9 E-Healthcare

E-healthcare is the use of digital data in support of healthcare at local and remote sites, and it involves electronic transmission, storage, and retrieval. It uses information technology, wearable devices, and smart phones to provide critical support in the healthcare delivery network.

10.9.1 E-Healthcare Functions

E-healthcare links patients, providers, payers, institutions, and the underlying processes. For example, data exchange between ambulatory care centers and hospital departments can significantly improve patient-access and cost effectiveness in the network. E-healthcare reduces the need to visit hospitals, saving time and money. It enables hospitals to focus on acute care patients who must be treated as inpatients. Electronic health records (EHR), smart cards, electronic order-entry, e-medicine, and digitized diagnostic – all contribute to improving the cost-effectiveness and performance of healthcare. As discussed earlier, EHR facilitates more effective treatment with improved knowledge of the patient's history and previous medical interventions. It can reduce medical errors through access to critical information, alerts, and guidance on best practices. It can also improve the cost-effectiveness of nursing, and administration. E-healthcare apps have proven useful in disease prevention, diagnosis, patient management, and care-provision.

Organizations need to build capabilities of different types that include content, connectivity, computer applications, commerce, and care (Schlachter 2004). *Content* provides information on diseases, drugs, therapies, research databases, and other clinical data. It may also contain information on healthcare providers – accreditations, and performance ratings. Different types of databases such as consumer data, genomic data, test images, and device-specific data are being built and integrated.

Connectivity capability includes technology such as the internet, electronic data interchange (EDI), mobile access, and satellite links. It must ensure data security and integrity, virus protection, and confidentiality. Connectivity can also be provided through telephone and fax lines.

Computer applications are a set of software that *run* database access, multimedia, virtual communities, transaction processing, monitoring, scheduling, access protocols, and clinical decisions. New applications need to be created for updating the database of evidence-based clinical knowledge, analyzing data for generating insights on performance and cost and identifying new trends.

Commerce includes platforms such as portals for online buying and selling, which provide clinical and business transactions capability for treatments, lab tests, invoicing, ordering, and payments. Online care is the capability to provide interactions between physicians, as diagnosis and treatments often require ability to extract expert knowledge available within specialist domains.

Adoption of e-healthcare is slow, its technological capability notwithstanding. The primary reason is the lack of motivation on the part of physicians, perceiving it as a threat to their income. They are also threatened by a perceived loss of autonomy and a sense of "being monitored" by the payers. In addition, medical practitioners are not sufficiently exposed to information technology and are not comfortable using it.

10.9.2 Wearable Technology

Wearable sensors can provide medical monitoring, memory enhancement, control of home appliances, medical data access, and rapid response in emergency situations. Continuous monitoring of patients facilitates early detection of emergencies and triggers appropriate services for people with cognitive and physical disabilities. The devices can monitor human physiological data to detect health conditions. Households where both parents work can benefit from high-quality care services for infants.

These devices can track and monitor doctors and patients inside a hospital as well, where patients may wear sensors for specific functions such as heart rate or blood pressure detection. They can help in drug administration in hospitals by attaching sensors to patients to identify their allergies and required medication through the EHR, and help minimize the side effects of drugs. For example, patients with chronic conditions can be monitored remotely using the eCAALYX system (Mosa et al. 2012). It receives data from sensors in a *smart garment* worn by the subject, transmits it to a monitoring server, and can identify signs of respiratory infections based on established medical knowledge.

Wearable device technology must have the capability to sense and capture both patient-movement data and physiological data, communicate the data to a remote center, and extract relevant inferences through data analysis. Advances in sensor technology, microelectronics, telecommunication, GPS tracking system, and data analysis are now making this possible.

10.9.3 Smart Phone Applications

The smart phone is a relatively new technology that combines mobile communication, graphics, and computation in a handheld device. It supports several modes of communication including voice calling, video calling, text messaging, email messaging, image messaging, and conferencing using the wireless access protocol.

Mobile connectivity enables providers to use them in local and remote clinical settings (Mosa et al. 2012). It can provide access to databases on disease-diagnostic guidelines, drug references, literature search, and medical education materials at the point of care, 24/7. It facilitates emergency rooms in remote locations to stay connected; they can access online consultation, telemedicine, and obtain help in trouble shooting. With its graphic capability, it should soon be possible to provide online consultation in real time surgery.

Smart phone apps are increasingly being used in performing simple medical exams such as vision test. Another example is helping diabetic patients. A smart phone can be used for sending patient data to the Diabeo system that automatically adjusts a patient's insulin dosage based on his/her condition (Mosa et al. 2012). Similarly, a smart phone connected to a GPS can send ECG rate, walking speed, heart rate, and patient location, to a server for real-time monitoring. Pulmonary patients can follow exercise instructions on their smart phones, which are based on patient heart and pulse rates and exercise time.

The wide adoption of smart phones by individuals emphasizes the opportunity mobile devices have through patient oriented applications; the full potential is yet to be exploited. There is a need for developing standards so that smart phone applications can be used seamlessly and integrated with EHR and patient monitoring systems.

10.10 Restructuring Healthcare Delivery Network

Patients seek high quality healthcare at a low price. Providers maintain that high quality comes at a high cost. Thus, patients and providers agree on the clinical side but not on the economics. While health insurance minimizes the impact of demand shocks, it does not address the quality vs. cost issue. The answer lies in technological and managerial innovations. While technological innovations in treatments and medicines are improving the quality of service; value-chain innovations in patient pathways, bundled pricing, and outcome emphasis are helping to cut waste from the system and reduce cost while setting appropriate incentives.

As a consequence, healthcare is mutating in different ways: extending the span of the delivery network, serving segmented markets, leveraging connectivity, building excellence in treatments, inculcating organizational responsibility, enabling patient-pulled services, offering different payment structures, improving the pharmaceutical supply chain, and altering the ownership structure, as shown in Table 10.4.

Table 10.4 Restructuring healthcare delivery network

Delivery network characteristics	Restructuring providers and processes
Extended span of delivery network	*Health driven lifestyles* causing wellness centers to be included in the network
	Acupuncture, homeopathy and other *non traditional providers* are being added to the network
	Minor surgery and other expertise are *transitioning upstream* in the value chain causing new clusters of ambulatory care and other services
	Traditional hospitals becoming *specialized* and smaller in size
Segmented market	Price leaders provide *standardized simple services* at low prices
	Concierge practice, and retail clinics for *consumer convenience*
	Outcome based providers deliver quick diagnosis, noninvasive treatment, and fast recovery
Leveraging connectivity	Electronic health records for *information transparency*
	Online transactions for ordering, invoicing, and payments
	Telemedicine for *remote consultation*
	Wearable devices – *patient monitoring*
	Smart phones and applications – *mobile remote connectivity*
Centers of excellence	*Clinical expertise* – organ transplantation, stem cell based treatment
	Personalized medicine enabled by Genomics
Organizational responsibility	ACO – federal structure of providers with *incentive alignment*
	Directly *accountable* for all services
	Providers may choose from *payment structures* that offer flexibility in risk exposure

(continued)

Table 10.4 (continued)

Delivery network characteristics	Restructuring providers and processes
Consumer-driven services	Consumers *participate* in treatment decisions
	Decreases *moral hazard*, and unnecessary procedures by some providers
	Supply chain for real time demand – *JIT, and VMI*
Payment structure	Multiple *pricing schemes* – fee for service (FFS), bundled payment, capitation, resource utilization, pay for performance (P4P)
	Transitioning from FFS to bundled payment or P4P
Pharmaceutical and medical device supply chains	Niche drugs – *restructuring the pharmaceutical supply chain* for small batch production, shorter patent life
	Generics – large batch production
	Transition from selling to wholesalers in large quantities to selling to many diverse customers in small batches
	Health safety *regulations define devices marketability*
	Convergence of devices, drugs, and diagnostics
Altering ownership structure	*Regulating insurance* companies as in Germany, and in the Affordable Care Act
	Government *ownership of providers* as in UK and Canada

References

Abelson, R., & Creswell, J. (2012, September 24). U.S. warning to hospitals on Medicare bill abuses. *New York Times*.

Adams, J., Bakalar, R., Boroch, M., Knecht, K., Mounib, E., & Stuart, N. (2008). *Healthcare 2015 and care delivery*. Somers: Institute for Business Value, IBM Global Services.

Black, N., & Gruen, R. (2005). *Understanding health services*. Maidenhead: Open University Press.

Bohmer, R., & Lee, T. (2009). The shifting mission of health care delivery organizations. *The New England Journal of Medicine*, *361*, 551–553. http://www.ascor.com.br/PDF/Literatura/Shifting%20mission%20health.pdf

CAM. (2005). *Complementary and alternative medicine in the United States*. Washington, DC: Institute of Medicine of the National Academies, National Academies Press. http://www.nap.edu/catalog.php?record_id=11182

Carlson, P., & Russ, E. (2007). *Health care industry trends*. http://www.workforcesolutionsalamo.org/pdf/NCEE%20Health%20Care%20Industry%20Trends.pdf

CHWS. (2006). *The impact of the aging population on the health workforce in the United States: Summary of key findings*. School of Public Health, University at Albany. http://www.albany.edu/news/pdf_files/impact_of_aging_excerpt.pdf

Cooper, W., Seiford, L., & Tone, K. (1999). *Data envelopment analysis*. Boston: Kluwer.

Deloitte. (2010). *Accountable care organizations: A new model for sustainable innovation*. Deloitte Center for Health Solutions. http://www.deloitte.com/assets/Dcom-UnitedStates/Local%20Assets/Documents/US_CHS_AccountableCareOrganizations_041910.pdf

DHCFP. (2011). *DHCFP report on bundled payments*, Vol. 1. Massachusetts Division of Health Care Finance and Policy. http://www.mass.gov/dhcfp

Edelson, E. (2011). Disease prevention programs: Worth the investment. *Health Day*. http://health.usnews.com/health-news/diet-fitness/diabetes/articles/2008/07/17/disease-prevention-programs-worth-the-investment

Edwards, N., Wyatt, S., & McKee, M. (2004). *Configuring the hospital for the 21st century*. Geneva: The European Observatory on Health Systems and Policies, World Health Organization. http://www.euro.who.int/__data/assets/pdf_file/0008/108845/E84697.pdf

Elhauge, E. (2010). *Why we should care about health care fragmentation and how to fix it*. New York: Oxford University Press. http://www.law.harvard.edu/faculty/elhauge/pdf/Elhauge%20The%20Fragmentation%20of%20US%20Health%20Care%20–%20Introductory%20Chpt.pdf

Field, R. (2008). Why is health care regulation so complex? *P&T, 33*, 607–608. http://www.ncbi.nlm.nih.gov/pmc/articles/PMC2730786/pdf/ptj33_10p607.pdf

Frenkel, M., & Borkan, J. (2003). An approach for integrating complementary: Alternative medicine into primary care. *Family Practice, 20*, 324–332. Oxford University Press. http://fampra.oxfordjournals.org/content/20/3/324.full.pdf+html

Friend, S., Swanick, M., & Arlington, S. (2011). *Pharma 2020: Supplying the future – Which path will you take?* Price Waterhouse Company. http://www.pwc.com/en_GX/gx/pharma-life-sciences/pharma-2020/pharma-2020-supplying-the-future.jhtml

Fuchs, V. (1978). The supply of surgeons and the demand for operations. *Journal of Human Resources, 13*, 121–133.

Harper, P. (2002). A framework for operational modeling of hospital resources. *Health Care Management Science, 5*, 165–173.

Health Affairs. (2010). *Accountable care organizations*. Health Policy Brief, Project HOPE – The People-to-People Health Foundation Inc. http://www.rwjf.org/files/research/66449.pdf

Herrick, D. (2007). *Medical tourism: Global competition in health care* (NCPA Policy Report No. 304). ISBN #1-56808-178-2. http://www.ncpa.org/pdfs/st304.pdf

Hu, S., Foster, T., & Kieffaber, A. (2005). Pharmacogenomics and personalized medicine: Mapping of future value creation. *BioTechniques, 39*. doi:10.2144/000112048. http://www.bionest.com/publications/Bionest-PM_Future_Value_Creation.pdf

IBM. (2003). *Strategies for clinical process redesign and implementing an ambulatory electronic medical record*. IBM Corporation. http://nchica.org/Past/07/Presentations/Snow.pdf

Keckley, P. (2008). *Retail clinics*. Deloitte Center for Health Solutions. http://www.deloitte.com/assets/Dcom-UnitedStates/Local%20Assets/Documents/us_chs_RetailClinics_230708(1).pdf

Keeling, D., & Schrader, U. (2012). *Operations for the executive suite: Opening new horizons for current and future pharma leaders*. McKinsey and Company. New York, USA

Koester, L., & Rash, K. (2005). *Building supply chain capabilities in the pharmaceutical industry. Part 2: Winning supply chain capabilities*. United Parcel Service of America, Inc. http://www.ups-scs.com/solutions/white_papers/wp_pharma2.pdf

Langenbrunner, J., Cashin, C., & O'Dougherty, S. (2009). *Designing and implementing health care provider payment systems*. Washington, DC: International Bank for Reconstruction and Development, World Bank. http://siteresources.worldbank.org/HEALTHNUTRITIONANDPOPULATION/Resources/Peer-Reviewed-Publications/ProviderPaymentHowTo.pdf

Lantronix. (2010). *Medical devices landscape*. Chicago: HIMSS Analytics. http://www.himssanalytics.org/docs/medicaldevices_landscape.pdf

Larew, M. (2011). *Healthcare medical devices*. Henry Fund Research, University of Iowa, Tippie School of Management. http://tippie.uiowa.edu/henry/reports11/medical_devices.pdf

LeCuyer, N., & Singhal, S. (2007). *Overhauling the US health care payment system*. McKinsey and Company. http://aclaimtransactions.com/public/docs/McKinsey-OverhaulingHealthCarePayments062807.pdf

McKesson. (2010). *Position paper: Bundled payment*. McKesson Health Solutions. http://www.mckesson.com/static_files/McKesson.com/MHS/Documents/Issue-Paper-Bundled-Payment0210.pdf

Miller, H. (2009). *How to create accountable care organizations.* Center for Healthcare Quality and Payment Reform. http://www.chqpr.org/downloads/HowtoCreateAccountableCareOrganizations.pdf

Mitre Corp. (2012). *Information technology for bundled payment.* http://www.ahcancal.org/facility_operations/hit/Documents/IT_Bundled_Payment.pdf

Mosa, M., Yoo, I., & Sheets, L. (2012). A systematic review of healthcare applications for smartphones. In *BMC medical informatics and decision making.* doi:10.1186/1472-6947-12-67. http://www.biomedcentral.com/content/pdf/1472-6947-12-67.pdf

Nahata, B., Ostaszewski, K., & Sahoo, P. (2005). Rising health care expenditures: A demand-side analysis. *Journal of Insurance Issues, 28*, 88–102.

Orszag, P. (2008). *Technological change and the growth of health care spending.* Congressional Budget Office, The Congress of the United States. http://www.cbo.gov/sites/default/files/cbofiles/ftpdocs/89xx/doc8947/01-31-techhealth.pdf

Pharmaceutical Research. (2007). *Drug discovery and development.* Innovation.org. http://www.innovation.org/drug_discovery/objects/pdf/RD_Brochure.pdf

Porter, M. (2012). *Value-based health care delivery.* HBS Healthcare Initiative Club. http://www.hbs.edu/healthcare/pdf/2012%2003%2007%20SUT%20HCI%20presentation.pdf

PWC. (2011). *Pharma 2020: Executive summary.* http://www.pwc.com/en_GX/gx/pharma-life-sciences/pdf/pharma-exec-summary.pdf

Rechel, B., Wright, S., Edwards, N., Dowdeswell, B., & McKee, M. (2009). *Investing in hospitals of the future.* World Health Organization and the European Observatory on Health Systems and Policies. http://www.euro.who.int/__data/assets/pdf_file/0009/98406/E92354.pdf

Rechel, B., Wright, S., Barlowc, J., & McKeea, M. (2010). Hospital capacity planning: From measuring stocks to modeling flows. *Bulletin of the World Health Organization, 88*, 632–636. http://ukpmc.ac.uk/articles/PMC2908974/pdf/BLT.09.073361.pdf

Rehan, M., Prashar, S., & Tang, Q. (2009). *Destination 2025: Focus on the future of the medical device industry.* Deloitte Development LLC. http://biobusinessalliance.org/Files/531111_d2025_whitepaper_medical_device_smaller_secured.pdf

Reinhardt, U. (1993). Reforming the health care system: The universal dilemma. *American Journal of Law and Medicine, 19*, 21–36.

Sakai, T., & Devita, M. (2009). Rapid response system. *Journal of Anesthesia, 23*, 403–408. doi:10.1007/s00540-009-0778-8. http://xa.yimg.com/kq/groups/16749867/304043226/name/Rapid+response+system.pdf

Schlachter, C. (2004). *New business models for E-healthcare and the role of trust.* Department of Informatics, University of Zurich. http://www.ifi.uzh.ch/archive/mastertheses/DA_Arbeiten_2004/Schlachter_Christoph.pdf

Schwarting, D., Bitar, J., Arya, Y., & Pfeiffer, T. (2011). *The transformative hospital supply chain: Balancing costs with quality.* Booz and Company. http://www.booz.com/media/uploads/BoozCo-Transformative-Hospital-Supply-Chain.pdf

Skoskiewicz, M., Burwell, B., & Bard, D. (2012). *New buying behavior provides value creation opportunities for medical device manufacturers.* Chicago: Marakon, Charles River Associates. http://www.marakon.com/downloads/Marakon-Commentary-Value%20Creation%20Opportunities%20for%20Medical%20Device%20Manufacturers.pdf

Spooner, J. (2007, January–February). Concierge medicine: Origins, growth, controversies, and implications to Medicare. *Medicare Patient Management*, 26–31. http://www.medicarepatientmanagement.com/issues/02-01/MPM02-01_Concierge.pdf

Thorpe, K., & Ogden, L. (2008). *What accounts for the rise in health care spending?* Institute for Advanced Policy Solutions, Emory University. http://www.emory.edu/policysolutions/pdfs/riseinhealthspending.pdf

Index

A.K. Chakravarty, *Supply Chain Transformation*, Springer Texts in Business and Economics, DOI 10.1007/978-3-642-41911-9,

The manufacturer's authorised representative in the EU is Springer Nature Customer Service Centre GmbH, Europaplatz 3, 69115 Heidelberg, Germany. If you have any concerns regarding our products, please contact ProductSafety@springernature.com

Printed and bound by CPI Group (UK) Ltd, Croydon, CR0 4YY
30/09/2026
02225602-0013